Automobile Mechanical and Electrical Systems

The third edition of *Automobile Mechanical and Electrical Systems* concentrates on core technologies to provide the essential information required to understand how different vehicle systems work. It gives a complete overview of the components and workings of a vehicle from the engine through to the chassis and electronics. It also explains the necessary tools and equipment needed in effective car maintenance and repair, and relevant safety procedures are included throughout.

Designed to make learning easier, this book contains:

▷ Photographs, flow charts and quick reference tables
▷ Detailed diagrams and clear descriptions that simplify the more complicated topics and aid revision
▷ Useful features throughout, including definitions, key facts and 'safety first' considerations.

In full colour and with support materials from the author's website (www.automotive-technology.org), this is the guide no student enrolled on an automotive maintenance and repair course should be without.

Tom Denton is a Fellow of the Institute of the Motor Industry and a Member of the Institute of Road Transport Engineers and of the Society of Automotive Engineers. He has written over 30 textbooks, along with support materials, and world-leading eLearning courses.

Hayley Pells is an experienced MOT tester, owner/operator of award winning Avia Sports Cars Ltd. She is a versatile automotive writer, regularly published since 2017. Hayley joined Tom as co-author for the second edition of *Automotive Technician Training* in 2020.

Automobile Mechanical and Electrical Systems

Third Edition

Tom Denton and Hayley Pells

Routledge
Taylor & Francis Group

LONDON AND NEW YORK

Third edition published 2023
by Routledge
2 Park Square, Milton Park, Abingdon, Oxon OX14 4RN

and by Routledge
605 Third Avenue, New York, NY 10158

Routledge is an imprint of the Taylor & Francis Group, an informa business

© 2023 Tom Denton and Hayley Pells

Second edition published by Butterworth-Heinemann 2011
First edition published by Routledge 2014

British Library Cataloguing-in-Publication Data
A catalogue record for this book is available from the British Library

Library of Congress Cataloging-in-Publication Data
A catalog record for this book has been requested

ISBN: 978-1-032-28909-0 (hbk)
ISBN: 978-1-032-28908-3 (pbk)
ISBN: 978-1-003-29906-6 (ebk)

DOI: 10.1201/9781003299066

Typeset in Univers
by Apex CoVantage, LLC

Contents

Preface

This third edition of Automobile Mechanical and Electrical Systems is aimed at students who are taking a course in automotive technology or are interested in learning for themselves about this fascinating industry.

I am delighted to be partnering on this book with my friend and colleague Hayley Pells, a multi award winning owner of a highly respected service and repair business, Avia Autos, in Bridgend, South Wales.

Comments, suggestions, and feedback are always welcome at my website: www.tomdenton.org. On this site, you will also find lots of **free** online resources to help with your studies. Several other books are available too:

▷ Automobile Electrical and Electronic Systems
▷ Automobile Advanced Fault Diagnosis
▷ Electric and Hybrid Vehicles
▷ Alternative Fuel Vehicles
▷ Automated Driving and Driver Assistance Systems.

We never stop learning, so I hope you find automotive technology as interesting as we still do.

About the authors

Tom Denton has been researching and writing best-selling automotive textbooks for over 30 years. His published work is endorsed by all leading professional organizations and used by automotive students across the world. He has taught college students at all levels, and always helped them achieve the best results. Tom was also a staff tutor and an associate lecturer for the Open University.

His postgraduate education in all aspects of technology and education, as well as many years of practical experience, has given him a broad base from which to approach and teach this interesting, yet challenging subject.

As a Fellow of the Institute of the Motor Industry, a Member of the Institute of Road Transport Engineers and the Society of Automotive Engineers, he keeps in contact with the latest technologies and innovations across all aspects of this fascinating industry.

As well as publishing over 40 textbooks, Tom has created amazing support materials and eLearning courses.

Hayley Pells joins her friend and inspiration Tom Denton as a co-author for this third edition. She has found that her work within the automotive industry has proved how the skills learned at the beginning of a career can be transferred to numerous aspects of work and personal life.

Hayley runs her own workshop, which you can see at www.aviaautos.com. She communicates with motorists as well as business clients; offering insight and information about the environment and how her team continues to learn and engage with further education.

Good luck with your journey, and may this learning enable you to increase your opportunity to enjoy our sector.

Acknowledgements

Over the years many people have helped in the production of my books. I am therefore very grateful to the following companies who provided information and/or permission to reproduce photographs and/or diagrams:

AA
AC Delco
ACEA
Alpine Audio Systems
Audi
Autologic Data Systems
BMW UK
Brembo brakes
C&K Components
Citroën UK
Clarion Car Audio
Continental
CuiCAR
Delphi Media
Eberspaecher
Fluke Instruments UK
Flybrid systems
Ford Motor Company
FreeScale Electronics
General Motors
GenRad
HaloIPT (Qualcomm)
Hella
HEVT
Honda

Hyundai
Institute of the Motor Industry
Jaguar Cars
Kavlico
Loctite
Lucas UK
LucasVarity
Mahle
Mazda
McLaren Electronic Systems
Mennekes
Mercedes
Mitsubishi
Most Corporation
NGK Plugs
Nissan
Oak Ridge National Labs
Peugeot
Philips
PicoTech/PicoScope
Pierburg
Pioneer Radio
Porsche
Renesas
Robert Bosch Gmbh/Media

Rolec
Rover Cars
Saab Media
Scandmec
SMSC
Snap-on Tools
Society of Motor Manufacturers
and Traders (SMMT)
Sofanou
Sun Electric
T&M Auto-Electrical
Tesla Motors
Thrust SSC Land Speed Team
Toyota
Tracker
Tula
Unipart Group
Valeo
Vauxhall
Varta
VDO Instruments
Volkswagen
Volvo Media
Wikimedia
ZF Servomatic

If I have used any information, or mentioned a company name that is not listed here, please accept my apologies and let me know so it can be rectified as soon as possible.

CHAPTER 1

Overview and introduction

1.1 Vehicle categories

1.1.1 Layouts

This section is a general introduction to the car with an internal combustion engine (ICE) as a whole. Many unusual designs have been tried over the years, but commonly they tend to be rectangular with a wheel at each corner! High voltage (HV) car layouts are covered in **Electric and Hybrid Vehicles**, which have further ways of delivering power. The focus for this book will concentrate on ICE layouts and we can categorize them in different ways such as:

▷ front engine driving the front wheels
▷ front engine driving the rear wheels
▷ front engine driving all four wheels
▷ rear engine driving the rear wheels
▷ mid-engine driving the rear wheels
▷ mid-engine driving all four wheels.

Common abbreviations for these layouts are given in Table 1.1.

A common layout for a standard car is the front-engine, front-wheel drive vehicle. This is because a design with the engine at the front driving the front wheels has a number of advantages:

▷ It provides protection in the case of a front-end collision.
▷ Engine cooling is easier because of the air flow.
▷ Cornering can be better if the weight is at the front.
▷ Front-wheel drive adds further advantages if the engine is mounted sideways-on (transversely).
▷ There is more room in the passenger compartment.
▷ The power unit can be made as a complete unit.
▷ The drive acts in the same direction as the steered wheels are pointing.

Rear-wheel drive from a front engine was the method used for many years. Some manufacturers have

Figure 1.1 Front-engine, front-wheel drive

Figure 1.2 Front-engine, rear-wheel drive

Automobile Mechanical and Electrical Systems. 978-0-415-72578-1 © 2018 Tom Denton.
Published by Taylor & Francis. All rights reserved.

DOI: 10.1201/9781003299066-1

Figure 1.3 Rear-engine, rear-wheel drive

Figure 1.4 Mid-engine, rear-wheel drive

Table 1.1 Common abbreviations

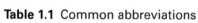

FWD	Front-wheel drive
RWD	Rear-wheel drive
AWD	All-wheel drive
4WD	Four-wheel drive

continued its use, BMW for example. A long propeller shaft from the gearbox to the final drive, which is part of the rear axle, is the main feature. The propshaft has universal joints to allow for suspension movement. This layout has some advantages:

▶ Weight transfers to the rear driving wheels when accelerating.
▶ Complicated constant velocity joints, such as used by front-wheel drive vehicles, are not needed.

Four-wheel drive combines all the good points mentioned above but makes the vehicle more complicated and therefore expensive. The main difference with four-wheel drive is that an extra gearbox known as a transfer box is needed to link the front- and rear-wheel drive.

The rear engine design has not been very popular but it was used for the bestselling car of all time: the VW Beetle. The advantages are that weight is placed on the rear wheels, giving good grip, and the power unit and drive can be all one assembly. One downside is that less room is available for luggage in the front. The biggest problem is that handling is affected because of less weight on the steered wheels. Flat-type engines are the most common choice for this type of vehicle.

Fitting the engine in the mid-position of a car has one major disadvantage: it takes up space inside the vehicle. This makes it impractical for most 'normal' vehicles. However, the distribution of weight is very

good, which makes it the choice of high-performance vehicle designers. A good example is the Ferrari Testarossa. Mid-engine is the term used to describe any vehicle where the engine is between the axles, even if it is not in the middle.

Key fact

In a mid-engine car the distribution of weight is very good.

1.1.2 Types and sizes

Vehicles are also categorized by type and size as in Table 1.2.

1.1.3 Body design

Types of light vehicle can range from small two-seat sports cars to large people carriers or sports utility vehicles (SUVs). Also included in the range are light commercial vehicles such as vans and pick-up trucks. It is hard to categorize a car exactly as there are several agreed systems in several different countries. Figures 1.5–1.13 show a number of different body types.

1.1.4 Chassis type and body panels

The vehicle chassis can be of two main types: separate or integrated. Separate chassis are usually used on heavier vehicles. The integrated type, often called monocoque, is used for almost all cars. The two main types are shown here in Figures 1.14 and 1.15.

Key fact

A vehicle chassis can be of two main types: separate or integrated.

Table 1.2 Common abbreviations and definitions

LV	Light vehicles (light vans and cars) with an MAM of up to 3500 kg, no more than eight passenger seats. Vehicles weighing between 3500 kg and 7500 kg are considered mid-sized
LGV	A large goods vehicle, known formerly and still in common use, as a heavy goods vehicle or HGV. LGV is the EU term for trucks or lorries with an MAM of over 3500 kg
PCV	A passenger-carrying vehicle or a bus, known formerly as omnibus, multibus or autobus, is a road vehicle designed to carry passengers. The most common type is the single-decker, with larger loads carried by double-decker and articulated buses, and smaller loads carried by minibuses. A luxury, long-distance bus is usually called a coach
LCV	Light commercial vehicle; the formal term in the EU for goods vehicles with an MAM of up to 3.5 tonnes. In general language, this kind of vehicle is usually called a van
MAM	Maximum allowed mass; EU: European Union

Figure 1.5 Saloon car.

Figure 1.6 Estate car.

Figure 1.7 Hatchback.

Figure 1.8 Coupé.

Figure 1.9 Convertible.

Figure 1.10 Modified cars.

Figure 1.11 Light van.

Figure 1.12 Pick-up truck.

Figure 1.13 Sports utility vehicle (SUV).

Figure 1.14 Multi purpose vehicle (MPV).

Figure 1.15 Leisure vehicles.

Figure 1.16 Body components (clockwise from the left): 1, bonnet (hood); 2, windscreen; 3, roof; 4, tailgate; 5, post; 6, rear quarter; 7, sill; 8, door; 9, front wing; 10, front bumper/trim

Figure 1.17 Ford engine.

Most vehicles are made of a number of separate panels. Figure 1.16 shows a car with the main panel or other body component named.

1.1.5 Main systems

No matter how we categorize them, all vehicle designs (including HV) have similar major components and these operate in much the same way. The four main areas of a vehicle are the motor, electrical, chassis, and transmission systems.

1.1.5.1 Motor

This area consists of the motor itself. For ICE this includes fuel, ignition, air supply, and exhaust systems (Fig. 1.17). In the ICE, a fuel–air mixture enters through an inlet manifold and is fired in each cylinder in turn. The resulting expanding gases push on pistons and connecting rods which are on cranks, just like a cyclist's legs driving the pedals, and this makes a crankshaft rotate. The pulses of power from each piston are smoothed out by a heavy flywheel. Power leaves the engine through the flywheel, which is fitted on the end of the crankshaft, and passes to the clutch. The spent gases leave via the exhaust system.

1.1.5.2 Electrical

Both ICE and HV vehicles use a 12V electrical system to cover many aspects such as lighting, wipers and instrumentation. A key component for the ICE is the alternator (Fig. 1.18), which, driven by the engine, produces electricity to run the electrical systems and charge the battery. A starter motor takes energy from the battery to crank over and start the ICE. Electrical components are controlled by a range of switches. Electronic systems use sensors to sense conditions and actuators to control a variety of things – in fact, on modern vehicles, almost everything.

1.1.5.3 Chassis

This area is made up of the braking, steering and suspension systems as well as the wheels and tyres. Hydraulic pressure is used to activate the brakes to slow down or stop the vehicle. Rotating discs are gripped between pads of friction lining (Fig. 1.19). The handbrake uses a mechanical linkage to operate parking brakes. Both front wheels are linked mechanically and must turn together to provide steering control. The most common method is to use a rack and pinion. The steering wheel is linked to the pinion and as this is turned it moves the rack to and fro, which in turn moves the wheels. Tyres also absorb some road shock and play a very important part in road holding. Most of the remaining shocks and vibrations are absorbed by springs in the driver and passenger seats. The springs can be coil type and are used in conjunction with a damper to stop them oscillating (bouncing up and down too much).

Figure 1.18 A modern alternator (Source: Denso)

Figure 1.19 Disc brakes and part of the suspension system.

1.1.5.4 Transmission

In this area, the clutch allows the driver to disconnect drive from the engine and move the vehicle off from rest. The engine flywheel and clutch cover are bolted together so the clutch always rotates with the engine, and when the clutch pedal is raised drive is passed to the gearbox. A gearbox is needed because an engine produces power only when turning quite quickly. The gearbox allows the driver to keep the engine at its best speed. When the gearbox is in neutral, power does not leave it. A final drive assembly and differential connect the drive to the wheels via axles or driveshafts. The differential allows the driveshafts and hence the wheels to rotate at different speeds when the vehicle is cornering.

> **Key fact**
>
> Some modern cars may now appear to be 'automatic', but in fact have two manual gearboxes working together!

1.1.6 Summary

The layout of a vehicle, such as where the engine is fitted and which wheels are driven, varies, as do body styles and shapes. However, the technologies used in the four main areas of a vehicle are similar no matter how it is described. These are:

- the power system (traditional ICE or HV)
- the 12V electrical system
- the chassis system
- the transmission system.

Figure 1.20 Self-adjusting clutch inspection.

These areas are covered in detail and make up the four main technology chapters of this book, but first, let's look at the wider picture of the motor industry.

1.2 The motor industry

1.2.1 Introduction

This section will outline some of the jobs that are open to you in the motor trade and help you understand more about the different types of business and how they operate.

It is easy to think that the operation of a business does not matter to you. However, I would strongly suggest we should all be interested in the whole business in which we are working. This does not mean to interfere in areas we do not understand. It means we should understand that all parts of the business are important. For example, when you complete a job, enter all the parts used so the person who writes the invoice knows what to charge.

The motor trade offers lots of opportunities for those who are willing to work hard and move forwards. There are many different types of job and you will find one to suit you with a little patience and study. To help you get started, some of the words and phrases in common use are listed in Table 1.3.

1.2.2 Types of motor vehicle companies

Motor vehicle companies can range from the very small one-person business to very large main dealers (Figs 1.21–1.26). The systems used by each will be different but the requirements are the same.

A system should be in place to ensure the level of service provided by the company meets the needs

Table 1.3 Motor industry words and phrases

Customer	The individuals or companies that spend their money at your place of work. This is where your wages come from
Job card/Job sheet	A document for recording, among other things, work required, work done, parts used and the time taken. Also known as a job sheet, it can be a printed sheet or a digital solution
Invoice	A description of the parts and services supplied with a demand for payment from the customer
Company system	A set way in which things work in one particular company. Most motor vehicle company systems will follow similar rules, but will all be a little different
Estimate	An estimate is an educated guess whereas a quote or contract is legally binding – it is important to understand the difference
Image	This is the impression given by the company to existing and potential customers. Not all companies will want to project the same image and may have policies about their digital presence
Warranty	An intention that if within an agreed time a problem occurs with the supplied goods or service, it will be rectified free of charge by the supplier
Recording system	An agreed system within a company so that all details of what is requested and/or carried out are recorded. The job card is one of the main parts of this system
Accreditation	Official recognition of status, for example 'Techsafe' is the official accreditation from the IMI to work on HV systems
Approved repairer	This can normally mean two things. The first is where a particular garage or bodyshop is used by an insurance company to carry out accident repair work. In some cases, however, general repair shops may be approved to carry out warranty work or servicing work by a particular vehicle or component manufacturer
After sales	This is a general term that applies to all aspects of a main dealer that are involved with looking after a customer's car, after it has been sold to them by the sales department. The service and repair workshop is the best example

Figure 1.21 Main dealer (Source: harry_nl, Flickr)

of the customer. Table 1.4 shows how diverse the trade is.

1.2.3 Company structure

A large motor vehicle company will probably be made up of at least the following departments:

- reception
- workshop
- bodyshop and paint shop
- parts department
- tyre bay
- MOT bay
- valeting
- new and second user car sales
- office support
- management
- cleaning and general duties.

Figure 1.22 Well-known chain independent workshop (Source: Halfords)

Figure 1.23 An independent garage and car sales business (Source: Cleevely EV)

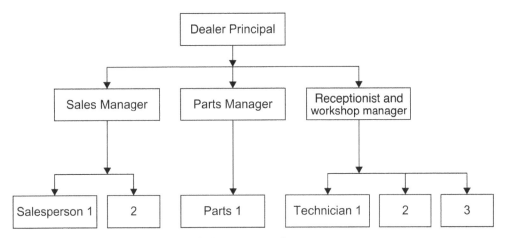

```
                        ┌─────────────────┐
                        │ Dealer Principal │
                        └─────────────────┘
                                 │
        ┌────────────────────────┼────────────────────────┐
        ▼                        ▼                        ▼
┌───────────────┐        ┌───────────────┐        ┌─────────────────────┐
│ Sales Manager │        │ Parts Manager │        │ Receptionist and     │
│               │        │               │        │ workshop manager     │
└───────────────┘        └───────────────┘        └─────────────────────┘
        │                        │                        │
   ┌────┴────┐                   │              ┌─────────┼─────────┐
   ▼         ▼                   ▼              ▼         ▼         ▼
┌──────────┐┌────┐       ┌───────────┐   ┌────────────┐┌────┐┌────┐
│Salesperson││ 2  │       │  Parts 1  │   │Technician 1││ 2  ││ 3  │
│    1     ││    │       │           │   │            ││    ││    │
└──────────┘└────┘       └───────────┘   └────────────┘└────┘└────┘
```

Figure 1.24 One way in which a company could be structured

Figure 1.25 A highstreet motor factor (Source: GSF)

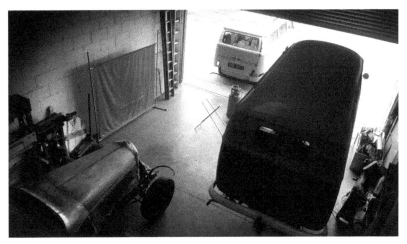

Figure 1.26 Small specialist companies can operate out of tiny premises

Table 1.4 Automotive trades

Mobile mechanics	Servicing and repairs at the owner's home or business. Usually a one-person company
Bodywork repairers and painters	Specialists in body repair and paintwork, a type of repair in this part of the sector can be termed small, medium area repair technology (SMART) and can be performed as a mobile service
Valeters	These companies specialize in valeting, which should be thought of as much more involved than getting the car washed. Specialist equipment and products are used and proper training is essential
Fuel stations	These may be owned by an oil company or be independent. Some also do vehicle repair work
Specialized repairers	Auto-electrical, air conditioning, automatic transmission and in-car entertainment systems are just some examples, but these repairers could also be independents specializing in a particular marque
General repair workshop or independent repairer	Servicing and repairs of most types of vehicles not linked to a specific manufacturer. Often this will be a small business, maybe employing two or three people. However, there are some very large independent repairers
Parts supply	Many companies now supply a wide range of parts. Many will deliver to the workshop
Fast-fit	Supplying and fitting of exhausts, tyres, radiators, batteries, clutches, brakes and windscreens
Fleet operator (with workshop)	Many large operators such as rental companies will operate their own workshops. A large company that has lots of cars, used by sales reps for example, may also have their own workshop and technicians
Non-franchised dealer	Main activity is the servicing and repairs of a wide range of vehicles, with some sales
Main dealer or franchised dealer	Usually franchised to one manufacturer, these companies hold a stock of vehicles and parts. The main dealer will be able to carry out all repairs to their own type of vehicle as they hold all of the parts and special tools. They also have access to the latest information specific to their franchise (e.g. Ford or Citroën). A 'franchise' means that the company has had to pay to become associated with a particular manufacturer but is then guaranteed a certain amount of work and that there will be no other similar dealers within a certain distance
Multi-franchised dealer	This type of dealer is just like the one above, except they hold more than one franchise (e.g. Volvo and Saab)
Breakdown services	The best known breakdown services in the UK are operated by the AA and the RAC. Others, e.g. National Breakdown and many independent garages, also offer roadside repair and recovery services
Motorists' shops	Often described as motorist discount centres or similar, these companies provide parts and materials to amateurs but in some cases also offer a service to those in the trade

Each area will employ one or a number of people. If you work in a very small garage you may have to be all of these people at once. In a large garage it is important that these different areas communicate with each other to ensure that a good service is provided to the customer. The main departments are explained further in the following sections.

> **Key fact**
>
> In a large organization it is important that different areas communicate with each other.

1.2.4 **Role of a franchised dealer**

The role of a franchised dealer (one contracted to a manufacturer) is to supply local:

- new and used franchised vehicles
- franchise parts and accessories
- repair and servicing facilities for franchise vehicles.

The dealer will also liaise with the vehicle manufacturer. This can include managing car recalls and manufacturer warranty claims.

1.2.5 **Reception and booking systems**

The reception, whether in a large or small company, is often the point of first contact with customers. It is very important therefore to get this bit right. The reception should be staffed by pleasant and qualified people. The purpose of a reception and booking system within a company can be best explained by following through a typical enquiry.

1 The customer enters reception area and is greeted in an appropriate way.
2 Attention is given to the customer to find out what is required. (Let's assume the car is difficult to start, in this case.)
3 Further questions can be used to determine the particular problem, bearing in mind the knowledge of vehicles the customer may, or may not have. (For example, is the problem worse when the weather is cold?)
4 Details are recorded securely about the customer, the vehicle and the nature of the problem. If the customer is new a record card

can be started, or one can be continued for an existing customer.

5 An explanation of expected costs is given as appropriate. An agreement to spend only a set amount, after which the customer will be contacted, is a common approach.

6 The date and time when the work will be carried out can now be agreed. This depends on workshop time availability and when is convenient for the customer. Managing expectation by giving realistic timescales is important for all involved.

7 The customer is thanked for visiting. If the vehicle is to be left at that time, the keys should be labelled and stored securely.

8 Follow-up for the customer can also be carried out by those in reception, such as workshop updates or completion of work.

More organizations are now automating more of these processes, such as allowing customers to book themselves in using a website. However there is likely to be a need for this to be a human role for some time!

1.2.6 Parts department

The parts department is the area where parts are kept and/or ordered. This will vary quite a lot between different companies. Large main dealers will have a very large stock of parts for their range of vehicles. They will have a parts manager and in some cases several other staff. In some very small garages the parts department will be a few shelves where popular items such as filters and brake pads are kept.

Even though the two examples given above are rather different in scale the basic principles are the same and can be summed up very briefly as follows:

▷ A set level of parts or stock is decided upon.
▷ Parts are stored so they can be easily found.
▷ A reordering system should be used to maintain the stock.

Security is important as most parts cost a lot of money. When parts are collected from the parts department or area, they will be used in one of three ways:

▷ for direct sale to a customer
▷ as part of a job
▷ for use on company vehicles.

In the first case, an invoice or a bill will be produced. In the second case, the parts will be entered on the customer's job record. The third case may also have a job card; if not, some other record must be kept. In all three cases keeping a record of parts used will allow them to be reordered if necessary. If parts are ordered and delivered by an external supplier, again they must be recorded.

> **Key fact**
> Security is important as most parts cost a lot of money.

1.2.7 Estimating costs and times

When a customer brings their car to a garage for work to be carried out, quite understandably he or she will want to know two things:

▷ How much will it cost?
▷ When will the car be ready?

In some cases, such as for a full service, this is quite easy as the company will have a set charge and by experience will know it takes a set time. For other types of job this is more difficult.

Most major manufacturers supply information to their dealers about standard times for jobs. These assume a skilled technician with all the necessary tools. For independent garages other publications are available. These give agreed standard times for all the most common tasks, on all popular makes of vehicle. To work out the cost of a job, you look up the required time and multiply it by the company's hourly rate. Don't forget that the cost of parts will also need to be included.

1.2.8 Job cards and systems

The job card (Fig. 1.27) is a vital part of the workshop system in a motor vehicle company. Many companies now dispense with the 'paper' altogether and use computer systems. These allow very fast, easy and accurate communication. Whether handwritten job cards or IT systems are used, the principle is the same and consists of a number of important stages. This is often described as the four-part job card system:

▷ **Reception** – Customers' details and requirements are entered on the job card or computer screen.
▷ **Workshop control** – Jobs are allocated to the appropriate technician using a loading sheet or again via the computer.
▷ **Parts department** – Parts used are added to the computer or job card.
▷ **Accounts** – Invoices are prepared from the information on the job card. Digital systems may automatically produce the invoice when the job is completed.

When a digital system is used each terminal will pass information to all the others. With job cards either the same card must be carried to each stage or copies are kept in each area. The different copies are collected and combined to produce the invoice.

11

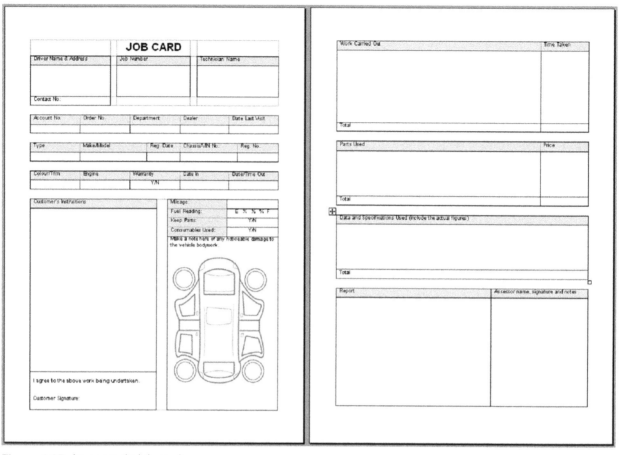

Figure 1.27 An example job card

1.2.9 Invoicing

As part of the contract made with a customer, an invoice for the work carried out is issued. The main parts of an invoice are as follows:

▶ **Labour charges** – The cost of doing the work; usually the time spent times the hourly rate.
▶ **Parts** – The retail price of the parts or as agreed.
▶ **Sundries** – Some companies add a small sundry charge to cover consumable items like nuts and bolts, cable ties.
▶ **Data charge** – Some workshops add the cost of accessing manufacturer's data.
▶ **MOT (Ministry of Transport) test** – If appropriate. This is separated because VAT is not charged on MOTs.
▶ **VAT (value added tax)** – Is charged at the current rate, if the company is registered.

> ### Definition
> **Contract:** In law, a contract is a binding legal agreement that is enforceable in a court of law or by binding arbitration. That is to say, a contract is an exchange of promises with a specific remedy if broken.

Hourly rates vary quite a lot between different garages. The hourly rate charged by the company has to pay for a lot more than your wages – hence it will be much higher than your hourly rate! Just take a look round in any good workshop: as well as the rent for the premises, some of the equipment can cost tens of thousands of pounds. The money has to come from somewhere.

1.2.10 Warranties

When a vehicle is sold a warranty is given, meaning that it is fit for the purpose for which it was sold. Further to this, the manufacturer will repair the vehicle at no cost to the customer if a problem develops within a set time. For most vehicles this is twelve months, but some periods are now longer. The term generally used for this is 'guarantee'. Quite often manufacturers advertise their guarantee as a selling point. Different lengths of time might apply for different parts of the repair.

> ### Key fact
> A warranty means that a vehicle is fit for the purpose for which it was sold.

It is also possible to have a warranty on a used vehicle or an extended warranty on a new vehicle. These often involve a separate payment to an insurance company. This type of warranty can be quite good but a number of exclusions and requirements may apply. Some examples are listed:

- Regular servicing must be carried out by an approved repairer.
- Only recommended parts must be used.
- Wear and tear is not included.
- Any work done must be authorized.
- Only recognized repairers may be used in some cases.

The question of authorization before work is carried out is very important for the garage to understand. Work carried out without proper authorization will not be paid for. If a customer returns a car within the warranty period then a set procedure must be followed:

1 Confirm that the work is within the terms of the warranty.
2 Get authorization if over an agreed limit.
3 Retain all parts replaced for inspection.
4 Produce an invoice which relates to standard or agreed times.

Often in the larger garages one person will be responsible for making warranty claims.

1.2.11 Digital workshop system

A managed workflow is a vital part of the workshop system in a motor vehicle company. Many companies may dispense with paper altogether and use digital systems. These can be expensive but allow fast, easy and accurate communication. Whether job cards, digital systems or even a combination are used, the principle is the same and consists of several important stages. This is often described as a four-part job card system:

1 **Reception** – Customers' details and requirements are collected and/or checked; this information could be stored in a paper or digital system or even both.
2 **Workshop control** – Jobs are allocated to the appropriate technician using a loading sheet or again via a digital solution.
3 **Parts department** – Parts used are added to the workshop information.
4 **Accounts** – Invoices are prepared from the information on the job card. Digital systems may automatically produce the invoice when the job is completed.

There are several digital based workshop management systems available. Some are specifically designed

for main dealers, some for the smaller independent company. In this part, I will outline a system called Garage Hive (www.garagehive.co.uk). This system has three specific strands:

- workshop management
- reporting and business intelligence
- finance and accountancy.

The workshop management built into Garage Hive has all the tools to run your workshop. A live dynamic work schedule, paperless jobcards, integrated Autodata repair times, technician time management, stock control and much more. You can also make informed decisions with the business intelligence and reporting application. View workshop profitability, technician efficiencies, margins and much more on all of your devices. The system offers a fully integrated accountancy system. Alternatively you can take advantage of the integrations with Xero, Sage and Quickbooks.

1.3 Working safely

1.3.1 Introduction

Health and safety law is designed to protect you.

In the UK the Health and Safety Executive (HSE) is the enforcement and legislative body set up by the government. The HSE has a very helpful website where you can get all the latest information, including a document specially developed for the motor industry. The address is: www.hse.gov.uk.

Similarly, in the USA, with the Occupational Safety and Health Act of 1970, Congress created the Occupational Safety and Health Administration (OSHA) to ensure safe and healthful working conditions for working men and women by setting and enforcing standards and by providing training, outreach,

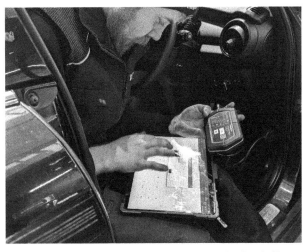

Figure 1.28 Creating a new vehicle health check file

13

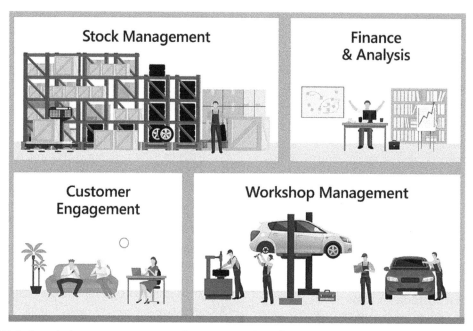

Figure 1.29 Digital workshop systems can perform many roles (Source: Garage Hive)

Figure 1.30 UK Health and Safety Executive (HSE) logo

Figure 1.31 USA OSHA logo

Safe Work Australia represents a genuine partnership between governments, unions and industry. Together they work towards the goal of reducing death, injury and disease in the workplace. Their website contains some useful content at: www.ascc.gov.au or http://safeworkaustralia.gov.au.

Now back to the UK's HSE. The emphasis is on preventing death, injury and ill-health in Britain's workplaces. However, the HSE does have the authority to come down hard on people who put others at risk, particularly where there is deliberate flouting of the law. Since 2009, HSE has published new versions of its approved health and safety poster and leaflet. The new versions are modern, eye-catching and easy to read. They are set out in simple terms, using numbered lists of basic points, and what employers and workers must do, and tell you what to do if there is a problem.

education and assistance. The OSHA has a good website at: www.osha.gov.

In Australia there is the office of the Australian Safety and Compensation Council (ASCC), in the Department of Employment and Workplace Relations. Safe Work Australia is an Australian Government statutory agency established in 2009, with the primary responsibility of improving work health and safety and workers' compensation arrangements across Australia. The agency is jointly funded by the Commonwealth, state and territory governments facilitated through an intergovernmental agreement signed in July 2008.

Safety first
Systems are quite similar, but always check the legislation and law in the country you are working in.

Figure 1.32 Safe Work Australia (SWA) logo

From 5 April 2014, employers must use the 2009 HSE poster and leaflet and ensure the contacts are up to date. Employers must display a poster or give a leaflet to workers. This is in the form of a pocket card that is better suited to the workplace.

Employers have a legal duty under the Health and Safety Information for Employees Regulations (HSIER) to display the poster in a prominent position in each workplace or provide each worker with a copy of the equivalent leaflet outlining British health and safety laws (Figs 1.33 and 1.34).

Employers must meet certain criteria, but health and safety is the responsibility of everyone in the workplace. The reason for the poster and the leaflets is to make everybody aware of this.

Figure 1.33 Health and Safety Law poster. (Source: HSE)

HSE

Health and Safety Executive

Health and Safety Law
What you need to know
Large print

All workers have a right to work in places where risks to their health and safety are properly controlled. Health and safety is about stopping you getting hurt at work or ill through work. Your employer is responsible for health and safety, but you must help.

What employers must do for you

1 Decide what could harm you in your job and the precautions to stop it. This is part of risk assessment.

2 In a way you can understand, explain how risks will be controlled and tell you who is responsible for this.

3 Consult and work with you and your health and safety representatives in protecting everyone from harm in the workplace.

4 Free of charge, give you the health and safety training you need to do your job.

5 Free of charge, provide you with any equipment and protective clothing you need, and ensure it is properly looked after.

Figure 1.34 Health and Safety Law leaflet (page 1). (Source: HSE)

Figure 1.35 Exhaust extraction isn't needed for this car!

Safety first

Health and safety is the responsibility of everyone in the workplace.

1.3.2 The key UK regulations and laws

There are a number of rules and regulations you need to be aware of. Check the details for the country in which you work. Table 1.5 lists some important areas for the UK.

1.3.3 Health and safety law: what you need to know

All workers have a right to work in places where risks to their health and safety are properly controlled (Fig. 1.34). Health and safety is about stopping you getting hurt at work or ill through work. Your employer is responsible for health and safety, but you must help. This section is taken from the HSE leaflet for employees.

What employers must do for you:

1 Decide what could harm you in your job and the precautions to stop it. This is part of risk assessment.

Figure 1.36 Use appropriate PPE for the job

2 In a way you can understand, explain how risks will be controlled and tell you who is responsible for this.

3 Consult and work with you and your health and safety representatives in protecting everyone from harm in the workplace.

4 Free of charge, give you the health and safety training you need to do your job.

5 Free of charge, provide you with any equipment and protective clothing you need, and ensure it is properly looked after.

6 Provide toilets, washing facilities and drinking water.

7 Provide adequate first-aid facilities.

8 Report injuries, diseases and dangerous incidents at work.

9 Have insurance that covers you in case you get hurt at work or ill through work. Display a hard copy or electronic copy of the current insurance certificate where you can easily read it.

10 Work with any other employers or contractors sharing the workplace or providing employees (such as agency workers), so that everyone's health and safety is protected.

What you must do:

1 Follow the training you have received when using any work items your employer has given you.

2 Take reasonable care of your own and other people's health and safety.

3 Co-operate with your employer on health and safety.

4 Tell someone (your employer, supervisor or health and safety representative) if you think the work or inadequate precautions are putting anyone's health and safety at serious risk.

17

Table 1.5 UK regulations and laws

Health and Safety Executive (HSE)	The HSE is the national independent watchdog for work-related health, safety and illness in the UK. It is an independent regulator and acts in the public interest to reduce work-related death and serious injury across Great Britain's workplaces. Other countries have similar organizations
Health and Safety at Work etc. Act 1974 (HASAW)	HASAW is the primary piece of legislation covering occupational health and safety in the UK. The HSE is responsible for enforcing the Act and a number of other Acts and Statutory Instruments relevant to the working environment
Control of Substances Hazardous to Health (COSHH)	This law requires employers to control substances that are hazardous to health, such as solvents
Reporting of Injuries, Diseases and Dangerous Occurrences Regulations 1995 (RIDDOR)	RIDDOR place a legal duty on employers, self-employed people and people in control of premises to report work-related deaths, major injuries or over-three-day injuries, work-related diseases and dangerous occurrences (near-miss accidents)
Provision and Use of Work Equipment Regulations 1998 (PUWER)	In general terms, PUWER requires that equipment provided for use at work is: • suitable for the intended use • safe for use, maintained in a safe condition and, in certain circumstances, inspected to ensure this remains the case • used only by people who have received adequate information, instruction and training • accompanied by suitable safety measures, e.g. protective devices, markings, warnings
Lifting Operations and Lifting Equipment Regulations 1998 (LOLER)	In general, LOLER requires that any lifting equipment used at work for lifting or lowering loads is: • strong and stable enough for particular use and marked to indicate safe working loads • positioned and installed to minimize any risks • used safely, i.e. the work is planned, organized and performed by competent people • subject to ongoing thorough examination and, where appropriate, inspection by competent people
Health and safety audit	Monitoring provides the information to let you or your employer review activities and decide how to improve performance. Audits, by company staff or outsiders, complement monitoring activities by looking to see whether your company policy, organization and systems are actually achieving the right results
Risk management and assessment	A risk assessment is simply a careful examination of what, in your work, could cause harm to people. This is done so that you and your company can decide whether you have taken enough precautions or should do more to prevent harm. Workers and others have a right to be protected from harm caused by a failure to take reasonable control measures. It is a legal requirement to assess the risks in the workplace, so you or your employer must put plans in place to control risks. How to assess the risks in your workplace: • Identify the hazards. • Decide who might be harmed and how. • Evaluate the risks and decide on precautions. • Record your findings and implement them. • Review your assessment and update if necessary.
Personal protective equipment (PPE)	PPE is defined in the Regulations as 'all equipment (including clothing affording protection against the weather) which is intended to be worn or held by a person at work and which protects him against one or more risks to his health or safety', e.g. safety helmets, gloves, eye protection, high-visibility clothing, safety footwear and safety harnesses. Hearing protection and respiratory protective equipment provided for most work situations are not covered by these Regulations because other regulations apply to them. However, these items need to be compatible with any other PPE provided

1.3.4 Personal protective equipment (PPE)

Personal protective equipment (PPE) such as safety clothing is very important to protect you (Figs 1.36 and 1.37). Some people think it clever or tough not to use protection. They are very sad and will die or be injured long before you! Some things are obvious; for example, when holding a hot or sharp exhaust you would likely be burnt or cut. Other things such as breathing in brake dust, or working in a noisy area, do

Figure 1.37 Protective clothing for welding

not produce immediately noticeable effects but could affect you later in life.

Safety first

Some people think it clever or tough not to use protection. They are very sad and will die or be injured long before you!

Safety first

Breathing in brake dust, or working in a noisy area, does not produce immediately noticeable effects but could affect you later in life.

Fortunately, the risks to workers are now quite well understood and we can protect ourselves before it is too late.

Table 1.6 lists a number of items classed as PPE together with suggested uses. You will see that the use of most items involves plain common sense.

1.3.5 Identifying and reducing hazards

Working in a motor vehicle workshop is a dangerous occupation, if you do not take care. The most important thing is to be aware of the hazards and then it is easy to avoid the danger. The hazards in a workshop are from two particular sources: you and your surroundings.

Hazards due to you may be caused by:

- carelessness – particularly while moving vehicles
- intoxication – badly affects your ability to react to dangerous situations
- tiredness or sickness – affects your ability to think and work safely
- messing about – most accidents are caused by people fooling about
- not using safety equipment – you have a duty to yourself and others to use safety equipment
- inexperience – or lack of supervision: if in doubt – ask.

The surroundings in which you work may have:

- bad ventilation
- poor lighting
- noise

Table 1.6 Examples of personal protective equipment (PPE)

Equipment	Notes	Suggested or examples where used
Ear defenders	Must meet appropriate standards	When working in noisy areas
Face mask	For individual personal use only	In dusty conditions. When cleaning brakes or preparing bodywork
High-visibility clothing	Fluorescent colours such as yellow or orange	Working in traffic, such as when on a breakdown
Leather apron	Should be replaced if it is holed or worn thin	When welding or working with very hot items
Leather gloves	Should be replaced when they become holed or worn thin	When welding or working with very hot items and also if handling sharp metalwork
Overalls/workshop uniform	Should be kept clean and be flameproof if used for welding	These should be worn at all times to protect your clothes and skin.
Rubber or plastic apron	Replace if holed	Use if you do a lot of work with battery acid or with strong solvents
Rubber or plastic gloves	Replace if holed	Gloves must always be worn when using degreasing equipment. There are special gloves for HV work
Safety shoes or boots	Strong toe caps are recommended	When working in any workshop with heavy equipment
Safety goggles	Keep the lenses clean and prevent scratches	Always use goggles when grinding or when at any risk of eye contamination. Cheap plastic goggles are much easier to come by than new eyes
Safety helmet	Must be to current standards	When working under a vehicle, in some cases
Welding goggles or welding mask	Check the goggles are suitable for the type of welding. Gas welding goggles are NOT good enough when arc welding	Wear welding goggles or use a mask even if you are only assisting by holding something

SAFETY DATA SHEET (REGULATION (EC) n° 1907/2006 - REACH) SPECIFIC 504 00 507 00 5W30 DE - 59110-1	Version 1.1 (27-05-2019) - Page 1/7

SAFETY DATA SHEET

MOTUL

(REACH regulation (EC) n° 1907/2006 - n° 2015/830)

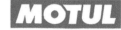

SECTION 1 : IDENTIFICATION OF THE SUBSTANCE/MIXTURE AND OF THE COMPANY/UNDERTAKING

1.1. Product identifier

Product name : SPECIFIC 504 00 507 00 5W30 DE
Product code : 59110-1

1.2. Relevant identified uses of the substance or mixture and uses advised against

4-stroke engine lubricant

1.3. Details of the supplier of the safety data sheet

Registered company name : MOTUL
Address : 119, Boulevard Felix Faure. 93300 AUBERVILLIERS CEDEX FRANCE
Telephone : 33.1.48.11.70.00. Fax: 33.1.48.33.28.79. Telex: .
Email : motul_hse@motul.fr

1.4. Emergency telephone number : +44 (0) 1235 239 670.

Association/Organisation : ORFILA.

Other emergency numbers

BRAZIL : +55 11 3197 5891 / COLOMBIA : +57 1 508 7337 / ARGENTINA : +54 11 5984 3690 / CHILE : +562 2582 9336
Ireland : +353 1 8092566
UNITED STATES: 001 866 928 0789 / CANADA: 001 800 579 7421 / MEXICO : +52 55 5004 8763 / MIDDLE EAST - AFRICA : +44 1235 239671
24 hours a day, 7 days a week

SECTION 2 : HAZARDS IDENTIFICATION

2.1. Classification of the substance or mixture

In compliance with EC regulation No. 1272/2008 and its amendments.

This mixture does not present a physical hazard. Refer to the recommendations regarding the other products present on the site.
This mixture does not present a health hazard with the exception of possible occupational exposure thresholds (see paragraphs 3 and 8).
This mixture does not present an environmental hazard. No known or foreseeable environmental damage under standard conditions of use.

2.2. Label elements

In compliance with EC regulation No. 1272/2008 and its amendments.

No labelling requirements for this mixture.

2.3. Other hazards

The mixture does not contain substances classified as 'Substances of Very High Concern' (SVHC) >= 0.1% published by the European CHemicals Agency (ECHA) under article 57 of REACH: http://echa.europa.eu/fr/candidate-list-table
The mixture fulfils neither the PBT nor the vPvB criteria for mixtures in accordance with annexe XIII of the REACH regulations EC 1907/2006.

SECTION 3 : COMPOSITION/INFORMATION ON INGREDIENTS

3.2. Mixtures

Composition :

Identification	(EC) 1272/2008	Note	%
CAS: 64742-54-7 EC: 265-157-1 DISTILLATES (PETROLEUM), HYDROTREATED HEAVY PARAFFINIC	GHS08 Dgr Asp. Tox. 1, H304	L	25 <= x % < 50
CAS: 64742-54-7 EC: 265-157-1 DISTILLATES (PETROLEUM), HYDROTREATED HEAVY PARAFFINIC		L	10 <= x % < 25
CAS: 64742-54-7 EC: 265-157-1 REACH: 01-2119484627-25 DISTILLATES (PETROLEUM),	GHS08 Dgr Asp. Tox. 1, H304		10 <= x % < 25

Made under licence of European Label System, Software of INFODYNE (http://www.infodyne.fr)

Figure 1.38 Example of a safety data sheet (Source: Motul)

Figure 1.39 Using an engine crane.

- dangerous substances stored incorrectly
- broken or worn tools and equipment
- faulty machinery
- slippery floors
- untidy benches and floors
- unguarded machinery
- unguarded pits.

Table 1.7 lists some of the hazards you will come across in a vehicle workshop. Also listed are some associated risks, together with ways they can be reduced. This is called risk management. An example of a safety data sheet is given in Figure 1.38.

1.3.6 Moving loads

Injuries in a workshop are often due to incorrect lifting or moving of heavy loads. In motor vehicle workshops, heavy and large components, such as engines and gearboxes, can cause injury when being removed and refitted. A few simple precautions will prevent you from injuring yourself, or others.

- Never try to lift anything beyond your capability – get a mate to help. The amount you can safely lift will vary but any more than you feel comfortable with, you should get help.
- Whenever possible use an engine crane, a transmission jack or a trolley jack (Fig. 1.39).
- Lift correctly, using the legs and keeping your back straight.
- When moving heavy loads on a trolley, get help and position yourself so you will not be run over if you lose control.

The ideal option in all cases is to avoid manual handling where possible.

Safety first

Use lifting equipment where there is risk of injury when moving loads manually.

1.3.7 Vehicle safety

Vehicle safety and the associated regulations can be very complicated. However, for our purposes we can consider the issue across two main areas: construction and operation of the vehicle.

1.3.7.1 Construction of the vehicle

Before a vehicle can be constructed a prototype has to be submitted for type approval. When awarded this means the vehicle has passed very stringent tests and that it meets all current safety requirements. Different countries have different systems, which means some modifications to a car may be necessary if it is imported or exported. The European Union (EU) has published many directives that each member country must incorporate into its own legislation. This has helped to standardize many aspects. In the UK the Road Vehicles (Construction & Use) Regulations 1986 is the act that ensures certain standards are met. If you become involved in modifying a vehicle, e.g. for import, you may need to refer to the details of this act. Other countries have similar legislation.

Many other laws exist relating to the motor vehicle and the environment. These are about emissions and pollution. Environmental laws change quite often and it is important to keep up to date.

The Department of Transport states that all vehicles over three years old must undergo a safety check that ensures the vehicle continues to meet the current legislation. First set up by the Ministry of Transport, it continues to be known as the MOT test. This test now includes checks relating to environmental laws.

1.3.7.2 Driving and operating the vehicle

To drive a vehicle on the road you must have an appropriate driving licence and insurance, and the vehicle must be taxed and must be in safe working order.

1.3.8 Safety procedures

When you know the set procedures to be followed, it is easier to look after yourself, your workshop and your workmates. You should know:

21

Table 1.7 Hazards and actions to reduce risk

Hazard	Risks	Action
Power tools	Damage to the vehicle or personal injury	Understand how to use the equipment and wear suitable protective clothing, e.g. gloves and goggles
Working under a car on the ramp	1 The vehicle could roll or be driven off the end 2 You could bang your head on hard or sharp objects when working under the car	1 Ensure you use wheel chocks 2 Set the ramp at the best working height; wear protection if appropriate
Working under a car on a jack	The vehicle could fall on top of you	The correct axle stands should be used and positioned in a secure place
Compressed air	Damage to sensitive organs such as ears or eyes. Death, if air is forced through the skin into your bloodstream	Do not fool around with compressed air. A safety nozzle prevents excessive air forces
Dirty hands and skin	Oil, fuel and other contaminants can cause serious health problems, ranging from dermatitis to skin cancer	Use gloves or a good quality barrier cream and wash your hands regularly. Do not allow dirt to transfer to other parts of your body. Good overalls should be worn at all times
Exhaust fumes	Poisonous gases such as carbon monoxide can kill. The other gases can restrict breathing and cause sore throats, and can cause cancer	Only allow running engines in very well-ventilated areas or use an exhaust extraction system
Engine crane	Injury or damage can be caused if the engine swings and falls off	Ensure the crane is strong enough: do not exceed its safe working load (SWL). Secure the engine with good quality sling straps and keep the engine near to the floor when moving across the workshop
Cleaning brakes	Brake dust is dangerous to health	Only clean with proper brake cleaner
Fuel	Fire or explosion	Keep all fuels away from sources of ignition. Do not smoke when working on a vehicle
Degreaser solvent	Damage to skin or to sensitive components	Wear proper gloves and make sure the solvent will not affect the items you are washing
Spillage such as oil	Easy to slip over or fall and be injured	Clean up spills as they happen and use absorbent granules
Battery electrolyte	Dangerous to your skin and in particular your eyes. It will also rot your clothes	Wear protective clothing and take extreme care
Welding a vehicle	The obvious risks are burns, fire and heat damage, but electric welders such as a MIG welder can damage sensitive electronic systems. Fire watch should continue for at least 30 minutes after welding is completed	Have fire extinguishers handy, remove combustible materials such as carpets and ensure fuel pipes are nowhere near. The battery earth lead must also be disconnected. Wear gloves and suitable protective clothing such as a leather jacket
Electric hand tools	The same risk as for power tools, but also the danger of electric shock, particularly in damp or wet conditions. This can be fatal	Do not use electric tools when damp or wet. Electrical equipment should be inspected regularly by a competent person
Driving over a pit	Driving into the pit	The pit should be covered, or you should have another person help guide you and drive very slowly
Broken tools	Personal injury or damage to the car. For example, a file without a handle could stab into your wrist or a faulty ratchet could slip	All tools should be kept in good order at all times. This will also make the work easier
Cleaning fluids	Skin damage or eye damage	Wear gloves and eye protection and also be aware of exactly what precautions are needed by referring to the safety data information. (Figure 1.38 shows an example)

- who does what during an emergency
- the fire procedure for your workplace
- about different types of fire extinguisher and their uses
- the procedure for reporting an accident.

If an accident does occur in your workplace the first bit of advice is: keep calm and don't panic! The HASAW states that for companies above a certain size:

- first aid equipment must be available
- employers should display simple first aid instructions
- fully trained first aiders must be employed.

In your own workplace you should know about the above three points. A guide to how to react if you come across a serious accident is given in Table 1.8.

Figure 1.40 Fire triangle. (Source: Wikimedia)

Safety first

Keep calm and don't panic!

1.3.9 Fire

Accidents involving fire are very serious. As well as you or a workmate calling the fire brigade (do not assume it has been done), three simple rules will help you to know what to do:

1 Get safe yourself, contact the emergency services – and shout FIRE!
2 Help others to get to safety if it does not put you or others at risk.
3 Wait somewhere safe nearby, inform the fire service of anyone trapped inside.

Of course, far better than the above situation is not to let a fire start in the first place.

Safety first

A fire is prevented or extinguished by removing any side of the fire triangle.

The fire triangle or combustion triangle is a simple model for understanding the ingredients necessary for most fires (Fig. 1.40). The triangle illustrates that a fire requires three elements: heat, fuel and an oxidizing agent (usually oxygen from the air). The fire is prevented or extinguished by removing any one of them. A fire naturally occurs when the elements are combined in the right mixture.

Without sufficient heat, a fire cannot start or continue. Heat can be removed by the application of a substance that reduces the amount of heat available to the fire reaction. This is often water, which requires heat to change from water to steam. Introducing sufficient quantities and types of powder or gas in the flame also reduces the amount of heat available for the fire reaction. Turning off the electricity in an electrical fire removes the ignition source.

Without fuel, a fire will stop. Fuel can be removed naturally, as where the fire has consumed all the burnable fuel, or manually, by mechanically or chemically removing the fuel from the fire. The fire goes out because a lower concentration of fuel vapour

Table 1.8 Actions in the case of an accident

Action	Notes
Assess the situation	Stay calm: a few seconds to think is important
Remove the danger	If the person was working with a machine, turn it off. If someone is electrocuted, switch off the power or use a rescue pole. Even if you are unable to help with the injury you can stop it getting worse
Get help	If you are not trained in first aid, get someone who is and phone for an ambulance
Stay with the casualty	If you can do nothing else, the casualty can be helped if you stay. Say that help is on the way and be ready to assist. You may need to guide the ambulance
Report the accident	All accidents must be reported by law (see RIDDOR). This is a record so that steps can be taken to prevent the accident happening again. Also, if the injured person claims compensation, underhanded companies could deny the accident happened
Learn first aid	First Aid is a useful skill that anyone can learn. Training providers can be found on the HSE website

Figure 1.41 Fire tetrahedron. (Source: Wikimedia)

in the flame leads to a decrease in energy release and a lower temperature. Removing the fuel therefore decreases the heat.

Without enough oxygen, a fire cannot start or continue. With a decreased oxygen concentration, the combustion process slows. In most cases, there is plenty of air left when the fire goes out, so this is commonly not a major factor.

The fire tetrahedron is an addition to the fire triangle (Fig. 1.41). It adds the requirement for the presence of the chemical reaction which is the process of fire. For example, the suppression effect of a Halon extinguisher is due to its interference in the fire chemical inhibition. Note that Halon extinguishers are only now allowed in certain situations and are illegal for normal use.

Combustion is the chemical reaction that feeds a fire more heat and allows it to continue. When the fire involves burning metals like magnesium (known as a class D fire), it becomes even more important to consider the energy release. The metals react faster with water than with oxygen and thereby more energy is released. Putting water on such a fire makes it worse. Carbon dioxide (CO_2) extinguishers are ineffective against certain metals such as titanium. Therefore, inert agents (e.g. dry sand) must be used to break the chain reaction of metallic combustion. In the same way, as soon as we remove

one out of the three elements of the triangle, the fire stops.

If a fire does happen your workplace should have a set procedure so you will know:

▶ how the alarm is raised
▶ what the alarm sounds like
▶ what to do when you hear the alarm
▶ your escape route from the building
▶ where to go to assemble
▶ who is responsible for calling the fire brigade.

There are a number of different types of fire, as shown in Table 1.9.

If it is safe to do so you should try to put out a small fire. Extinguishers and a fire blanket should be provided. Remember, if you remove one side of the fire triangle, the fire will go out. If you put enough water on a fire it will cool down and go out. However, spraying water on an electrical circuit could kill you. Spraying water on a petroleum fire could spread it about and make the problem far worse. This means that a number of different fire extinguishers are needed. Internationally there are several accepted classification methods for hand-held fire extinguishers. Each classification is useful in fighting fires with a particular group of fuel. Tables 1.10–1.12 show the differences: study the one that relates to your situation.

Safety first

Study the table on fire extinguishers that relates to your situation.

In Australia (Table 1.10), yellow (Halon) fire extinguishers are illegal to own or use on a fire, unless an essential use exemption has been granted.

There is no official standard in the USA for the colour of fire extinguishers, though they are typically red, except for class D extinguishers, which are usually yellow, and water extinguishers, which are usually silver, or, if water mist types, white (Table 1.11). Extinguishers are marked with pictograms depicting

Table 1.9 Classification of fires

European/Australian/Asian	American	Fuel/heat source
Class A	Class A	Ordinary combustibles
Class B	Class B	Flammable liquids
Class C		Flammable gases
Class D	Class D	Combustible metals
Class E	Class C	Electrical equipment
Class F	Class K	Cooking oil or fat

Table 1.10 Australia: fire extinguishers

Type	Pre-1997		Current	Suitable for use on fire class[a]					
Water	Solid red	▨	Solid red	A					
Foam	Solid blue	▨	Red with a blue band	A	B				
Dry chemical (powder)	Red with a white band		Red with a white band	A	B	C		E	
Carbon dioxide	Red with a black band	■	Red with a black band	(A)	B		D		F
Vaporizing liquid (non-Halon clean agents)	Not yet in use		Red with a yellow band	A	B	C		E	
Halon	Solid yellow		No longer produced	A	B			E	
Wet chemical	Solid oatmeal		Red with an oatmeal band	A					F

[a]Parentheses denote sometimes applicable.

Table 1.11 USA: fire extinguishers

Fire class	Geometric symbol		Pictogram	Intended use
A	Green triangle	▨	Garbage can and wood pile burning	Ordinary solid combustibles
B	Red square	▨	Fuel container and burning puddle	Flammable liquids and gases
C	Blue circle	▨	Electric plug and burning outlet	Energized electrical equipment
D	Yellow decagon (star)		Burning gear and bearing	Combustible metals
K	Black hexagon	■	Pan burning	Cooking oils and fats

the types of fires that the extinguisher is approved to fight. In the past, extinguishers were marked with coloured geometric symbols. Some extinguishers still use both symbols.

Fire extinguishers in the UK, and throughout Europe, are red (Fig. 1.42), with a band or circle of a second colour covering between 5 and 10% of the surface area of the extinguisher to indicate its contents (Table 1.12). Prior to 1997, the entire body of the fire extinguisher was colour coded.

In the UK the use of Halon gas is now prohibited except under certain situations such as on aircraft and by the military and police.

1.3.10 Clean working environment

There are three main reasons for keeping your workshop and equipment clean and tidy:

▷ It makes the workshop a safer place to work.
▷ It makes it a better place to work.
▷ It gives a better image to your customers.

Servicing and fixing motor vehicles can be a dirty job, but if you clean up after any job then you will find your workshop a much more pleasant place in which to work.

▷ The workshop and floor should be uncluttered and clean to prevent accidents and fires as well as maintaining the general appearance.
▷ Your workspace reflects your ability as a technician. A tidy workspace equals a tidy mind equals a tidy job equals a tidy wage when you are qualified.

Figure 1.42 CO_2 and water extinguishers and information posters

▷ Hand tools should be kept clean as you are working. You will pay a lot of money for your tools; look after them and they will look after you in the long term.
▷ Large equipment should only be cleaned by a trained person or a person under supervision.

25

Table 1.12 UK and Europe: fire extinguishers

Type	Old code		BS EN 3 colour code	Suitable for use on fire class[a]				
Water	Signal red		Signal red	A				
Foam	Cream		Red with a cream panel above the operating instructions	A	B			
Dry powder	French blue		Red with a blue panel above the operating instructions	(A)	B	C	E	
Carbon dioxide (CO$_2$)	Black		Red with a black panel above the operating instructions		B		E	
Wet chemical	Not yet in use		Red with a canary yellow panel above the operating instructions	A	(B)			F
Class D powder	French blue		Red with a blue panel above the operating instructions				D	
Halon 1211/BCF	Emerald green		No longer in general use	A	B		E	

[a]Parentheses denote sometimes applicable.

Table 1.13 Cleaning materials

Material	Purpose	Notes
Detergents	Mixed with water for washing vehicles, etc. Also used in steam cleaners for engine washing, etc.	Some industrial detergents are very strong and should not be allowed in contact with your skin
Solvents	To wash away and dissolve grease and oil, etc. The best example is the liquid in the degreaser or parts washer which all workshops will have	NEVER use solvents such as thinners or fuel because they are highly inflammable. Suitable PPE should be used, e.g. gloves They may attack your skin Many are flammable The vapour given off can be dangerous Serious problems if splashed into eyes Read the label
Absorbent granules	To mop up oil and other types of spills. They soak up the spillage after a short time and can then be swept up	Most granules are a chalk or clay type material that has been dried out

Obvious precautions are to ensure that equipment cannot be operated while you are working on it and only using appropriate cleaning methods. For example, would you use a bucket of water or a brush to clean down an electric pillar drill? I hope you answered 'a brush'!

In motor vehicle workshops many different cleaning operations are carried out. This means a number of different materials are required. It is not possible to mention every brand name here, but the materials are split into three different types in Table 1.13. It is important to note that the manufacturer's instructions printed on the container must be followed at all times.

1.3.11 Signage

A key safety aspect is first to identify hazards and then to remove them or, if this is not possible, reduce the risk as much as possible and bring the hazard to everyone's attention. This is usually done by using signs or markings. Signs used to mark hazards are often as shown in Tables 1.14 and 1.15.

Safety first

Identify hazards and then remove them. If this is not possible, make others aware of them using signs.

1.3.12 Environmental protection

Environmental protection is all about protecting the environment, on individual, organizational or governmental levels. Owing to the pressures of population and technology the Earth's environment is being degraded, sometimes permanently. Activism by the environmental movement has created awareness of the various environmental issues. This has led to governments placing restraints on activities that cause environmental problems and producing regulations.

In a workshop these regulations relate to many items such as solvents used for cleaning or painting, fuels, oil and many other items. Disposal methods must not breach current regulations and, in many cases, only licensed contractors can dispose of certain materials. Failure to comply can result in heavy penalties. Make sure you are aware of your local regulations, as these can change.

Table 1.14 Hazard, mandatory and warning signs

Function	Example	Background colour	Foreground colour	Sign
Hazard warning	Danger of electric shock	Yellow	Black	 Figure 1.43 Electricity
Mandatory	Use ear defenders when operating this machine	Blue	White	 Figure 1.44 Wear ear protection
Prohibition	Not drinking water	White	Red/black	 Figure 1.45 Not drinking water

Table 1.15 Other common signage

Function	Example	Background colour	Foreground colour	Sign
First aid (escape routes are a similar design)	Location of safety equipment such as first aid	Green	White	 Figure 1.46 First aid
Fire	Location of fire extinguishers	Red	White	 Figure 1.47 Extinguisher
Recycling	Recycling point	White	Green	 Figure 1.48 The three Rs of the environment

Safety first

Disposal methods must not breach current regulations.

Finally, let's consider the three Rs:

- **Reduce** the amount of the Earth's resources that we use.
- **Reuse** – Don't just bin it: could someone else make use of it?
- **Recycle** – Can the materials be made into something new?

And, as we are automotive technicians, maybe there is a fourth: **Repair**?

1.3.13 High voltage vehicles

This section covers some risks when working with electricity or electrical systems, together with suggestions for reducing them. This is known as a risk assessment. The diagnostic process is no different but avoid working on high voltage vehicles unless you are trained and have specialist equipment – stay safe.

Electric vehicles (pure or hybrid) use high voltage batteries so that energy can be delivered to a drive motor or returned to a battery pack in a very short time. Voltages of 400V are now common and some are up to 700V, so clearly there are electrical safety issues when working with these vehicles. High voltage means the classification of an electric component or circuit if its working voltage is >60V and ≤1500V DC or >30V and ≤1000V AC root mean square (rms). High voltages can also be present on vehicles without high voltage batteries such as headlamps that use ballast resisters for example.

EV batteries and motors have high electrical and magnetic potential that can severely injure or kill if not handled correctly. It is essential that you take note of all the warnings and recommended safety measures outlined by manufacturers. Any person with a heart pacemaker or any other electronic medical devices should not work on an EV motor since the magnetic effects could be dangerous. In addition, other medical devices such as intravenous insulin injectors or meters can be affected.

The electrical energy is conducted to or from the motor via thick orange wires connected to the battery. If these wires have to be disconnected, SWITCH OFF or DE-ENERGIZE the high voltage system. This will prevent the risk of electric shock or short-circuit of the high voltage system.

Electrolytes used in EV batteries can be harmful and it is important to remember that both acid and alkali solutions are used in this technology. Understanding before you start work what type of chemical is present will enable selecting the correct procedure when dealing with a spill.

The general advice about working on high voltage vehicles is split into four areas:

Before work: Turn OFF the ignition switch and remove the key.

Figure 1.49 High voltage (HV) battery pack (Source: Chevrolet Media)

▷ Switch OFF the battery module switch or de-energize the system.
▷ Wait for 5 minutes before performing any maintenance procedures on the system. This allows any storage capacitors to be discharged.
▷ Isolate system and stow key in a safe location only accessible to the technician.

During work: Always wear correct PPE when there is risk of touching a live component.

▷ Always use insulated tools when performing service procedures to the high voltage system. This precaution will prevent accidental short-circuits.
▷ Always use correct workshop equipment to define the work area, warn others of the dangers present and use a correct insulated rubber mat when appropriate.

Interruptions: When maintenance procedures must be interrupted while some high voltage components are uncovered or disassembled, make sure that:

▷ The key is only accessible with the working technician's knowledge.
▷ The battery module switch is switched off.
▷ No untrained persons have access to the area and prevent any unintended touching of the components.

After work: Before switching on or re-energizing the battery module after repairs have been completed, make sure that:

▷ All terminals have been tightened to the specified torque.
▷ No high voltage wires or terminals have been damaged or shorted to the body.
▷ The insulation resistance between each high voltage terminal, of the part you disassembled and the vehicle's body, has been checked.

> **Key fact**
> Understanding basic principles will mean that you will be a better technician.

1.4 Basic science, materials, mathematics and mechanics

1.4.1 Introduction

When you want to work on motor vehicles, it is easy to wonder why you should study maths, science, materials, electricity and other similar subjects. The answer is that understanding basic principles will mean that you will be a better technician because you know how things really work – and you will have the skills to figure out how something you have not seen before works.

Table 1.16 Useful terminology

SI units	A set of standard units so we all talk the same language. SI stands for 'Système International'. This is French for 'International System'!
Ratio	The amount of one thing compared to another, e.g. two to one is written as 2:1
Area (m²)	Amount of surface of anything, e.g. the surface area of a car roof would help you know how much paint would be needed to cover it
Volume (m³)	Capacity of an object, e.g. 1000 cc (cubic centimetres) or one litre of paint to do the job above
Mass (kg)	The quantity of matter in a body. Volume does not matter, e.g. which has the greater mass, a kilogram of lead or a kilogram of feathers? They both have the same mass, but have different volumes
Density (kg/m³)	A full paint tin has a greater mass than an empty tin, but the volumes are the same
Energy (J)	The ability to do work or the amount of work stored in something; e.g. petrol contains a lot of energy in chemical form
Force (N)	When you push an object it moves (if you can apply enough force)
Work (J)	Work is done when the force applied to an object makes it move. Work can also be said to be done when energy is converted from one form to another
Power (W)	The rate at which work can be done, e.g. energy used per second
Torque (Nm)	A turning force like a spanner turning a nut. A longer spanner needs less force
Velocity (m/s)	A scientific name for speed; e.g. the UK national velocity limit is 70 mph (not an SI unit!)
Acceleration (m/s²)	The rate at which velocity changes. If positive then the car, for example, will increase in speed. If negative (or deceleration) such as when braking, the car's speed decreases
Momentum (kg m/s)	The combination of the mass of a body and its velocity. A large goods vehicle has much greater momentum than a car at the same speed. It must have much better brakes or it will take a lot longer to stop
Friction (μ)	When one surface moves over another friction tries to stop the movement. It is interesting to note that without friction a moving object such as a car would not stop!
Heat (J)	This is a measure of the amount of energy in a body. Heat can only transfer from a higher to a lower temperature and this will be by conduction, convection or radiation
Temperature (°C)	A measure of how hot something is, but this must not be confused with the amount of heat energy
Pressure (N/m² or Pa)	This is a force per area; e.g. the old tyre pressure measurement for many cars was 28 psi (pounds per square inch). The better unit to get used to is the bar: the tyre pressure would be about 1.8 bar. The SI unit is the pascal or newtons per metre squared (Pa or N/m²). The pressure in this room is about 1 bar or 1 atmosphere or 100,000 Pa. It may be much more if you have been reading about science for a long time!
Centrifugal force (N)	If you swing a stone on a string round your head it tries to move outwards and you can feel the centrifugal force on the string. The faster you swing it the greater the force. When a car wheel is rotating very quickly a small imbalance in the tyre causes unequal centrifugal force and this makes the wheel wobble
Weight (N)	The mass of an object acted upon by the Earth's gravity gives it a weight. When you next go into outer space, you will find that your weight is zero, or in other words you are weightless. You still have the same mass, however. The word weight is often used incorrectly, but as gravity is the same all over the Earth it doesn't often make any difference
Centre of gravity	The point within an object at which it will balance. All the weight of an object such as a car can be said to act through the centre of gravity. If the force due to gravity and acceleration acting through this point falls outside the wheels of the car, the car will fall over!
Electricity	This is the movement of electrons known as a current flow in a conductor or a wire. Electricity is a very convenient way of transferring energy
Strength	This is hard to define because different materials are strong in different ways. A material can be strong by providing opposition to bending, tension, compression or shear force
Corrosion	Corrosion of materials is by a chemical process; e.g. if iron is left open to the air or water it rusts. The chemical process is that the iron reacts with oxygen in the air and turns into iron oxide (rust)
Machines	A machine is something that converts one form of energy into another; e.g. an alternator converts mechanical energy from the engine into electrical energy
Hydraulics	When fluids are used to do 'work' this is described as hydraulics. The braking system of a car is a good example
Oscillation	If you bounce a mass on a spring (a car on its suspension) it will move up and down (oscillate) until all the mechanical energy in the spring has been converted to another form (mostly heat due to friction). Dampers are used on a car to make this time as short as possible

Often, the words used to describe scientific principles can be confusing. Table 1.16 picks out the most important terminology and a simple explanation is given. Some of the terms are described in more detail in later sections.

1.4.2 Units

When I go into a café or a bar and I ask for a pint of beer or half a litre of coke I usually get what I want (Fig. 1.50). This is because I ask by using the correct units. When you blow up the tyres on a car you check the pressure in a book or on a chart and then look at the gauge. It will have the same units, and you can inflate the tyres to the correct pressure.

> **Definition**
>
> **SI:** SI stands for 'Système International' (often described as the metric system).

The easiest units to work with are called SI units, sometimes described as the metric system. Other systems are fine, of course, and whatever is in common use, or whatever is stated in manufacturer's data is what you should use. However, the basic SI units you will need to know are listed in Table 1.17.

Many other units in use are derived from the basic SI units. Some of them are combined and given new names (Table 1.18).

When dealing with some of these units or derived units, we need a way of describing very large or very small quantities. For example, I would not

say that I live 24,000 metres away from where I work. I would say I live 24 kilometres away, normally written as 24 km. The 'k' is known as a multiplier and in this case you will see it has the value of 1000.

Likewise, if setting a spark plug gap I could set it at 0.001 metres, or it might be easier to say 1 millimetre, normally written as 1 mm. The 'm' can be thought of as a divider which in this case is 1000 or a multiplier of 0.001. Common multipliers are listed in Table 1.19.

> **Definition**
>
> **Velocity:** Velocity = Distance travelled/Time taken ($v = d/t$)

1.4.3 Velocity and acceleration

Velocity is the speed of an object in a given direction. Velocity is a 'vector quantity', meaning that its direction is important as well as its speed. The velocity v of an object travelling in a fixed direction may be calculated by dividing the distance s it has travelled by the time taken t. It is expressed as miles per hour (mph) or m/s (metres per second).

Acceleration is the rate of change of velocity (how quickly speed is increasing or decreasing). It is usually measured in metres per second per second. Newton's second law of motion says that a body will accelerate only if it is acted upon by an outside force. The outside force on a car is either the accelerator to increase speed (accelerate) or the brakes to decrease speed (decelerate). It is usually expressed as metres per second per second or ms^{-2}.

Acceleration due to gravity is the acceleration of an object falling due to the Earth's gravity. The value used for gravitational acceleration or g is 9.806 ms^{-2} (10 ms^{-2} is usually near enough for our calculations).

The average acceleration a of an object travelling in a straight line over a time t may be calculated using the formula:

$$\text{Acceleration} = \text{Change of velocity/Time taken}$$

Or, if v is its final velocity and u its initial velocity:

$$a = (v - u)/t$$

A negative answer (less than zero, e.g. −5 ms^{-2}) would mean that the object is slowing down (decelerating).

Figure 1.50 Mmm!

Table 1.17 SI units

Unit	Abbreviation	Quantity	Example
metre	m	Length	The distance from one point to another
kilogram	kg	Mass	The quantity of matter which makes an object
second	s	Time	About 300 s to boil an egg!
ampere	A	Electric current	The flow rate of electricity through a wire
kelvin	K	Temperature	How hot the radiator of a car is
candela	cd	Luminous intensity	How brightly a headlight shines

Table 1.18 Derived SI units

Unit	Abbreviation	Quantity
joule	J	Energy
newton	N	Force
watt	W	Power
area	m^2	Square metres
volume	m^3	Cubic metres
torque	Nm	Newton metres
velocity	m/s or ms^{-1}	Metres per second
acceleration	m/s/s or ms^{-2}	Metres per second per second

Table 1.19 Common multipliers

Prefix	Symbol	Value	Long value
mega	M	10^6	1,000,000
kilo	k	10^3	1000
hecto	h	10^2	100
centi	c	10^{-2}	0.01
milli	m	10^{-3}	0.001
micro	μ	10^{-6}	0.000 001

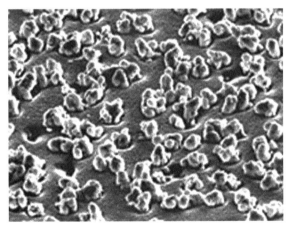

Figure 1.51 Surface of a smooth material magnified thousands of times

Key fact

The force that opposes the relative motion of two bodies in contact is known as friction.

1.4.4 Friction

The force that opposes the relative motion of two bodies in contact is known as friction. The coefficient of friction is the ratio of the force needed to achieve this motion to the force pressing the two bodies together.

For motor vehicle use friction is greatly reduced in some places by using lubricants such as oil and grease.

In other places friction is deliberately increased; for example, brake shoes, pads, drive belts and tyres.

Definition

Pressure: The SI unit of pressure is the pascal (Pa), equal to a pressure of 1 newton per square metre.

1.4.5 Pressure

In a fluid or gas, pressure is said to be the force that acts at right angles per unit surface area of something immersed in the fluid or gas. The SI unit of pressure is the pascal (Pa), equal to a pressure of 1 newton per square metre. In the atmosphere, the pressure decreases as you go higher, from about 101 kPa at sea level to zero, where the atmosphere dwindles into space. The other common units of pressure you will meet are the bar and psi. One bar (100 kPa) is atmospheric pressure, which is also about 14.7 psi (pounds per square inch).

Absolute pressure is measured from a perfect vacuum or zero pressure (Fig. 1.52). Gauge pressure is the difference between the measured pressure and atmospheric pressure. A tyre gauge works like this because it reads zero in atmospheric pressure. When we talk about a vacuum or a depression, what we really mean is a pressure less than atmospheric. It is best to use absolute pressure figures for discussing subjects such as the operation of an engine, or at least make sure you do not confuse the different figures.

31

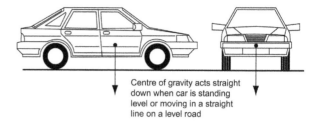

Definition

Absolute pressure: Absolute pressure is measured from a perfect vacuum or zero pressure.

Centre of gravity acts straight down when car is standing level or moving in a straight line on a level road

1.4.6 Centre of gravity or centre of mass

The centre of gravity (or mass) is a point in or near to an object about which it would turn if it could rotate freely. A symmetrical object, such as a cube or ball, has its centre of mass at its geometrical centre; a hollow object such as a beer glass may have its centre of gravity in the space inside it.

For an object such as a car to be stable, a perpendicular line down through its centre of gravity must run within the boundaries of its wheelbase (Fig. 1.53). If the car is tilted until this line falls outside the wheelbase, it will become unstable and fall over.

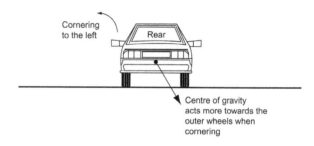

Cornering to the left

Rear

Centre of gravity acts more towards the outer wheels when cornering

Definition

Time period: For any vibration, the time for one complete oscillation.

1.4.7 Oscillation

An oscillation is one complete to-and-fro movement of a vibrating object or system (Fig. 1.54). For any vibration, the time for one complete oscillation is its time period.

The number of oscillations in one second is the frequency. In most mechanical systems in the car,

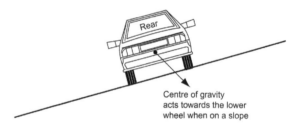

Rear

Centre of gravity acts towards the lower wheel when on a slope

Figure 1.53 Centre of gravity of a car

oscillations are damped down. The dampers (shock absorbers) fitted to the suspension help to prevent the springs oscillating.

Key fact

Energy cannot be destroyed, only converted to another form.

1.4.8 Energy, work and power

Energy can be thought of as the ability to do work or the amount of work stored up, and is measured in joules. When you have no energy it's hard to work! Energy cannot be destroyed, only converted to another form. It can be stored in a number of forms.

Most types of energy are listed here, together with an example (Fig. 1.55):

▷ kinetic or mechanical energy, e.g. the movement of an engine
▷ potential or position energy, e.g. when you lift a hammer its potential energy increases
▷ electrical energy, e.g. that made by an alternator
▷ chemical energy, e.g. stored in a battery
▷ heat energy, e.g. from burning a fuel
▷ nuclear energy – which is not yet used in motor vehicles, fortunately!

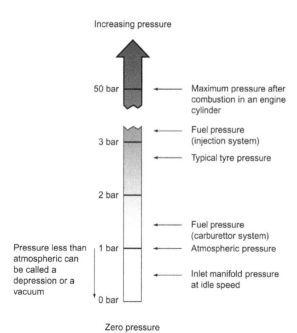

Increasing pressure

50 bar — Maximum pressure after combustion in an engine cylinder

3 bar — Fuel pressure (injection system)

— Typical tyre pressure

2 bar

— Fuel pressure (carburettor system)

Pressure less than atmospheric can be called a depression or a vacuum

1 bar — Atmospheric pressure

— Inlet manifold pressure at idle speed

0 bar

Zero pressure

Figure 1.52 Absolute pressure

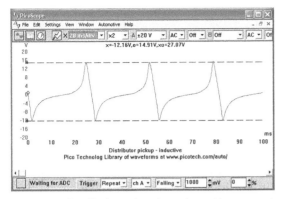

Figure 1.54 Oscillating signal produced by a crank sensor. (Source: PicoScope)

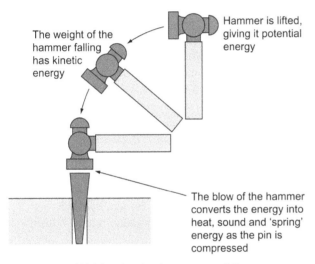

The weight of the hammer falling has kinetic energy

Hammer is lifted, giving it potential energy

The blow of the hammer converts the energy into heat, sound and 'spring' energy as the pin is compressed

Figure 1.55 Waiting for the hammer to fall

Power is the rate of doing work or converting energy. It is measured in watts. If the work done, or energy converted, is E joules in t seconds, then the power P is calculated by:

$$\text{Power} = \text{Work done}/\text{Time}$$
$$P = E/t$$

1.4.9 Force and torque

A force is thought of as any influence that tends to change the state of rest or the motion in a straight line of an object, just like braking force slows a vehicle down. If the body cannot move freely it will deform or bend. Force is a vector quantity, which means it must have both size and direction; its unit is the newton (N).

Torque is the turning effect of force on an object (Fig. 1.56). A car engine produces a torque at the wheels. Torque is measured by multiplying the force by its perpendicular distance away from the turning point; its unit therefore is the newton metre (Nm).

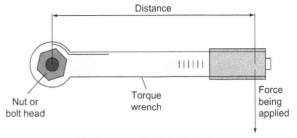

Distance

Nut or bolt head

Torque wrench

Force being applied

The torque applied is the force in newtons times the distance in metres

Figure 1.56 Torque wrench

1.4.10 Mass, weight and force

Mass is the quantity of matter in a body as measured by its resistance to movement. The SI system base unit of mass is the kilogram (kg). The mass of an object such as a car determines how much driving force is needed to produce acceleration. The mass also determines the force exerted on a body by gravity.

> **Key fact**
> The mass of an object determines how much driving force is needed to produce acceleration.

The force F, mass m and acceleration a (or g, if due to gravity) can be calculated using:

$$\text{Force} = \text{Mass} \times \text{Acceleration}$$
$$F = ma$$

Or

$$\text{Force (weight)} = \text{Mass} \times \text{Gravity}$$
$$F = mg$$

At a given place, equal masses experience equal gravity, which are known as the weights of the bodies. Masses can be compared by comparing the weights of bodies as long as they are at the same place (Fig. 1.57).

1.4.11 Volume and density

Density is a measure of the compactness of a substance; it is measured in kilograms per cubic metre (kg/m^3 or kg m^{-3}). The density D of a mass m, occupying a volume V, is given by:

$$\text{Density} = \text{Mass}/\text{Volume}$$
$$D = m/V$$

Relative density is the ratio of the density of one substance to another. This is useful for testing older

33

Figure 1.57 Mass on the Earth and Moon remains the same but the weight will change

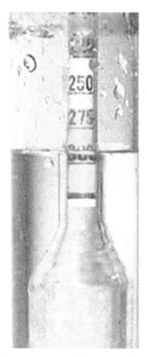

Figure 1.58 Measuring relative density with a hydrometer

types of battery by comparing the density of the electrolyte to that of water (Fig. 1.58). It is sometimes described as specific gravity.

Definition

Relative density: The ratio of the density of one substance to another.

1.4.12 Heat and temperature

Heat is a form of energy possessed by a substance by virtue of the vibrating movement or kinetic energy of

its molecules or atoms. Heat only flows from a higher temperature to a lower temperature. Its effect on a substance may be simply to raise its temperature, or to cause it to expand. Solids can melt, liquids vaporize and gases if confined will increase in pressure. This is much like ice, water, steam and steam pressure in a boiler.

Quantities of heat are usually measured in units of energy, such as joules (J). The specific heat capacity of a substance is the ratio of the amount of heat energy required to raise the temperature of a given mass of the substance through a given range of temperature, to the heat required to raise the temperature of an equal mass of water through the same range. This is useful for comparing materials.

Heat energy is transferred by conduction, convection and radiation. Conduction is the passing of heat along a medium to neighbouring parts. For example, the whole length of a metal rod becomes hot when one end is held in the flame of a welding torch. Convection is the transmission of heat through a liquid or gas in currents, for example when the air in a car is warmed by the heater matrix and blower. Radiation is heat transfer by infrared rays. It can pass through a vacuum and travels at the speed as light. For example, you can feel radiated heat from a vehicle headlight just in front of the glass.

Definition

Heat: Heat is a form of energy.

1.4.13 Percentages

1.4.13.1 Example 1

If a data book says 30% antifreeze and the cooling system holds 8 litres, how much antifreeze should you add? (Fig. 1.59)

30% means 30/100, which cancels to 3/10
3/10 × 8 = 24/10 = 2.4 litres

1.4.13.2 Example 2

If your normal pay rate is £10 per hour, how much will you get if you are given a 22% rise?

22% means 22/100
22/100 × £10 = £2.20

Your new pay rate is £10 + £2.20p = £12.20 per hour

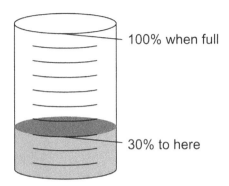

Figure 1.59 Percentage

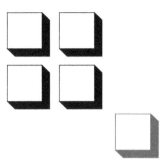

Figure 1.60 The red block is a quarter of the four black ones

1.4.14 Fractions

1.4.14.1 Example 1

If your normal pay rate is £5 per hour, how much will you get if your pay increases by a quarter for overtime? (Fig. 1.60)

Time and a quarter means 1¼ × your normal rate
1¼ × £5 = 5/4 × 5/1 = 25/4 = 6.25

Your overtime pay rate is £6.25 per hour

1.4.14.2 Example 2

If a heater blower circuit has a 2 Ω and a 3 Ω resistor connected in parallel by the speed control switch, what is the combined resistance?

The formula is: $1/R_T = 1/R_1 + 1/R_2$
Which means $1/R_T = 1/2 + 1/3$

To add fractions the bottom numbers must be the same:

$1/R_T = 3/6 + 2/6$
$1/R_T = 5/6$ $R_T = 6/5 = 1.2$ Ω

1.4.15 Ratios

1.4.15.1 Example

If the maximum speed an alternator can run at is 15,000 rpm and the top speed of the engine is 6000 rpm, why is the pulley ratio 2.5:1? (Fig. 1.61)

15,000/6000 = 2.5 (the ratio of the speeds)

Therefore, the alternator can never be driven too fast.

1.4.16 Areas

1.4.16.1 Example

If the car roof is 1.2 m long and 1.1 m wide and the aerosol says it will cover 1.5 m², will there be enough paint? (Fig. 1.62)

The area is 1.2 × 1.1 = 1.32m²

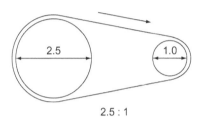

Figure 1.61 Pulley ratio

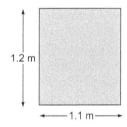

Figure 1.62 Car roof

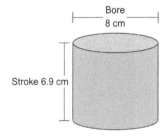

Figure 1.63 Cylinder

So yes, you have got enough paint (for one coat)!

1.4.17 Volumes

1.4.17.1 Example

If the bore of a four-cylinder engine is 8 cm and the stroke (distance from bottom dead centre to top dead centre) is 6.9 cm, what is the capacity of the engine? (Fig. 1.63)

The volume of a regular solid is the area multiplied by the height. For a cylinder, the area is πr^2, so the

35

volume must be $\pi r^2 h$ (r, the radius, is half the bore diameter; h is the stroke):

$$V = 3.14 \times 4 \times 4 \times 6.9 = 346.66 \text{ (now} \times 4 \text{ cylinders)}$$
$$= 1386.62 \text{ cc}$$

This engine would be called a 1400 cc or a 1.4 litre engine.

1.4.18 Indices

1.4.18.1 Example

A current flow of 1 ampere means that 6,000,000,000,000,000,000 electrons pass a point in one second! It is much easier to write 6×10^{18}, this simply means 6 with 18 zeros after it.

This quantity of electrons is known as a coulomb. It is about enough electricity to work a heavy-duty starter motor for about 0.001 seconds. This could be written as 10^{-3} seconds, where the –3 means moving the decimal point 3 places to the left (dividing by 1000).

1.4.19 Drawings

Drawings are an ideal way to pass on important information (Figs 1.64 and 1.65). Many manufacturers provide information in the form of drawings. It is essential that you can interpret the details you need and are not put off by the amount of information presented.

To help you do this, some simple standards are used relating to the type of line:

- continuous thick line ———— visible outlines
- continuous thin line ———— projection, dimension and hatching lines
- short thin dashes - - - - - - hidden details
- long thin chain - — - — - centre lines
- long chain — - — - — cutting planes

There are many types of drawing methods and some of these are outlined in Table 1.20. There are many

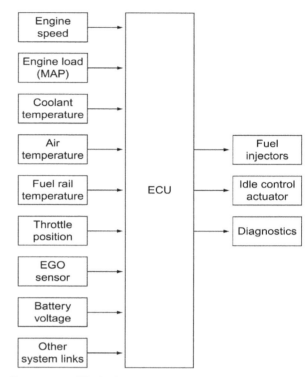

Figure 1.65 Block diagram representing a fuel injection system. MAP, manifold absolute pressure; EGO, exhaust gas oxygen; ECU, electronic control unit

sources of information that use drawings; some are outlined in Table 1.21.

Drawings can be produced in a number of ways. The most ideal for engineering drawings are as follows.

- **Orthographic** projections show three elevations, usually a front, plan and end view.
- **Pictorial** projections, such as isometric and oblique, show a representation of what the item looks like. The isometric view is used often in workshop manuals to show the arrangement of a complicated system.

> **Definition**
>
> **Machine:** A simple mechanical machine is a device that allows a small force to overcome a larger force.
>
> A machine is a device that converts one form of energy to another.

1.4.20 Mechanical machines

A simple mechanical machine is a device that allows a small force to overcome a larger force. There are only three basic machines (Fig. 1.66):

- ramps – such as a wedge

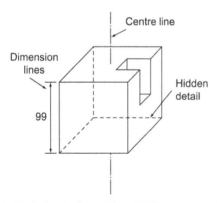

Figure 1.64 A three-dimensional (3D) representation (oblique projection)

Table 1.20 Terminology used on drawings

Sections	When the inside details of an object are important it is often convenient to show this by sectioning or cutting the object in a suitable place. A sectioned view of a brake cylinder is a good example
Dimensions	Lines with arrowheads in some cases simply show the size of the object. These are used more for drawings used to make an item than to pass on information for repair
Tolerances and limits	Dimensions can never be completely accurate, although they are very close in some cases. A good example of a tolerance or limit on a motor vehicle drawing is the bore of a cylinder given as 70 ± 0.05 mm
Fits	Two types of 'fit' can be used. Clearance fit is where, say, a pin is slightly smaller than and therefore slides into a hole. Interference fit is where the pin is very slightly too large and needs pressing into the hole
Projection	A term used to describe the way an object is drawn. You can imagine it as if projected on to a screen from different angles
Line diagram	A simplified diagram showing only the most basic of information
Block diagram	Complicated systems can be simplified by representing, say, the fuel system as one block and the engine as another, and so on
Exploded diagram	This is often used in workshop manuals. It shows a collection of components spread apart to show their details and suggest their original positions

Table 1.21 Sources of drawings and other information

Technical bulletins	Sheets of information sent from manufacturers to the dealers outlining the latest repair information
Parts books	The pictures or drawings in parts books are useful for repair procedures as they often show all the component parts of an object
Textbooks	This one is the best, of course, but a number of other good books are available
Workshop manuals	These are the traditional source of detailed information on specific vehicles and systems
Digital	Subscription-based data is available from manufacturers and specialist providers
CD and DVD	One DVD can hold a massive amount of information (equivalent to many workshop manuals), which can be accessed quickly and easily by a computer
Online databases	Many sources of information are now available online. This means that with a computer and modem you can access remote databases

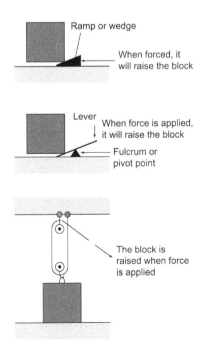

Figure 1.66 The three basic types of machine

- levers – such as a lever!
- wheels and axles – such as pulleys and a belt.

All other machines are combinations of these three. This is a good way of making any complicated machine easier to understand. The main features of a machine are:

- mechanical advantage, which is the ratio of load to effort (think of a car jack)
- velocity ratio, which is the velocity input compared to the velocity output (think of a car gear box)
- efficiency, which is the work output divided by the work input as a percentage.

In a perfect machine, with no friction, the efficiency would be 100%. All practical machines have efficiencies of less than 100% – otherwise perpetual motion would be possible!

1.4.21 Gears

Gears are toothed wheels (with lots of small levers) that transmit the turning movement of one shaft to another. Gear wheels may be used in pairs or in threes if both shafts need to turn in the same direction. The gear ratio, which is the ratio of the number of teeth on the two wheels, determines:

▷ **the torque ratio** – the turning force on the output shaft compared with the turning force on the input shaft
▷ **the speed ratio** – the speed of the output shaft compared with the speed of the input shaft.

Gears with the ratio 2:1 (say 20 teeth input and 10 teeth output) will have an output twice the speed and half the torque of the input.

A common type of gear for parallel shafts is the spur gear, with straight teeth parallel to the shaft axis. The helical gear (most common in car gear boxes) has teeth cut at an angle in a corkscrew shape.

1.4.22 Hydraulics

Hydraulics means using the properties of liquids to transmit pressure and movement (Fig. 1.68). The best known machines of this type are the hydraulic press and the hydraulic jack. The principle of pressurized liquid and increasing mechanical efficiency is also ideal for use on vehicle braking systems.

A basic hydraulic system consists of two liquid-connected pistons in cylinders, one of narrow bore, one of large bore. A force applied to the narrow piston applies a certain pressure to the liquid, which is transmitted to the larger piston.

Because the area of this piston is larger, the force exerted on it is larger. The original force has been increased, although the smaller piston has to move a greater distance to move the larger piston only a small

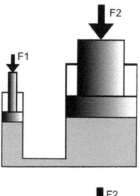
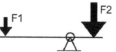

Figure 1.68 A hydraulic system like this one is sometimes called a liquid lever

Figure 1.69 Cast light alloy throttle body

distance. Mechanical advantage is gained in force, but movement is lost.

Definition

Hydraulics: Using the properties of liquids to transmit pressure and movement.

1.4.23 Materials and properties

Different materials are used in different places on motor vehicles because of their properties. For example, cast iron is a fairly obvious choice for use as an exhaust manifold because of the very high temperatures. Plastic would probably melt so would not be used. However, perhaps aluminium could be used instead? This time it is not obvious and more thought is required to decide the most suitable material.

Figure 1.67 Remember when using a hydraulic jack to use axle stands.

Table 1.22 Comparison of material properties

Material	Ease of shaping	Strength	Resistance to heat	Electrical resistance	Corrosion resistance	Cost	Typical motor vehicle uses
Copper	2	3	2	1	3	4	Wires and electrical parts
Aluminium	2	3	2	1	3	4	Cylinder heads
Steel	3	2	1	1	4	3	Body panels and exhausts
Cast iron	3	2	1	1	4	3	Manifolds and engine blocks
Platinum	3	1	1	1	2	5	Spark plug tips
Soft plastic	1	5	5	5	1	1	Electrical insulators
Hard plastic	1	4	4	5	1	1	Interiors and some engine components
Glass	3	5	2	5	1	2	Screens and windows
Rubber	2	4	5	5	3	2	Tyres and hoses
Ceramics	4	4	1	5	1	4	Spark plug insulators

1 = best; 5 = worst.

Figure 1.70 Corrosion can cause component failure (Source: K.Baker)

In Table 1.22 are listed several types of material together with important properties. As a very rough guide these are given as a number from 1 (best) to 5 (worst) in a kind of league table. This makes the table easier to use for comparing one material with another. Note that this table is just to help you compare properties; the league table positions are only rough estimates and will vary with different examples of the same material.

Corrosion is the eating away and eventual destruction of metals and alloys (mixtures of metals) by chemical action (Fig. 1.70). Most metals corrode eventually, but the rusting of ordinary iron and steel is the most common form of corrosion. Rusting of iron or steel takes place in damp air, when the iron combines with oxygen and water to form a brown deposit known as rust. Higher temperatures make this reaction work more quickly. Salty road and air conditions make car bodies rust more quickly.

Material properties are listed in Table 1.23. Some materials other than metals corrode or perish over a period of time, for example rubber-based materials. Plastics have a great advantage because some appear to last for ever!

1.5 Tools and equipment

1.5.1 Hand tools

Using hand tools is something you will learn by experience, but an important first step is to understand the purpose of the common types (Fig. 1.71). This section therefore starts by listing some of the more

Table 1.23 Terms used to describe materials

Property	Explanation
Hardness	The ability to withstand indentation (marking)
Softness	Can be easily indented
Toughness	The ability to resist fracture
Brittleness	Breaks or shatters under shock loads (impact)
Ductility	Plastic (deforms and stays that way) under tension or stretching
Malleability	Plastic under compression (squeezing)
Plasticity	The ability to retain a deformation after a load is removed
Elasticity	The ability to return to its original shape when a deforming load is removed
Strength	The ability to withstand a load without breaking

39

Figure 1.71 Ratchet open ended spanner

popular tools, with examples of their use, and ends with some general advice and instructions.

Practise until you understand the use and purpose of the tools listed in Table 1.24 when working on vehicles.

General advice and instructions for the use of hand tools (taken from advice by Snap-on):

▶ Only use a tool for its intended purpose.
▶ Always use the correct size tool for the job you are doing.
▶ Pull a spanner or wrench rather than pushing whenever possible.
▶ Do not use a file, or similar, without a handle.
▶ Keep all tools clean and replace them in a suitable box or cabinet.
▶ Do not use a screwdriver as a pry bar.
▶ Look after your tools and they will look after you!

1.5.2 Test equipment

Removing, refitting and adjusting components to ensure the vehicle system operates within specification is a summary of almost all the work you will be doing. The use, care, calibration and storage of

Table 1.24 Hand tools

Hand tool	Example uses and/or notes
Adjustable spanner (wrench)	An ideal standby tool and useful for holding one end of a nut and bolt
Open-ended spanner	Use for nuts and bolts where access is limited or a ring spanner cannot be used
Ring spanner	The best tool for holding hexagon bolts or nuts. If fitted correctly it will not slip and damage both you and the bolt head
Torque wrench	Essential for correct tightening of fixings. The wrench can be set in most cases to 'click' when the required torque has been reached. Digital wrenches can be preprogrammed and some can provide a log of settings used
Socket wrench	Often contains a ratchet to make operation far easier
Hexagon socket spanner	Sockets are ideal for many jobs where a spanner cannot be used. In many cases a socket is quicker and easier than a spanner. Extensions and swivel joints are also available to help reach that awkward bolt
Impact gun	These tools are great for speeding up your work but it is easy to damage components because it is very powerful. Only special, extra strong, high-quality sockets should be used. The tool can be battery or air powered
Battery ratchet wrench	Often a smaller drive than the impact gun, they can speed up a job – but also the damage if ill used!
Screwdrivers	Learn not to confuse the different types, pozidriv, phillips, crosshead, and blade. The wrong type will slip and damage will occur
Torx®	Similar to a hexagon tool like an Allen key but with further flutes cut in the side. It can transmit good torque
Special-purpose wrenches	Many different types are available. As an example, mole grips are very useful tools as they hold like pliers but can lock in position
Pliers	Used for gripping and pulling or bending. They are available in a wide variety of sizes, from snipe nose for electrical work, to engineers' pliers for larger jobs such as fitting split pins
Lever	Used to apply a very large force to a small area. If you remember this you will realize how if incorrectly applied, it is easy to damage a component
Hammer	Anybody can hit something with a hammer, but exactly how hard and where is a great skill to learn!

test equipment are therefore very important. In this sense, 'test equipment' means:

▷ measuring equipment, e.g. a micrometer
▷ hand instruments, e.g. a spring balance
▷ electrical meters, e.g. a digital multimeter (DMM) (Fig. 1.72) or an oscilloscope.

Figure 1.72 Digital multimeter in use

Figure 1.73 Battery maintenance system

Figure 1.74 Suspension arm lever

The operation and care of this equipment will vary with different types. Therefore, you should always read the manufacturer's instructions carefully before use, or if you have a problem. The following list sets out good general guidelines:

▷ Follow the manufacturer's instructions – at all times.
▷ Handle with care – do not drop and keep the instrument in its box.
▷ Ensure regular calibration – check for accuracy.
▷ Understand how to interpret results – if in doubt, ask!

My favourite piece of test equipment is the PicoScope (Fig. 1.75). This is an oscilloscope that works through a laptop or computer. It will test all engine management systems and other electrical and electronic devices. Check out www.picoauto.com for more information.

1.5.3 Workshop equipment

In addition to hand tools and test equipment, most workshops will also have a range of equipment for lifting and supporting as well as electrical or air-operated tools. Table 1.25 provides some examples of common workshop equipment together with typical uses.

> **Key fact**
> An oscilloscope draws a graph of voltage against time.

1.6 Workshop bench skills

1.6.1 Introduction

As well as the obvious skills such as knowledge of the systems and the ability to use normal hand tools for vehicle repairs, bench fitting and in some cases machining skills are also essential.

Figure 1.75 Automotive PicoScope. (Source: Pico Tech Media)

41

Figure 1.76 Digital tyre tread depth gauge

Figure 1.77 Gas welding

Figure 1.78 Wheel clamps for ADAS calibration or wheel alignment

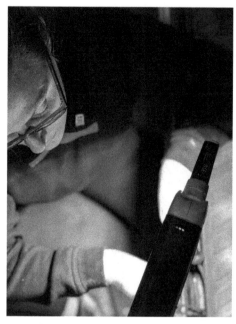

Figure 1.79 Inspection lamp

This usually involves metal cutting operations but it can involve other materials such as wood and plastics. In this sense, the word cutting is a very general term and can refer to:

▶ sawing
▶ drilling
▶ filing
▶ tapping
▶ machining.

1.6.2 Fitting and machining

Fitting and machining skills may be needed to complete a particular job (Fig. 1.80). In the context of an automotive engineer, the term 'fitting' is often used as a general description of hand skills usually used on a work bench or similar, to construct an item that cannot be easily purchased – for example, a support bracket for a modified exhaust or a spacer plate to allow the connection of an accessory of some type, such as additional lights.

Machinists usually work to very small tolerances, for example ±0.1 mm, and deal with all aspects of shaping and cutting. The operations most often carried out by machinists are milling, drilling, turning and grinding. To carry out fitting or machining operations you should be familiar with:

▶ measuring tools, e.g. a micrometer
▶ hand tools, as found in a standard tool kit
▶ machine tools, e.g. a bench drill
▶ work holders, e.g. a vice
▶ tool holders, e.g. the chuck of a drill
▶ cutting tools, e.g. saws and files.

Table 1.25 Examples of workshop equipment

Equipment	Common use
Ramp or hoist	Used for raising a vehicle off the floor. Other designs include four-post and scissor types where the mechanism is built into the workshop floor
Jack and axle stands	A trolley jack is used for raising part of a vehicle such as the front or one corner or side. It should always be positioned under suitable jacking points, or axle or suspension mountings. When raised, stands must always be used in case the seals in the jack fail, causing the vehicle to drop
Impact wrench	A battery or air gun is commonly used in workshops. Note that when replacing wheel fixings it is essential to use a torque wrench
Electric drill	The electric drill is just one example of electric power tools used for automotive repair. Note that it should never be used in wet or damp conditions
Parts washer	A number of companies will supply a parts washer and change the fluid it contains at regular intervals
Steam cleaner	Steam cleaners can be used to remove protective wax from new vehicles as well as to clean grease, oil and road deposits from cars in use. They are supplied with electricity, water and a fuel to run a heater, so caution is necessary
Electric welder	A number of forms of welding are used in repair shops. The two most common are metal inert gas (MIG) and manual metal arc (MMA)
Induction heater	Using induction, heat can be applied to precise areas in a safe manner. This is more economic than gas heating techniques. The applied heat expands the threads that are gripping the bolt, not the nut itself
Engine crane	A crane of some type is essential for removing the engine on most vehicles. It usually consists of two legs with wheels that go under the front of the car and a jib that is operated by a hydraulic ram. Chains or straps are used to connect to or wrap around the engine
Transmission jack	On many vehicles the transmission is removed from underneath. The car is supported on a lift and then the transmission jack is rolled underneath.

Figure 1.80 Machines such as lathes require maintenance, and supervision when learning at a young age! (no lathes were harmed taking this staged picture)

Figure 1.81 Surface grinder, used here to reface a cylinder head

Key fact

Files have forward-facing cutting teeth that cut best when pushed over the workpiece.

1.6.3 Filing and grinding

Filing and grinding is the process of removing material when manufacturing something; it is used mostly for finishing operations. Filing can be used on a wide range of materials as a finishing process. Emery paper may be considered as a filing tool.

Grinders are power tools that can speed up the process of filing, but they can also speed up the ability to make mistakes!

Files come in a wide variety of sizes, shapes, cuts and tooth configurations. The most common cross-sections of a file are: flat, round, half-round, triangular and square. The cut of the file refers to how fine its teeth are. They are described, from roughest to smoothest, as: rough, middle, bastard, second cut, smooth and dead smooth. Most files have teeth on all faces, but some flat files have teeth only on one

43

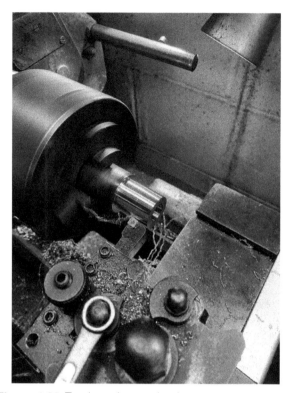

Figure 1.82 Turning a bronze bush

Figure 1.83 Grinding a crank shaft

Figure 1.84 Plasma cutting

face or edge, so that the file can work against another edge without causing damage.

1.6.4 Drilling

Drilling is a cutting process that uses a drill bit to cut or enlarge a hole in a solid material. The drill bit cuts by applying pressure and rotation to the workpiece, which forms chips at the cutting edge. The flutes remove these chips.

In use, drill bits have a tendency to 'walk' if not held very steadily. This can be minimized by keeping the drill perpendicular to the work surface. This walking or slipping across the surface can be prevented by making a centring mark before drilling. This is most often done by centre punching. If a large hole is needed, then centre drilling with a smaller bit may be necessary.

Drill bits used for metalworking will also work in wood. However, they tend to chip or break the wood, particularly at the exit of the hole. Some materials such as plastics have a tendency to heat up during the drilling process. This heat can make the material expand, resulting in a hole that is smaller than the drill bit used.

1.6.5 Cutting

A hacksaw is a fine-tooth saw with a blade under tension in a frame. Hand-held hacksaws consist of a metal arch with a handle, usually a pistol grip, with

pins for attaching a narrow disposable blade. A screw or another mechanism is used to put the blade under tension. The blade can be mounted with the teeth facing toward or away from the handle, resulting in cutting action on either the push or pull stroke. The push stroke is most common.

Blades are available in standardized lengths, usually 10 or 12 inches (15 or 30 cm) for a standard hacksaw. Junior hacksaws are usually half this size. Powered hacksaws may use large blades in a range of sizes.

Cutting can also be performed using an angle grinder fitted with a cutting disc, or an electric reciprocating saw, which can be very useful for removing items such as old exhaust systems. Some workshops may also have access to a plasma cutter for heavier work.

1.6.6 Thread cutting

Taps and dies are cutting tools used to create screw threads. A tap is used to cut the female part of the mating pair (e.g. a nut) and a die is used to cut the male portion (e.g. a screw). Cutting threads using a tap is called tapping, and using a die is called threading. Both tools can also be used to clean a thread, in a process known as chasing. The use of a suitable lubricant is recommended for most threading operations.

A tap cuts a thread on the inside surface of a hole, creating a female surface that functions like a nut. The three basic types are:

▷ The bottoming tap has a continuous cutting edge with almost no taper, which allows it to cut threads to the bottom of a blind hole.
▷ The intermediate tap, second tap or plug tap has tapered cutting edges, which assist in aligning and starting it into an untapped hole.
▷ The taper tap is similar to a plug tap but has a longer taper, which results in a more gradual cutting action.

Key fact

A lubricant is recommended for most threading operations.

The process of tapping begins with drilling and slightly countersinking a hole. The diameter of the hole is determined using a drill and tap size chart.

A T-shaped handle is used to rotate the tap. This is often turned in steps of one turn clockwise and about a quarter turn back. This helps to break off the chips, which avoids jamming. With hard materials, it is common to start with a taper tap, because the shallower cut reduces the amount of torque required to make the threads. If threads are to be cut to the bottom of a blind hole, the taper

Figure 1.85 Specialist thread repair set

Figure 1.86 Sticky tape can be an effective temporary repair to get a vehicle to the workshop

tap is followed by an intermediate (plug) tap and a bottoming tap.

The die cuts a thread on a cylindrical rod, which creates a male threaded piece that functions like a bolt. The rod is usually just less than the required diameter of the thread and is machined with a taper. This allows the die to start cutting the rod gently, before it cuts enough thread to pull itself along. Adjusting screws on some types of die allow them to be closed or opened slightly to allow small variations in size. Split dies can be adjusted by screws in the die holder. The action used to cut the thread is similar to that used when tapping.

Die nuts have no split for resizing and are made from a hexagonal bar so that a wrench or spanner can be used to turn them. Die nuts are used to clean up existing threads and should not be used to cut new threads.

1.6.7 Joining

It is very important for the correct methods of joining to be used in the construction and repair of a modern motor vehicle. Joining can cover many aspects, ranging from simple nuts and bolts to very modern and sophisticated adhesives.

The choice of a joining method for a repair will depend on the original method used as well as consideration of the cost and strength required. Table 1.26 shows some typical joining methods which include the use of gaskets in some cases.

Figure 1.87 A selection of joining or fastening components

Figure 1.88 Pop rivets

Table 1.26 Joining methods

Joining method	Examples of uses	Notes
Pins, dowels and keys	Clutch pressure plate to the flywheel	Used for strength and alignment in conjunction with nuts or bolts in most cases
Riveting	Some brake shoe linings	Involves metal pegs which are deformed to make the joint. Figure 1.88 shows some pop rivets, which are a popular repair component
Compression fitting	Wheel bearings	Often also called an interference fit. The part to be fitted is slightly too large or small, as appropriate, and therefore pressure has to be used to make the part fit
Shrinking	Flywheel ring gear	The ring gear is heated to make it expand and then fitted in position. As it cools it contracts and holds firmly in place
Adhesives	Body panels and sound deadening	Adhesive or glue is very popular as it is often cheap, quick, easy to apply and waterproof. Also, when two items are bonded together the whole structure becomes stronger
Nuts, screws, washers and bolts	Just about everything!	Metric sizes are now most common but many other sizes and thread patterns are available. This is a very convenient and strong fixing method. Figure 1.87 shows how varied the different types are
Welding	Exhaust pipes and boxes	There are several methods of welding, oxyacetylene and MIG being the most common. The principle is simple, in that the parts to be joined are melted so they mix together and then set in position
Brazing	Some body panels	Brazing involves using high temperatures to melt brass which forms the join between two metal components
Soldering	Electrical connections	Solder is made from lead and tin. It is melted with an electric iron to make it flow into the joint
Clips, clamps and ties	Hoses, cables, etc.	Hose clips, for example, are designed to secure a hose to, say, the radiator and prevent it leaking

Examples of use and useful notes are also given in the table.

Methods of joining are described as permanent or non-permanent. The best example of permanent joining is any form of welding. An example of non-permanent would be nuts and bolts. In simple terms, the permanent methods would mean that some damage would occur if the joint had to be undone.

Definition

Joining methods: Methods of joining are described as permanent or non-permanent.

1.6.8 **Nuts and bolts**

The nut and bolt is by far the most common method of joining two components together. Figure 1.89 shows some common nuts and bolts. The head of the bolt is

Figure 1.89 Nuts and bolts

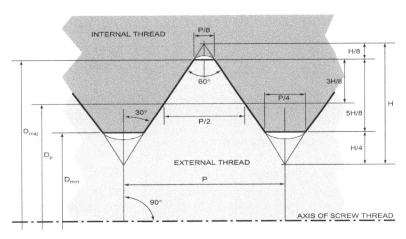

Figure 1.90 ISO metric thread profile

Figure 1.91 Tig welding aluminium

usually a hexagon, but an Allen socket or a Torx® drive or a number of other designs can be used. Smaller bolts can have a screwdriver type head such as a slot, cross, Phillips, Pozidriv or some other design.

The material used to make a nut or bolt depends on the application. For example, sump bolts will be basic mild steel, whereas long through bolts on some engines are made from sophisticated high tensile steel so that they will stretch. The size of the nut and bolt will depend on the size of components to be secured. Thread sizes used to be a problem, but fortunately now most nuts and bolts are metric. Figure 1.90 shows a metric thread profile.

Metric nuts and bolts are described as in the following example:

$$M10 \times 1.5$$

The M is metric, the 10 is the bolt diameter and the 1.5 is the pitch of the thread.

When joining with nuts and bolts it is common to find flat washers and in many cases some type of locking device. Metric threads are quite good at locking in position as they are, but for safety, extra devices are often used. Vibration is the main cause of bolts coming loose, as well as their not being tightened to the correct torque in the first place.

Key fact

Vibration is the main cause of bolts coming loose.

Another common method of securing threads is to use a locking compound such as Loctite®. This is in effect an adhesive which sticks the threads together. When the correct compound is applied with care, it is a very secure way of preventing important components from working loose.

1.6.9 Adhesives

A very wide range of adhesives is used in today's automotive industry. The number of applications is increasing daily and tending to replace older methods such as welding. There are too many types of adhesives to cover here but most of the basic requirements are the same. It is very important to note, however, that manufacturers' instructions must always be followed. This is because of the following:

▷ Many adhesives give off toxic fumes and must be used with care.
▷ Most types are highly flammable.
▷ Adhesives are often designed for a specific application.

47

Adhesives also have a number of important terms associated with them:

- **Cleanliness** – Surfaces to be joined must be clean.
- **Curing** – The process of setting, often described as 'going off'.
- **Wetting** – The adhesive spreads evenly and fully over the surface.
- **Thermosetting** – Heat is required to cure the adhesive.
- **Thermoplastic** – Melts when heated.
- **Contact adhesive** – Makes a strong joint as soon as contact is made.
- **'Super glue'** – Cyanoacrylate adhesive that bonds suitable materials in seconds, including skin – take care!

> **Safety first**
>
> Manufacturers' instructions must always be followed when working with adhesives.

Adhesives have many advantages, which is why they are becoming more widely used. These include:

- even stress distribution over the whole surface
- waterproof
- good for joining delicate materials
- no distortion when joining
- a wide variety of materials can be joined
- a neat, clean join can be made with little practice.

As a final point in relation to adhesives, the importance of choosing the correct type for the job in hand must be stressed. For example, an adhesive designed to bond plastic will not work when joining rubber to metal. And don't forget, if the surfaces to be joined are not clean you will make a very good job of bonding dirt to dirt instead of what you intended!

Figure 1.92 Soldering an electronic circuit

1.6.10 Soldering

Soft soldering is a process used to join materials such as steel, brass, tin or copper. It involves melting a mixture of lead and tin to act as the bond. A common example of a soldered joint is the electrical connection between the stator and diode pack in an alternator. Figure 1.92 shows this process using the most common heat source, which is an electric soldering iron.

The process of soldering is as follows:

1 Prepare the surfaces to be joined by cleaning and using emery cloth or wire wool as appropriate.
2 Add a flux to prevent the surfaces becoming dirty with oxide when heated, or use a solder with a flux core.
3 Apply heat to the joint and add solder so it runs into the joint.
4 Complete the process as quickly as possible to prevent heat damage.
5 Use a heat sink if necessary.

Soldering, in common with many other things, is easy after some practice; take time to do this in your workshop. Note that some materials such as aluminium cannot be soldered by ordinary methods.

1.6.11 Brazing

Brazing is a similar process to soldering except a higher temperature is needed and different filler is used. The materials to be joined are heated to red heat and the filler rod (bronze brass or similar), after being dipped in flux, is applied to the joint. The heat from the materials is enough to melt the rod and it flows into the gap making a good, strong, but slightly flexible, joint. Dissimilar metals such as brass and steel can also be joined and less heat is required than when fusion welding. Brazing is only used on a few areas of the vehicle body.

1.6.12 Welding

Welding is a method of joining metals by applying heat, combined with pressure in some cases. A filler rod of a similar metal is often used. The welding process joins metals by melting them, fusing the melted areas and then solidifying the joined area to form a very strong bond. Welding technology is widely used in the automotive industry.

The principal processes used today are gas and arc welding, in which the heat from a gas flame or an electric arc melts the faces to be joined. Figure 1.93 shows a welding process in action.

Several welding processes are used:

- Gas welding uses a mixture of acetylene and oxygen which burns at a very high temperature. This

Figure 1.93 MIG welding process

is used to melt the host metal with the addition of a filler rod if required (OA or oxyacetylene).

▷ Shielded metal-arc welding uses an electric arc between an electrode and the work to be joined; the electrode has a coating that decomposes to protect the weld area from contamination and the rod melts to form filler metal (MMA or manual metal arc).

▷ Gas-shielded arc welding produces a welded joint under a protective gas (MIG or metal inert gas, or TIG or tungsten inert gas).

▷ Arc welding produces a welded joint within an active gas (MAG or metal active gas).

▷ Resistance welding is a method in which the weld is formed by a combination of pressure and resistance heating from an electric current (spot welding).

Other, specialized types of welding include laser-beam welding, which makes use of the intensive heat produced by a light beam to melt and join the metals, and ultrasonic welding, which creates a bond through the application of high-frequency vibration while the parts to be joined are held under pressure.

1.6.13 Shrinking

When parts are to be fitted by shrinking they first have to be heated so they expand, or cooled so they contract. In both cases the component to be fitted must be made to an exact size. If parts fitted in this way are to be removed, it is usual to destroy them in the process. For example, a flywheel ring gear has to be cut through with a hacksaw to remove it.

For a hot shrink fitting, the part will have a smaller internal diameter than the one on which it is to be fitted. It is important not to overheat the components

or damage will occur. An oven is best, but a welding torch may be used with great care. When the component has been heated and therefore expanded, it is placed in position at once. It will then cool and make a good tight joint.

Cold shrinking is very similar except the component to be fitted is made very slightly larger than the hole in which it is to be fitted. A cylinder head valve insert is one example. The process is the opposite of hot shrinking. The component is cooled so it contracts, after which it is placed in position where it warms back up and expands, making a secure joint. Cold shrinking is normally a specialist job, but it is possible to buy aerosols of carbon dioxide under pressure which can be used to make a component very cold (dry ice).

1.6.14 Compression fitting

Many parts are fitted by compression or pressure. Bearings are the most common example (Fig. 1.94). The key to compression fitting is an interference fit. This means that the component, say a bearing, is very slightly larger than the hole in which it is to be fitted. Pressure is therefore used to force the bearing into place. Suspension bushes are often also fitted in this way.

The secret is to apply the force in a way that does not make the components go together at an incorrect angle. They must be fitted true to each other.

1.6.15 Riveting

Riveting is a method of joining metal plates, fabric to metal or brake linings to the shoes. A metal pin called a rivet, which has a head at one end, is inserted into matching holes in two overlapping parts. The other end is struck and formed into another head, holding the parts together. This is the basic principle of riveting, but many variations are possible.

Figure 1.88 shows some pop rivets, which are one of the most common for motor vehicle repair. These are hollow

Figure 1.94 This bearing on a gearbox shaft is held in place by compression

rivets that are already mounted on to a steel pin. The rivet is placed through the holes in the parts to be joined and a special rivet gun grips the pin and pulls it with great force. This causes the second rivet head to be formed and when the pin reaches a set tension it breaks off, leaving the rivet securely in place. The great advantage of this method is that you can work blind. In other words, you don't need access to the other side of the hole.

1.6.16 Gaskets

Gaskets are used to make a fluid- or pressure-tight seal between two component faces. The best example of this is the cylinder head gasket, which also has to withstand very high pressures and temperatures (Fig. 1.95). Gaskets are often used to make up for less than perfect surfaces and therefore act as a seal between the surfaces. Also, as temperature changes, the gasket can take up the difference in expansion between the two components. Gaskets are made from different materials depending on the task they have to perform (Table 1.27).

> **Key fact**
>
> Gaskets are used to make a fluid- or pressure-tight seal between two component faces.

The general rules for obtaining a good joint, with a gasket or otherwise, are as follows:

▷ cleanliness of the surfaces to be joined
▷ removal of burrs from the materials

Table 1.27 Gaskets and typical uses

Gasket material	Examples of where used
Paper or card	General purpose, such as thermostat housings
Fibre	General purpose
Cork	Earlier types of rocker covers
Rubber, often synthetic	Water pump sealing ring
Plastics, various types	Fuel pump to engine block
Copper, asbestos or similar	Exhaust flange – note safety issues of asbestos
Copper and aluminium	Head gaskets
Metal and fibre compounds, with metal composites	Head gaskets

▷ use of the correct materials
▷ following manufacturers' instructions (such as tighten to the correct torque in the correct sequence)
▷ safe working (this applies to everything you do).

1.6.17 Sealants

Many manufacturers are now specifying the use of sealants in place of traditional gaskets. The main reason for this is a better quality of joint. Liquid sealants, often known as instant gasket, are a type of liquid rubber that forms into a perfect gasket as the surfaces are mated together. The three major advantages of this technique are:

▷ It is easier to apply.
▷ A perfect seal is made with very little space being taken up.
▷ The adhesive bonding effect reduces fretting due to vibration, and hence the seal is less likely to leak.

Figure 1.96 shows a sealant being applied. A major advantage as far as the repair trade is concerned is that a good selection of jointing sealants means you can manufacture a gasket on the spot at any time!

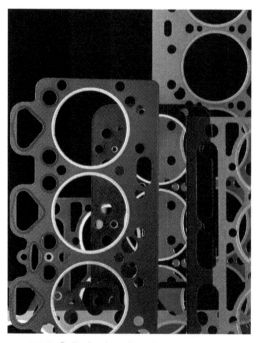

Figure 1.95 Cylinder head gaskets

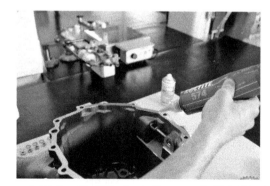

Figure 1.96 Loctite® Sealant. (Source: © 2010 Henkel AG & Co. KGaA, Düsseldorf. All rights reserved)

Figure 1.97 Oil seals. (Source: © 2005 Newsad Energy Company. All Rights Reserved)

Note the recommendations of the manufacturers, however, as only the correct material must be used.

1.6.18 Oil seals

The most common type of oil seal is the neoprene (synthetic rubber) radial lip seal. The seal is fitted into a recess and the soft lip rubs against the rotating component. The lip is held in place by a spring. Figure 1.97 shows this type of seal; note how the lip faces the oil such that any pressure will cause the lip to fit more tightly, rather than allow oil to be forced underneath.

1.7 Servicing and inspections

1.7.1 Introduction

It is important to carry out regular servicing and inspections of vehicles for a number of reasons:

- to ensure the vehicle stays in a safe condition
- to keep the vehicle operating within tolerances specified by the manufacturer and regulations
- to ensure the vehicle is reliable and to reduce downtime

- to maintain efficiency
- to extend components' and the vehicle's life
- to reduce running costs
- to keep the vehicle looking good and to limit damage from corrosion.
- to check for software updates and safety recalls.

To carry out servicing and inspections you should understand how the vehicle systems operate. It is also important to keep suitable records; these are often known as a vehicle's service history. Services and inspections of vehicles vary a little from one manufacturer to another. Servicing data and servicing requirement books are available, as well as the original manufacturer's information. This type of data should always be read carefully to ensure that all the required tasks are completed. Table 1.28 lists some important words and phrases relating to the servicing and inspection.

> **Key fact**
>
> Keep proper records when servicing; these are known as the vehicle's service history.

Clearly, it is important to keep a customer's vehicle in a clean condition (Fig. 1.98). This can be achieved by using:

- seat covers to keep the seats clean
- floor mats to protect the carpets from dirt
- steering wheel covers to keep greasy handprints off the wheel
- wing covers to keep the paintwork clean and to prevent damage.

1.7.2 Rules and regulations

The three main regulations that cover the repair and service of motor vehicles in the UK are:

- **Road Traffic Act** – this covers things like road signage and insurance requirements. It also covers issues relating to vehicle safety. For example, if a

Table 1.28 Servicing terminology

First service	The first service is commonly carried out by the main dealer, it is a safety inspection – but can also include software upgrades and an oil change
Distance-based services	10,000, 12,000 or 20,000 mile intervals are common distances, but manufacturers vary their recommendations. Most have specific requirements at set distances
Time-based services	For most light vehicles, distance-based services are best. Some vehicles though, run for long periods but do not cover great distances. In this case the servicing is carried out at set time intervals. This could be every six months, six weeks or after a set number of hours run
Inspection	The MOT test, which must be carried out each year after a light vehicle is older than three years, is a good example of an inspection. However, an inspection can be carried out at any time and should form part of most services
Records	A vital part of a service, to ensure all aspects are covered and to keep information available for future use
Customer contracts	When you make an offer to do a service, and the customer accepts the terms and agrees to pay, you have made a contract. Remember that this is legally enforceable by both parties

Figure 1.98 Bodywork protection in use during repairs

▶ define the work to be carried out
▶ record the work carried out
▶ record the time spent
▶ record materials consumed
▶ allow invoices to be prepared
▶ record stock that may need replacing
▶ form evidence in the event of an accident or customer complaint.

Table 1.30 is an example of a service sheet showing tasks carried out and at what service intervals. Please note once again that this list, while quite comprehensive, is not suitable for all vehicles, and the manufacturer's recommendations must always be followed. Some of the tasks are only appropriate for certain types of vehicle. The table also lists the work in a recommended order, including the use of a lift.

car suspension was modified it may become unsafe and not conform to the law.
▶ **Driver and Vehicle Standards Agency (DVSA) regulations** – the main one of these being the annual MOT test requirements.
▶ **Highway Code** – which all drivers must follow and forms part of the driving test.

Similar regulations are in place in other countries. The regulations are designed to improve safety.

Some of the main vehicle systems relating to safety are listed in Table 1.29, together with examples of the requirements. Note, though, that these are just examples and that specific data must be studied relating to specific vehicles.

1.7.3 Service sheets

Service sheets are used and records must be kept because they:

1.7.4 Road test

Assuming you are a qualified driver, or you are able to tell a driver what you want, then a road test is an excellent way of checking the operation of a vehicle. A checklist is again a useful reminder of what should be done. A typical road test following a service or inspection would be much as follows (but remember to check specific manufacturers' requirements):

▶ Fit trade plates to vehicle if necessary.
▶ Check operation of starter and inhibitor switch (automatic).
▶ Check operation of lights, horn(s), indicators, wipers and washers.
▶ Check indicators self-cancel.
▶ Check operation of all warning indicators.
▶ Check footbrake and handbrake.
▶ Check engine noise levels, performance and throttle operation.
▶ Check clutch for free play, slipping and judder.
▶ Check gear selection and noise levels in all gears.

Table 1.29 Safety inspection examples

Brakes	The footbrake must produce 50% of the vehicle weight braking force and the parking brake 16% (this assumes a modern dual-line braking system). The brakes must work evenly and show no signs of leaks
Exhaust	Should not leak, which could allow fumes into the vehicle, and should not be excessively noisy
Horn	Should be noisy! (up to a legal maximum level)
Lights	All lights should work and the headlights must be correctly adjusted
Number plates	Only the correct style and size must be fitted. The numbers and letters should also be correctly spaced and not altered (DAN 15H is right, DANISH is wrong)
Seat belts	All belts must be in good condition and work correctly
Speedometer	Should be accurate and illuminate when dark
Steering	All components must be secure and serviceable
Tyres	Correct tread depth is just one example, but assessment of wear can indicate if an alignment or repair is required
Windscreens and other glass	You should be able to see right through this one! No cracks allowed in the screen within the driver's vision

Table 1.30 Example service sheet

Driving vehicle into workshop	
Instrument gauges, warning/control lights and horn	Check operation
Washers, wipers	Check operation/adjust, if necessary
Inside vehicle	
Exterior and respective control lights; instrument cluster illumination	Check operation/condition
Service interval indicator	Reset after every oil change if applicable
Handbrake	Check operation/adjust, if necessary
Seat belts, buckles and stalks	Check operation/condition
Pollen filter	Renew
Warning vest	Check availability – if applicable
First aid kit	Check availability and expiry date – if applicable
Warning triangle	Check availability – if applicable
Outside vehicle	
Hood latch/safety catch and hinges	Check operation/grease
Road (MOT) test	Check regarding next road (MOT) test due date – if applicable
Emission test	Check regarding next emission test due – if applicable
Under bonnet (hood)	
Wiring, pipes, hoses, oil and fuel feed lines	Check for routing, damage, chafing and leaks
Engine, vacuum pump, heater and radiator	Check for damage and leaks
Coolant	Check antifreeze concentration: °C
Coolant expansion tank and washer reservoirs	Check/top up fluid levels as necessary – in case of abnormal fluid loss, a separate order is required to investigate and rectify
Power steering fluid	Check/top up fluid levels as necessary – in case of abnormal fluid loss, a separate order is required to investigate and rectify
Battery terminals	Clean, if necessary/grease
Battery	Visual check for leaks – in case of abnormal fluid loss, a separate order is required to investigate and rectify
Fuel filter	Drain water, if not renewed – diesel models (with drain facility)
Headlamp alignment	Check – adjust alignment, if necessary
Brake fluid	Check/top up fluid levels as necessary – in case of abnormal fluid loss, a separate order is required to investigate and rectify
Under vehicle	
Engine	Drain oil and renew oil filter
Steering, suspension linkages, ball joints, sideshaft joints, gaiters	Check for damage, wear, security and rubber deterioration
Engine, transmission	Check for damage and leaks
Pipes, hoses, wiring, oil and fuel feed lines, exhaust	Check for routing, damage, chafing and leaks
Underbody	Check condition of PVC coating
Tyres	Check wear and condition, especially at tyre wall, note tread depth: RF mm, LF mm, LR mm, RR mm, Spare mm
Brake system	With wheels off, check brake pads, discs and linings for wear, and check brake cylinders for condition: check rubber components for deterioration

- Check steering for noise, effort required, free play, wander and self-centring.
- Check suspension for noise, irregularity in ride and wheel imbalance.
- Check footbrake pedal effort, travel, braking efficiency, pulling and binding.
- Check speedometer for steady operation, noise and operation of mileage recorder.
- Check operation of all instruments.
- Check for abnormal body noises.
- Check operation of seat belts, including operation of inertia reels.
- Check handbrake ratchet and hold.
- Position car on lift.
- Recheck tension if drive belts have been renewed.
- Raise lift.
- Inspect engine and transmission for oil leaks.
- Check exhaust system for condition, leakage and security.
- Lower lift: drive vehicle off lift.
- Report on Road Test findings.
- Remove car protection kit.
- Ensure cleanliness of controls, door handles, etc.
- Remove trade plates if fitted.

1.7.5 Effects of incorrect adjustments

Table 1.31 lists a selection of possible incorrect adjustments, together with their effects on the operation of the vehicle. This is intended to be an exercise to help you see why correct adjustments are so important; not so you know how to do it wrong! You must also be able to make a record and tell a customer the effects, if you are unable to make the correct adjustments. This could be due to some parts being worn so that adjustment is not possible.

Remember though, anyone can mess with a vehicle and get it wrong. As a professional you will get it right, the customer and your company will be happy and it *will* affect your pay rates in years to come.

One of the problems that can arise after a vehicle has been serviced is when the customer expected a certain task to be completed, but it was not. For example, on a basic, an interim or even in some cases a full service, little or no work is carried out on the ignition system. This will not therefore rectify a misfire. It is important that the customer is aware of what will be done, as well as what was done to their vehicle. And if you notice a fault during a service, report it.

Key fact

It is important that a customer is aware of what will be done, as well as what was done to their vehicle during a service.

1.7.6 Maintenance and inspections

The purpose of routine maintenance is simple; it is to keep the vehicle in good working order and in a safe condition. Manufacturers specify intervals and set tasks that should be carried out at these times (Fig. 1.100). It is usually a condition of the warranty that a vehicle should be serviced according to the manufacturer's needs. The main purpose of regular inspection, therefore, is to check for the following:

- malfunction of systems and components
- damage and corrosion to structural and support regions
- leaks
- water ingress
- component and system wear and security.

Inspections are usually:

- audial – listening for problems
- visual – looking for problems
- tactile – feeling for problems
- olfactory – smelling for problems
- functional – checking that things work.

The main types of inspection, in addition to what is carried out when servicing, are:

- prework
- postwork
- pre-delivery inspection (PDI)
- used vehicle inspection
- special inspection (e.g. after an accident).

A prework inspection is used to find out what work needs to be carried out on a vehicle. Postwork inspections are done to make sure the repairs have been carried out correctly and that no other faults have been introduced.

A PDI is carried out on all new vehicles to check certain safety items and to, for example, remove any transport packaging such as suspension locks or similar. A used vehicle inspection is done to determine the safety and saleability of a vehicle as well as checking that everything works. After gaining experience you may be asked to carry out an inspection of a vehicle after an accident to check the condition of the brakes, for example.

Definition

PDI: Pre-delivery inspection.

Key fact

Sometimes you can smell a problem first, such as an overheated friction plate on a clutch.

In all cases, a recommended checklist should be used and careful records of your findings should be kept.

Table 1.31 Possible results of incorrect adjustments

Incorrect adjustment	Possible effects						
Brake	Excessive pedal and lever travel	Reduced braking efficiency	Unbalanced braking	Skidding and a serious accident			
Drive belts	Overheating	Battery recharge rate slow	Power steering problems	Air conditioning not operating			
Fuel system	Poor starting or non-start	Lack of power or hesitation	Uneven running and stalling	Popping back or backfiring	Running on or detonation	Heavy fuel usage	Fuel leaks and smells
Ignition	Poor starting or non-start	Lack of power	Hesitation	Exhaust emission	Running on		
Plug gaps	Poor starting or non-start	Lack of power	Hesitation	Uneven running	Misfiring	Exhaust emissions	
Steering system	Abnormal or uneven tyre wear	Heavy steering	Pulling to one side	Poor self-centring	Wandering	Steering wheel alignment	Excessive free play
Tyre pressures	Abnormal tyre wear	Heavy steering	Uneven braking	Heavy fuel usage	Reduced tyre lifetime		
Valve clearances	Lack of power	Uneven running	Misfiring	Excessive fuel usage	Exhaust emissions	Noise from valves or camshaft	

Figure 1.99 Boiling brake fluid to collect maintenance data

Working to timescales, or reporting to a supervisor that timescales cannot be met, is essential:

▶ When a customer books a car in for work to be done they expect it to be ready at the agreed time. Clearly, if this deadline cannot be met the customer needs to be informed.
▶ To make the running of a workshop efficient and profitable, a technician will have jobs allocated that will take a certain amount of time to complete. If for any reason this time cannot be met then action will need to be taken by the workshop manager or supervisor.

1.7.7 Information sources

The main sources of information are:

▶ workshop manuals
▶ technical bulletins
▶ servicing schedules
▶ job card instructions
▶ inspection records
▶ checklists
▶ online data and repair instructions
▶ MOT history
▶ Vehicle safety recalls.

All main manufacturers now allow online access to the full range of their data. This is accessed through special websites where you have to create an account and pay for the time.

Several companies supply information, such as Haynes Pro and Autodata.

Information and data relating to vehicles are available for carrying out many forms of diagnostic work. They used to come as a book or on CD/DVD. This is now online or as part of a package. This information is essential to ensure that you find the fault – as long as you have developed the diagnostic skills to go with it of course.

The type of information available is:

▶ engine testing and tuning
▶ servicing, repair processes, and labour times
▶ fuel and ignition systems
▶ circuit diagrams
▶ component location
▶ alignment data
▶ diagnostic routines
▶ air conditioning data.

There are some excellent packages that you can buy on subscription, or they are included with a diagnostic tool. Some have a pay-as-you-go option. Some example sources are:

▶ Bosch
▶ Snap-on
▶ Hella
▶ Delphi
▶ Thatcham
▶ AllData
▶ Autodata
▶ Haynes Pro
▶ Tech4Techs
▶ REPXPERT

Manufacturers' data (some of the above packages also access this) is highly recommended because you know you have the correct and latest information. This can be bought as required from almost all manufacturers. For example, Volkswagen data is available from: https://erwin.volkswagen. de. It is necessary to register on each site but then you can access the same information that is supplied to the main dealers. At the time of writing, the Volkswagen site allowed a range of payment options, but an example was one hour of full access for 8EUR (2022). This is more than enough time to find what you need for a specific repair or diagnostic job. Some workshops include this cost in their standard rate or add it as an extra on the customer's invoice.

1.8 Diagnostic techniques

1.8.1 Introduction and diagnostic process

Diagnostics or faultfinding is a fundamental part of an automotive technician's work. The subject of diagnostics does not relate to individual areas of the vehicle. If your knowledge of a vehicle system is at

Figure 1.100 Checking data from test equipment

Table 1.32 Useful words related to diagnostics

Symptom	The effect of a fault noticed by the driver, user or technician
Fault	The cause of a symptom/problem
Root cause	The cause of the fault
Diagnostics	The process of tracing a fault by means of its symptoms, applying knowledge and analysing test results
Knowledge	The understanding of a system that is required to diagnose faults
Logical procedure	A step-by-step method used to ensure nothing is missed
Concern, cause, correction	A reminder of the process starting from what the driver reports, to the correction of the problem
Report	A standard format for the presentation of results

a suitable level, the same logical process is used for diagnosing the fault, whatever the system.

The terminology included in the tables is provided to ensure we are talking the same language. These tables are provided as a simple reference source.

Where to stop? This is one of the most difficult skills to learn. It is also one of the most important. The secret is twofold:

1 Know your own limitations – it is not possible to be good at everything.
2 Leave circuits alone where you could cause more damage or even injury – for example air bag circuits.

Despite the best of intentions, a person new to diagnostics may not only fail to find the fault but introduce more faults into the system in the process. I would suggest you learn your own strengths and weaknesses; you may be confident and good at dealing with mechanical system problems but less so when electronics is involved. Of course you may be just the opposite of this.

Remember that diagnostic skill is in two parts – the knowledge of the system and the ability to apply diagnostics. If you do not yet fully understand a system – leave it alone until you do.

A key checklist – the six stages of fault diagnosis – is given in this list:

1 **Verify**: Is there actually a problem, can you confirm the symptoms?
2 **Collect**: Get further information about the problem, by observation and research.
3 **Evaluate**: Stop and think about the evidence.
4 **Test**: Carry out further tests in a logical sequence.
5 **Rectify**: Fix the problem.
6 **Check**: Make sure all systems now work correctly.

Here is a very simple example to illustrate the diagnostic process. The reported fault is excessive use of engine oil:

1 Question the motorist to find out how much oil is being used (is it excessive?) Check service history.
2 Examine the vehicle for oil leaks and blue smoke from the exhaust. Are there any service bulletins?
3 If leaks are found, the engine could still be burning oil but leaks would be a likely cause.
4 A compression test, if the results were acceptable, would indicate a leak to be the most likely fault. Clean down the engine and then run again to show a leak more easily.
5 Change a gasket or seal, etc.
6 Run through an inspection of the vehicle systems particularly associated with the engine. Double check the fault has been rectified and that you have not caused any further problems.

The knowledge needed for accurate diagnostics is in two parts:

1 understanding the system in which the problem exists
2 having the ability to apply a logical diagnostic routine

The knowledge requirement and use of diagnostic skills can be illustrated with a very simple example:

57

Table 1.33 General terminology

System	A collection of components that carry out a function
Efficiency	This is a simple measure of any system. It can be scientific. For example, if the power out of a system is less than the power put in, its percentage efficiency can be determined (P-out/P-in x 100%). This could, for example, be given as say 80%. In a less scientific example, a vehicle using more fuel than normal is said to be inefficient
Noise	Emanations of a sound from a system that is either simply unwanted or is not the normal sound that should be produced
Active	Any system that is in operation all the time (steering for example)
Passive	A system that waits for an event before it is activated (an air bag is a good example)
Short-circuit	An electrical conductor is touching something that it should not be (usually another conductor or the chassis)
Open circuit	A circuit that is broken (a switched-off switch is an open circuit)
High resistance	In relation to electricity, this is part of a circuit that has become more difficult for the electricity to get through. In a mechanical system a partially blocked pipe would have a resistance to the flow of fluid
Worn	This works better with further additions such as: 'Worn to excess', 'worn out of tolerance' or even, 'worn, but within tolerance'
Quote	Exact information on the price of a part or service. A quotation may often be considered to be legally binding
Estimate	A statement of the anticpated cost of a job (e.g. a service or repairs). An estimate is normally an approximation, and is not legally binding
Bad	Not good – and also not descriptive enough really…
Dodgy, knackered or @#%&*	Words often used to describe a system or component, but they mean nothing. Get used to describing things so that misunderstandings are eliminated

Figure 1.101 An obvious electrical fault

After connecting a hose pipe and turning on the tap, no water comes out of the end. Your knowledge of this system tells you that water should come out providing the tap is on, because the pressure from a tap pushes water through the pipe, and so on. This is where your diagnostic skills become essential. The following shows the required stages.

1 Confirm that no water is coming out by looking down the end of the pipe.
2 Check if water comes out of the other taps, or did it come out of this tap before you connected the hose?

3 Consider what this information tells you; for example, if the answer is 'Yes' the hose must be blocked or kinked.
4 Walk the length of the pipe looking for a kink.
5 Straighten out the hose.
6 Check that water now comes out and that no other problems have been created.

A very simple example, but the procedure you have just followed made the hose work and it is also guaranteed to find a fault in any system. It is easy to see how it works in connection with a hose pipe and I'm sure anybody could have found that fault (well most people anyway).

We have introduced the six-stage process of diagnostics, not so that it should always be used as a checklist but to illustrate how important it is to follow a process. Much more detail will be given later, in particular about stages 3 and 4. The purpose of this set process is to ensure that work is carried out in a set, logical way.

1.8.2 Electrical diagnostics

Electrical work is often the area that technicians find most difficult. Here is some important guidance.

Start all 'hands on' diagnostic routines with hand and eye checks. In other words look over the vehicle for obvious faults. For example, if the battery terminals are loose or corroded then put this right before carrying out complicated voltage readings. Here are some further suggestions that will at some point save you a lot of time:

▷ A misfire may be caused by a loose plug lead – it is easier to look for this than interpret the ignition waveforms on a scope.

▷ If the ABS warning light stays on look to see if the wheel speed sensor(s) are covered in mud or oil.

A test lamp is ideal for tracing faults in, say, a lighting circuit because it will cause a current to flow, which tests out high resistance connections. However, it is this same property that will damage delicate electronic circuits – so don't use it for any circuit that contains an electronic control unit (ECU).

A digital multimeter is ideal for all forms of testing. Most have an internal resistance in excess of 10 MΩ, which means that the current they draw is almost insignificant. An LED test lamp or a logic probe is also acceptable.

The following procedure is generic but with a little adaptation can be applied to any electrical system. Refer to manufacturer's recommendations if in any doubt.

Volt drop is a term used to describe the difference between two points in a circuit. In this way we can talk about a voltage drop across a battery (normally about 12.6 V) or the voltage drop across a closed switch (ideally 0 V but may be 0.1 or 0.2 V).

The first secret to volt drop testing is to remember that the sum of all volt drops around a circuit always adds up to the supply. The second secret is to ensure the circuit is switched on and operating – or at least the circuit should be 'trying to operate'.

In this picture this means that if the circuit is operating correctly, $V_1 + V_2 + V_3 = V_s$. When electrical testing therefore, and if the battery voltage is measured at say 12 V, a reading of less than 12 V at V_2 would indicate a volt drop between the terminals of V_1 and/or V_3. Likewise the correct operation of the switch, that is, it closes and makes a good connection, would be confirmed by a very low reading on V_1.

What is often described as a 'bad earth' (when what is meant is a high resistance to earth), could equally be determined by the reading on V_3. To further narrow the cause of a volt drop down, simply measure across a smaller area. The voltmeter V_4 for

Figure 1.102 A misfire caused by a failed injector

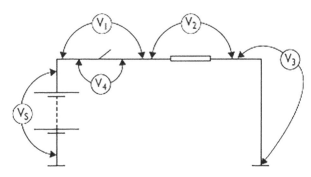

Figure 1.104 Volt drop testing

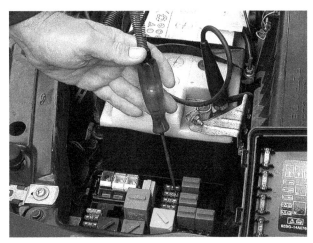

Figure 1.103 Test lamp in use

Figure 1.105 Headlight connections

59

example, would only assess the condition of the switch contacts.

A short-circuit to earth fault will normally blow a fuse – or burn out the wiring completely. To trace a short-circuit is very different to looking for a high resistance connection or an open circuit. The volt drop testing above will trace an open circuit or a high resistance connection. My preferred method of tracing a short, after looking for the obvious signs of trapped wires, is to connect a bulb or test lamp across the blown fuse and switch on the circuit. The bulb will light because on one side it is connected to the supply for the fuse and on the other side it is connected to earth via the short-circuit fault. Now disconnect small sections of the circuit one at a time until the test lamp goes out. This will indicate the particular circuit section that has shorted out.

On load means that a circuit is drawing a current; off load means it is not. One example where this may be an issue is when testing a starter circuit. Battery voltage may be 12 V (well 12.6 V) off load, but may be as low as 9 V when on load (cranking a cold engine perhaps).

A second example is the supply voltage to the positive terminal of an ignition coil via a high resistance connection (corroded switch terminal for example). With the ignition on and the vehicle not running, the reading will almost certainly be battery voltage because the ignition ECU switches off the primary circuit and no volt drop will show up. However, if the circuit were switched on (with a fused jumper lead if necessary) a lower reading would result showing up the fault.

The technique outlined here is known as 'black box faultfinding'. This is an excellent technique and can be applied to many vehicle systems from engine management and ABS to cruise control and instrumentation.

As most systems now revolve around an ECU, the ECU is considered to be a 'black box', in other words we know what it should do but the exact details of how it does it are less important.

Treating the ECU as a 'black box' allows us to ignore its complexity. The theory is that if all the sensors and associated wiring to the 'black box' are OK, all the output actuators and their wiring are OK and the supply/earth (ground) connections are OK, then the fault must be the 'black box'. Most ECUs are very reliable, however, and it is far more likely that the fault will be found in the inputs or outputs.

If you are testing for a supply (for example) at an ECU, then use the probes of your digital meter with care. Connecting to the back of the terminals will not

Figure 1.107 Electric window switches

Figure 1.108 Test the voltage by back probing a connector with care

Figure 1.106 Short-circuit testing

damage the connecting surfaces provided excessive force is not applied. Sometimes a pin clamped in the test lead's crocodile/alligator clip is ideal for connecting 'through' the insulation of a wire without having to disconnect it.

The key to electrical diagnostics, as with all other systems, is to work methodically and logically. And finally, remember to take care making test connections so that you do not cause more faults as you carry out tests!

1.8.3 Multimeters

There are lots of different options or settings available when using a multimeter, but the three most common

Figure 1.109 Voltage supply to a fuse box

Figure 1.110 Checking a simple resistor

measurements are: voltage (volts), resistance (ohms) and current (amps).

To measure voltage the meter is connected in parallel with the circuit. Most voltage measurements on a vehicle are DC. Remember to set the range of the meter (some are auto-ranging) and if in doubt, start with a higher range and work downwards.

To measure resistance the meter must be connected across (in parallel with) the component or circuit under test. However, the circuit must be switched off or isolated. If not, the meter will be damaged. Likewise, because an ohmmeter causes a current to flow there are some circuits, such as Hall effect sensors, that can be damaged by the meter.

Current can be measured in two ways:

1 Connect the meter in series with the circuit (in other words break the circuit and reconnect it through the meter).
2 Use an inductive amp clamp around the wire. This is a safer way to measure, but less accurate at low values.

Figure 1.111 Inductive ammeter clamp on a high voltage cable (measuring the current drawn by the EV cabin heater)

Figure 1.112 Ghost voltage caused by shaking the red lead

Voltmeters can display a 'ghost' voltage rather than zero when the leads are open circuit. In other words, if checking the voltage at an earth/chassis connection we would expect a 0V reading. However, the meter will also display zero before it is connected, so how do we know the reading is correct when it is connected? The answer is to shake the multimeter leads, a 'ghost' voltage will fluctuate, a real voltage will not!

1.8.4 Oscilloscopes

An oscilloscope (often shortened to scope) is an instrument used to display the amplitude and period of a signal, as well as its shape. Amplitude is its height, and period is the time over which it repeats. A scope draws a two-dimensional graph. Normally, voltage is displayed on the vertical, or Y-axis and time on the horizontal, or X-axis.

The settings for these can be changed and are usually described as volts per division (volts/div) and time per division (time/div). The graphs can be moved up and down on the screen with a control known as Y-shift. This means that the zero voltage position can, for example, be set to the middle of the screen. This means that the blue line in this image shows a DC voltage of 0V. The red line represents a DC voltage of about 12V because the volts per division (each small vertical square) in this case, is set at 5V. The green trace shows an AC signal that, assuming the volts/div is still set at 5V, has a peak-to-peak voltage of 20V.

Much more information is shown compared to other instruments such as multimeters or frequency meters. For example, when using a scope, you can determine the amplitude and shape, how much noise is present, and the frequency of a signal.

The frequency of a signal can be worked out from the time per division settings. In the image shown, the time/div (each small horizontal square), is set to 25ms. This means that the sinewave repeats every 100ms (four divisions). Because frequency is measured in cycles per second (hertz or Hz), we simply need to divide 1s (1000ms) by the time period, 100ms in this case, which means that this signal has a frequency of 10Hz.

As well as directly displaying electric signals, scopes can measure non-electrical values if appropriate transducers are used. Transducers change one kind of variable into another. For example, a pressure transducer (often described as a sensor), produces a voltage that is proportional to pressure.

The speed at which the trace moves across the screen is known as the time base, which can be adjusted either automatically or manually. The start of the trace moving across the screen is known as the trigger. This can be internal, such that it flies back and re-starts every two seconds for example, or it can be external, so it starts every time a fuel injector operates for example. The voltage from the item under test can either be amplified or attenuated (reduced), much like changing the scale on a voltmeter.

Now almost all automotive oscilloscopes are digital and use a computer screen to display signals. This also allows values such as the voltage and time base scales, and frequency to be shown on the screen. It also means the waveforms can be saved and shared. Waveforms gathered over a period of time can also be saved so that they can be replayed after, for example, a road test.

The number of traces that can be displayed on an oscilloscope at the same time is known as the number of channels. Most scopes designed for automotive use will have a minimum of two channels and more often four. This is so that different waveforms (traces, patterns, signals etc.) can be displayed at the same

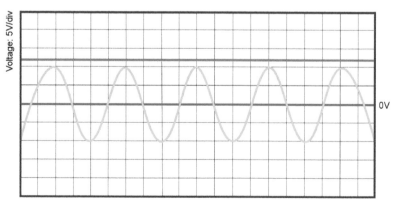

Figure 1.113 Oscilloscope screen

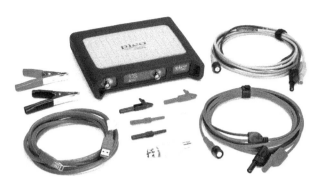

Figure 1.114 Pico 2-channel starter kit scope (Source: PicoScope, www.picoauto.com)

time, using different voltage scales if necessary, so that comparisons can be made.

There are a range of PicoScope automotive diagnostic kits. They integrate with software on a PC to measure and test virtually all of the electrical and electronic components and circuits in any modern vehicle. Common systems to measure include:

- ignition (primary and secondary)
- injectors and fuel pumps
- batteries, alternators and starter motors
- lambda, airflow, ABS and MAP sensors

- electronic throttle control
- CAN bus, LIN bus and FlexRay.

A two- or four-channel PicoScope is recommended for general workshop use. On these devices, a separate ground connection is used for each channel and the instrument is protected up to 200V.

Figure 1.115 PicoScope advanced kit (Source: PicoScope, www.picoauto.com)

Engine systems

2.1 Engine mechanical

2.1.1 Introduction and operating cycles

The modern motor vehicle engine burns a fuel to obtain power. The fuel is usually petrol (gasoline) or diesel, although liquid petroleum gas (LPG) and compressed natural gas (CNG) are sometimes used. Specialist fuels have been developed for racing car engines. Motor vehicle engines are known as internal combustion engines (ICE) because the energy from the combustion of the fuel, and the resulting pressure from expansion of the heated air and fuel charge, is applied directly to pistons inside closed cylinders in the engine. The term 'reciprocating piston engine' describes the movement of the pistons, which go up and down in the cylinders. The pistons are connected by a rod to a crankshaft to give a rotary output (Fig. 2.1).

> **Key fact**
> Specialist fuels and oils have been developed for racing car engines.

Fuel is metered into the engine together with an air charge for most petrol engines. However, most now use injectors that inject directly into the engine cylinder. In diesel engines, the fuel is injected into a compressed air charge in the combustion chamber. In order for the air and fuel to enter the engine and for the burnt or exhaust gases to leave the engine, a series of ports is connected to the combustion chambers (Figs 2.2 and 2.3). The combustion chambers are formed in the space above the pistons when they are at the top of the cylinders. Valves in the combustion chamber at the ends of the

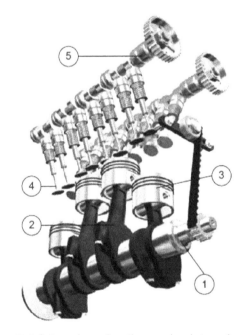

Figure 2.1 Internal combustion engine internal components: 1, crankshaft; 2, connecting rods; 3, pistons; 4, valves; 5, camshafts

ports control the air charge and exhaust gas movements into and out from the combustion chambers.

> **Definitions**
> **TDC:** top dead centre
> **BDC:** bottom dead centre
> **ICE:** internal combustion engine

The valves are described as 'poppet' valves and have a circular plate at right angles to a central stem that runs through a guide tube. The plate has a chamfered

Automobile Mechanical and Electrical Systems. 978-0-415-72578-1 © 2018 Tom Denton.
Published by Taylor & Francis. All rights reserved.

DOI: 10.1201/9781003299066-2

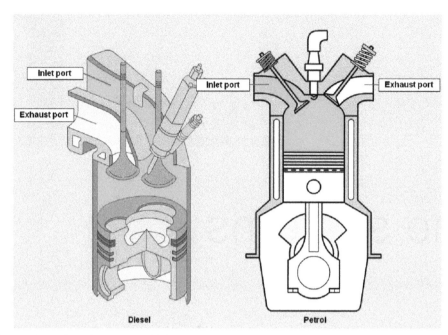

Figure 2.2 Internal combustion engine (ICE): pistons at bottom dead centre (BDC) before start of the compression strokes

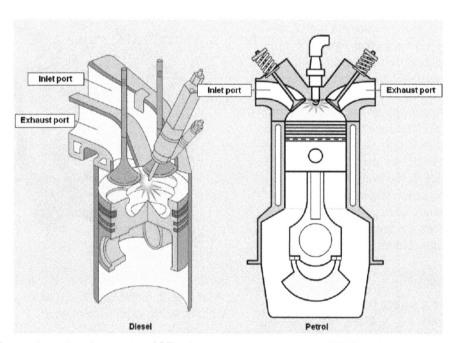

Figure 2.3 Internal combustion engine (ICE): pistons at top dead centre (TDC) before start of the combustion or power strokes

sealing face in contact with a matching sealing face in the port. The valve is opened by a rotating cam and associated linkage. It is closed and held closed by a coil spring. The opening and closing of the valves and the movement of the pistons in the cylinders follow a cycle of events called the 'four-stroke cycle' or the 'Otto cycle' after its originator (Figs 2.4–2.7).

Key fact

Combustion chambers are formed by the space above the pistons when they are at the top of the cylinders (TDC).

The first stroke of the four-stroke cycle is the induction or intake stroke. This occurs when the piston is

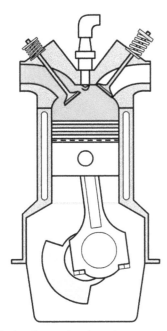

Figure 2.4 Induction or intake: piston moving down

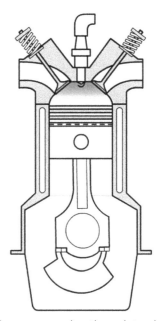

Figure 2.6 Power or combustion: piston is forced down

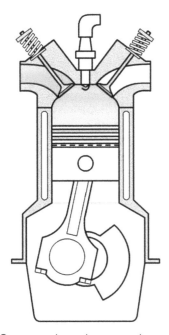

Figure 2.5 Compression: piston moving up

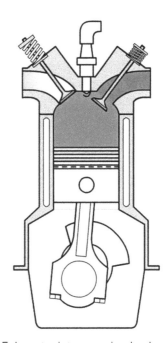

Figure 2.7 Exhaust: piston moving back up

moving down in the cylinder from top dead centre (TDC) to bottom dead centre (BDC) and the inlet valve is open. The movement of the pistons increases the volume of the cylinder so that air and fuel enter the engine.

Key fact

Towards the end of the compression stroke, the fuel is ignited and burns to give a large pressure rise.

The next stroke is the compression stroke, when the piston moves upwards in the cylinder. Both the inlet and exhaust valves are closed and the space in the cylinder above the piston is reduced. This causes the air and fuel charge to be compressed, which is necessary for clean and efficient combustion of the fuel.

Towards the end of the compression stroke, the fuel is ignited and burns to give a large pressure rise in the cylinder above the piston. This pressure rise forces the piston down in the cylinder on the combustion or power stroke.

67

Key fact

The four-stroke cycle is sometimes described as: suck, squeeze, bang, blow. . . .

Once the energy from the fuel has been used, the exhaust valve opens so that the waste gases can leave the engine through the exhaust port. To complete the exhausting of the burnt gases the piston moves upward in the cylinder. This final stroke is called the exhaust stroke.

The four-stroke cycle then repeats over and over again, as the engine runs. A heavy flywheel keeps the engine turning between power strokes.

On the induction stroke of a petrol engine (most types), air and petrol enter the cylinder so the inlet valve in the inlet port must be open. In diesel, and gasoline (petrol) direct injection (GDi) engines, only air enters the cylinder. A rotating cam on the camshaft provides a lifting movement when it runs in contact with a follower. A mechanical linkage is used to transfer the movement to the valve stem and the valve is lifted off its seat so that the inlet port is opened to the combustion chamber. The air and fuel charge or air charge can now enter the cylinder. The inlet valve begins to open shortly before the piston reaches TDC. The exhaust valve, which is operated by its own cam in the same way as the inlet valve, is beginning to close as the piston passes TDC at the end of the exhaust stroke. Valve overlap helps clear the remaining exhaust gases from the combustion chamber. The incoming air charge fills the combustion chamber as the last quantity of exhaust gas leaves through the exhaust port. This is known as 'scavenging'; it helps cool the combustion chamber by removing hot exhaust gases and gives a completely fresh air charge.

Key fact

Valve overlap helps clear the remaining exhaust gases from the combustion chamber.

The terms TDC and BDC are used to describe the position of the piston and crankshaft when the piston is at the end of a stroke and the axis of the piston and crankshaft bearing journals are in a straight line and at 0° (TDC) and 180° (BDC) of crankshaft revolution. To the abbreviations are added the letters 'A' to indicate degrees 'after' TDC or BDC and the letter 'B' to indicate 'before' TDC or BDC. See Figures 2.2 and 2.3.

Key fact

The camshaft rotates once for the two revolutions of the crankshaft during the four-stroke cycle.

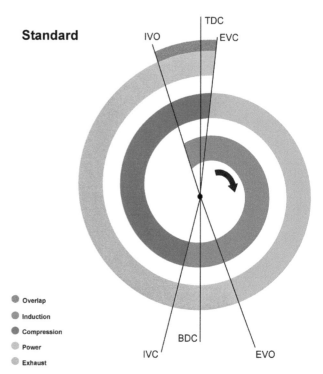

Figure 2.8 Spiral valve timing diagram for a standard four-stroke engine (TDC = Top dead centre, BDC = Bottom dead centre, IVO = Inlet valve opens, IVC = Inlet valve closes, EVO = Exhaust valve opens, EVC = Exhaust valve closes)

The camshaft (see Fig. 2.1) rotates once for the two revolutions of the crankshaft during the four-stroke cycle. The drive from the crankshaft to the camshaft has a 2:1 ratio produced by the numbers of teeth on the driven and driver gears. Rotational data for the camshaft is usually given as degrees of crankshaft rotation and this should to be considered in relation to the four-stroke cycle. The four-stroke cycle occurs over two full revolutions of the crankshaft, which is a 720° rotational movement.

2.1.1.1 Valve timing

In four-stroke engines the valve timing is controlled by a camshaft. It can be varied by modifying the camshaft, but it is a fixed mechanical setting. It is affected by the adjustment of the valve mechanism, and particularly by the valve clearance. Figure 2.8 shows a spiral type valve timing diagram. The colours represent the open positions of the valves. During the red part for example (compression), both valves are closed.

With fixed valve timing an engine has a period of valve overlap at the end of the exhaust stroke, when both the intake and exhaust valves are open (shown in grey). The intake valve is opened before the exhaust gases have completely left the cylinder, and because of their movement, this helps to draw in the new air/

fuel mixture. Ideally, the exhaust valve will close just as the new charge from the inlet valve reaches it.

When a valve opens before top dead centre (TDC) or bottom dead centre (BDC) is reached it is known as valve lead. Where it closes after TDC or BDC, it is called valve lag.

The timing of a fixed system, however, will always be a compromise for different engine operating speeds and load conditions. This is why variable valve timing (VVT) was developed.

All internal combustion engines have an induction, compression, expansion and exhaust process. For a four-stroke engine, each of these processes requires half an engine revolution, so the complete engine cycle takes two complete engine revolutions. That is, there is a working and a non-working (gas-exchange) revolution of the engine within the cycle. However, a two-stroke engine combines two of the processes in each half turn of the engine; thus, all processes are complete in one engine revolution and the engine has a power stroke with every revolution (Fig. 2.9). In order to operate, the two-stroke petrol engine uses the crankcase (piston underside) for induction of the fuel/air mixture and transfer into the cylinder via ports in the cylinder barrel.

On the upstroke, the piston moves upwards towards TDC, and fuel/air charge trapped in the cylinder space above the piston is compressed and, around TDC, ignited by a spark. This is the beginning of the power stroke. As the piston rises during the upstroke, the volume in the crankcase increases and atmospheric pressure forces the fresh fuel/air charge into the crankcase (under the piston).

On the downstroke the piston moves towards BDC as the power stroke begins; the expanding gases

force the piston down the bore, producing torque at the crankshaft via the connecting rod. At the same time, the crankcase volume decreases and the fuel/air mixture is compressed under the piston. As the piston approaches BDC, the transfer port connecting the cylinder volume to the crankcase volume is uncovered by the piston. On the opposing side of the cylinder, the exhaust port is also uncovered.

This allows the fresh charge in the crankcase volume to transfer and fill the cylinder volume, at the same time forcing the exhaust gases out of the cylinder via the exhaust port. The efficiency of this scavenging process is very dependent on the port exposure timing and the gas dynamics. Often the piston crown has a deflector to assist this process and to prevent losing fresh charge down the exhaust. Note that two-stroke gasoline engines are normally lubricated via the provision of an oil mist in the crankcase. This is provided by oil mixed in with the fuel/air (premixed or injected); hence the oil is burnt in the combustion process, which produces excessive hydrocarbon emissions.

Two-stroke engines are generally more powerful for a given displacement owing to the extra power stroke compared with a four-stroke engine, but the problem is that the expansion stroke is short and volumetric efficiency (how easy it is to get the gases in and out of the engine) is poor, so they are less efficient. In addition, exhaust emissions are higher than from a four-stroke engine.

Some large static diesel engines are often two-stroke types (Fig. 2.10). Note that all four operating processes are executed in one engine revolution (induction, compression, expansion and exhaust). The diesel engine requires a charge of air that is compressed to raise its temperature above the self-ignition point of the fuel. This air charge is supplied by an air pump or pressure charging device (turbo or supercharger). The pressurized

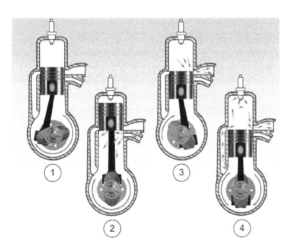

Figure 2.9 Two-stroke operating cycle: 1, compression; 2, induction; 3, combustion (power); 4, exhaust (transfer also taking place)

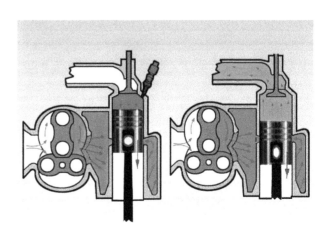

Figure 2.10 Diesel two-stroke cycle

air from this device passes into the combustion chamber via ports in the cylinder wall. The exhaust gases leave the combustion chamber via cam-operated poppet valves. The incoming charge forces the exhaust gases out via these valves (the cylinder scavenging process).

During the downwards movement of the piston, the hot expanding gases are forcing the piston down the bore, producing torque at the crankshaft. This is the expansion process. As the piston approaches BDC, the exhaust valve opens and the remaining pressure in the exhaust gas starts the evacuation of the gases in the cylinder via the open valves. As the piston moves further down to BDC, inlet ports are exposed around the bottom part of the cylinder bore, which allow the pressurized, fresh air charge from the air pump (or turbocharger) to fill the cylinder, evacuating the remaining exhaust gas via the valves and completing the exhaust and induction cycles.

At BDC, the cylinder contains a fresh air charge and the piston then begins to move up the cylinder bore. The inlet ports are closed off by the piston movement and the air charge is trapped and compressed by the decreasing volume in the cylinder. At a few degrees before TDC, the air temperature has risen owing to the compression process and fuel is injected directly into the combustion chamber, into the hot air charge, where it vaporizes, burns and generates thermal and pressure energy. This energy is converted to torque at the crankshaft via the piston, connecting rod and crankshaft during the downstroke.

Another variation on engine operation is the Wankel (the name of the inventor) or rotary engine (Fig. 2.11). This engine has been used in a limited number of passenger car applications. The engine uses a complex geometric rotor that moves within a specially shaped housing. The rotor is connected to the engine crankshaft and turns within the housing to create

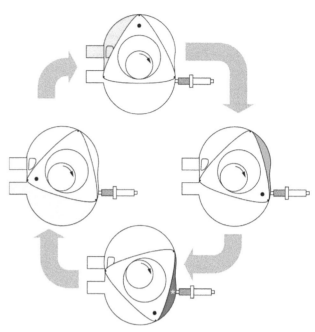

Figure 2.12 Rotary engine cycle: starting with the top image, induction, compression, power, exhaust. (Source: Wikimedia)

working chambers. These are exposed to inlet and exhaust ports to allow a fuel/air charge in, compress it and expand it (thus extracting work), then evacuate the waste gases and restart the cycle (Fig. 2.12). The rotor has special tips to provide a gas-tight seal between the working chambers. The movement of the rotor in this engine follows a path know as an epitrochoid.

No matter what design of engine, it has to be positioned in the vehicle. There are various configurations that manufacturers have used in the configuration of their vehicle powertrains. The engine can be front, mid or rear mounted and can be installed in-line (along the vehicle axis) or transverse (across the vehicle axis) (Fig. 2.13).

Definition

Epitrochoid: A roulette traced by a point attached to a circle rolling around the outside of a fixed circle.

Key fact

The mounting system has to isolate the vehicle from the engine vibrations.

The engine mounting system is important as it supports the weight of the engine in the vehicle. In addition, it counteracts the torque reaction under load conditions. The mounting system has to

Figure 2.11 Rotary engine. (Source: Mazda Media)

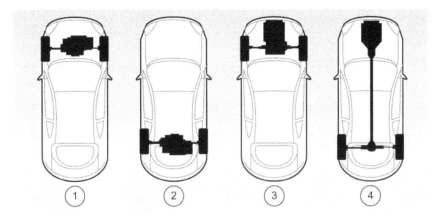

Figure 2.13 Typical positions for the engine: 1, front transverse engine FWD; 2, rear transverse engine RWD; 3, front longitudinal engine FWD; 4, front longitudinal engine RWD

Figure 2.14 Engine mountings

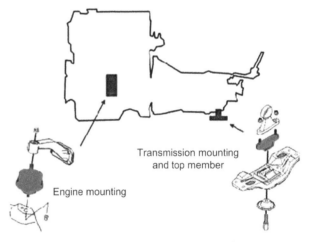

Figure 2.15 Typical engine mountings for front-engine RWD

isolate the vehicle from the engine vibrations. The engine mounts consist of steel plates with a rubber sandwich between to provide the vibration isolation (Fig. 2.14). The mountings have appropriate brackets and fittings to fix to the engine and vehicle frame.

For a front-engine, rear-drive powertrain layout, the engine mounts are often at the centre position of the engine side, approximately at the engine centre of gravity (Fig. 2.15). The engine mounts bear compression and shear forces in supporting the engine weight and torque. The rear of the engine is bolted to the transmission, which in turn is supported at the rear end via a rubber mounting system. This three-point mounting is very common for this powertrain configuration.

For a front-wheel drive, transverse powertrain layout (Fig. 2.16), the mounting system has to cope with the weight of the engine, plus the torque reaction of the wheel torque. The mounting system therefore includes

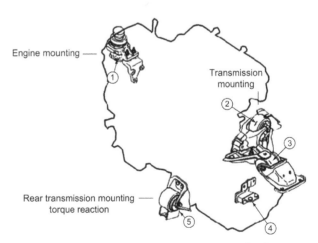

Figure 2.16 Five typical engine mountings for front-engine FWD

mountings to support weight and counteract torque separately. These are mounted at the top or bottom of the engine, respectively.

> **Definition**
>
> **Naturally aspirated engine:** An engine that does not use forced induction by a supercharger or turbocharger.

> **Definition**
>
> **Volumetric efficiency:** A measure of the efficiency of the cylinder charging process during the induction stroke.

2.1.2 Engine operating details

The air above the Earth's surface is like a fluid that exerts a pressure on all points around it because of the Earth's gravitational force pulling it down. This creates a pressure known as atmospheric pressure and is 101.325 kPa, 760 mmHg, 29.92 inches Hg, 14.696 psi or 1013.25 millibars (let's stick with approximately 1 bar or 15 psi).

A naturally aspirated engine (one that does not use forced induction by a supercharger or turbocharger) relies on atmospheric pressure to charge the cylinder with gas (air or air/fuel mixture) ready for the combustion process. As the piston moves down the cylinder (from TDC to BDC), the volume increases and this causes the pressure in the cylinder to reduce, becoming lower than atmospheric pressure. This creates a pressure difference between the inside and outside of the cylinder, and as a result the atmospheric pressure (the higher pressure) forces gases into the cylinder (where there

is lower pressure) until the pressure is balanced. Note that any restriction to the flow of gas will reduce the effectiveness of the cylinder charging process.

Volumetric efficiency is a measure of the efficiency of the cylinder charging process during the induction stroke. Theoretically, the cylinder should be completely filled with a mass of gas, but in practice this never happens owing to flow losses and inefficiencies. Therefore, the volumetric efficiency is a measure of the actual amount of gas induced compared to the theoretical amount (which is the mass required to completely fill the cylinder volume) and is expressed as a percentage (Fig. 2.17). It is calculated as:

(Actual mass of air/Theoretical mass of air) × 100%

The more efficiently the engine cylinders can fill with gas, the more air, or fuel/air, is available for the combustion process and this improves overall engine efficiency (Fig. 2.18). The process of getting gases into and out of the engine is known as 'aspiration' or 'engine breathing'.

Combustion in the engine cylinder takes place because of a chemical reaction between the carbon and hydrogen in the fuel and the oxygen in the air. This reaction releases energy from the fuel in the form of heat that generates pressure in the cylinder to force movement of the piston. To achieve efficient combustion, the quality of the fuel/air mixture is important; that is, how evenly mixed the fuel droplets are in the induced air. Movement of the air as it enters the cylinder is important for this process and

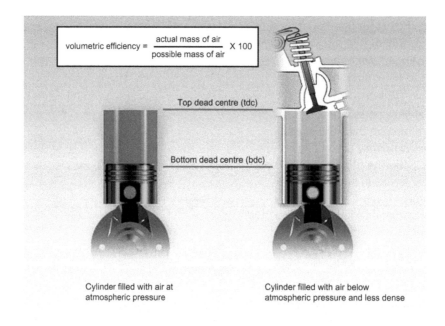

$$\text{volumetric efficiency} = \frac{\text{actual mass of air}}{\text{possible mass of air}} \times 100$$

Top dead centre (tdc)

Bottom dead centre (bdc)

Cylinder filled with air at atmospheric pressure

Cylinder filled with air below atmospheric pressure and less dense

Figure 2.17 Volumetric efficiency

Figure 2.18 This very efficient engine produces 98 g of CO_2/km. (Source: GM Media)

the requirements are different for petrol and diesel engines. The required air movements for each engine type are created by careful design of the components that form the inlet tract and combustion chamber.

Key fact

Combustion in the engine cylinder takes place because of a chemical reaction between the carbon and hydrogen in the fuel and the oxygen in the air.

The inlet valve opens and closes according to piston position and controls the incoming gas charge into the engine. It generally remains open for a small period after the piston has reached BDC (i.e. beyond the

end of the inlet stroke). This allows the energy of the moving gas column in the inlet tract to assist in the cylinder charging process, which helps to increase volumetric (and engine) efficiency.

After the combustion chamber has been charged with gas (air or fuel/air) during the induction stroke, the cylinder inlet and exhaust valves are both closed and seal the combustion chamber. The piston begins to rise in the cylinder, thus reducing the volume of the cylinder space and hence increasing the pressure of the trapped gas charge in the cylinder before combustion. The opening and closing of the valves is executed in sequence via the engine valve gear, synchronized with the four-stroke cycle and piston position.

It is important that the closed cylinder is sealed properly to maintain the appropriate pressures in the cylinder during the working cycle. Any losses in pressure would significantly reduce the efficiency of the engine. To seal the piston and bore, piston rings are fitted into radial groves near the top of the piston and provide a gas-tight seal between the moving piston and the cylinder bore. When the cylinder volume is reduced during the compression stroke, the trapped gas is compressed and the amount of compression is known as the compression ratio. Compressing the charge before combustion allows more oxygen or fuel/oxygen in the cylinder than would otherwise have been available without compression and this improves combustion efficiency. Most spark ignition engines have a compression ratio of 8:1 to 10:1. This means that the cylinder volume reduces by eight or ten times during the compression stroke (Fig. 2.19).

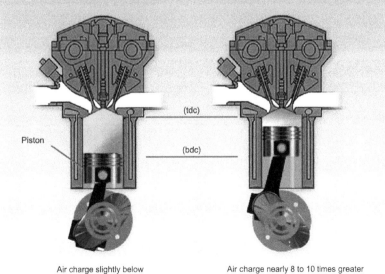

Piston

(tdc)

(bdc)

Air charge slightly below atmospheric pressure

Air charge nearly 8 to 10 times greater than atmospheric pressure

Figure 2.19 Compression

2.1.2.1 Spark ignition (SI)

During compression of the fuel/air mixture in a petrol engine, heat energy and kinetic energy (due to gas movement) are imparted into the mixture owing to the reducing volume and rising pressure. This creates a significant temperature increase and the magnitude of this increase depends upon the speed of the compression process and the amount of heat rejected to the surroundings (via the cylinder combustion space, walls, head, etc.). The temperature rise elevates to a point just below the self-ignition temperature of the fuel/air charge, which will combust at or above the flashpoint when ignited via an external source (i.e. the spark plug). Note that if the temperature of the mixture was too high, spontaneous self-ignition could occur and this would be a limiting factor for the maximum compression ratio in a petrol engine.

After compression of the inlet charge, combustion of the fuel creates heat and pressure energy, which is imparted on the piston to generate mechanical work. In a petrol engine, this process is initiated by the high-voltage arc at the spark plug electrodes in the cylinder.

Combustion in the cylinder of an engine is a chemical reaction process between carbon and hydrogen in the fuel and oxygen present in the induced air. The carbon and oxygen combine to form carbon dioxide (CO_2), and the hydrogen combines with oxygen to form water (H_2O). Nitrogen passes through the engine as long as the combustion chamber temperatures remain below critical limits.

If the combustion process is not efficient, incomplete combustion will result and this produces carbon monoxide (CO). If combustion chamber temperatures are high, oxides of nitrogen (NO_x) are produced. These are harmful pollutants and their emissions from motor vehicles are closely regulated and controlled by environmental protection agencies and bodies around the world.

The combustion process should occur in a rapid but controlled manner. The flame propagation and energy release in the cylinder should have a predictable, stable behaviour depending on the engine operating conditions. The timing of the spark ignition is critical to achieve appropriate energy release for maximum efficiency in the energy conversion process that takes place in the combustion chamber. The burn duration of the fuel varies according to engine conditions; therefore, the spark must be adjusted to occur at the correct time, according to these conditions, to obtain the optimum torque from the engine. The optimum spark advance for a given engine condition is known as minimum spark advance for best torque (MBT) (Fig. 2.20).

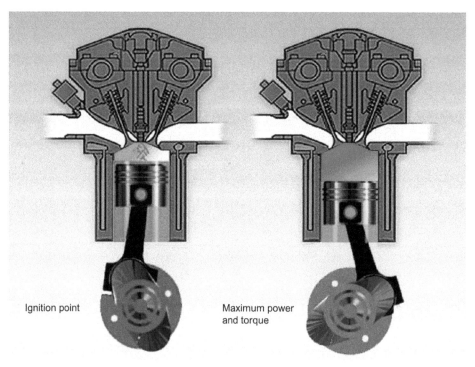

Ignition point   Maximum power and torque

Figure 2.20 Ignition timing for best torque

Definition

MBT: Minimum spark advance for best torque.

The quality of petrol (gasoline) is measured by a parameter called the octane rating, which gives an indication of the fuel's resistance to engine 'knock' or uncontrolled, spontaneous combustion, which causes engine damage. Fuels with a higher octane rating burn more slowly and in a more controlled manner, and hence have a greater resistance to knock. The octane rating of the fuel determines the limit of ignition advance for a given engine speed and load condition. Therefore, it is particularly important to operate the engine on the correct fuel, to prevent damage to the engine due to knocking.

A chemically correct air and fuel ratio mixture must exist to ensure that sufficient oxygen is present to completely combust all of the fuel. This is known as mixture strength and is the ratio of air mass to fuel mass (Fig. 2.21). For petrol, the correct ratio is approximately 14.7 air mass to 1 part fuel mass. If more air is present then the mixture strength is known as 'weak'. If less than a 14.7 air/fuel ratio, then the mixture strength is known as 'rich'. Weak and rich mixtures are less than optimum for the engine, although under certain conditions the mixture strength is adjusted by the engine control system according to demand. For example, for full power a slightly rich mixture is needed and this is provided when the engine is at full throttle. Extended running on rich or weak mixtures reduces engine efficiency and can cause damage to the engine and its subsystems.

Definition

Stoichiometric ratio: For petrol, the correct ratio is approximately 14.7 air mass to 1 part fuel by mass.

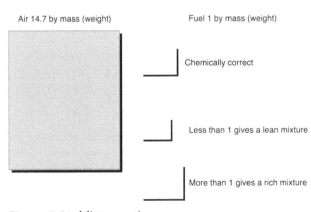

Figure 2.21 Mixture ratio

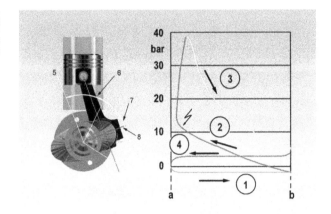

Figure 2.22 Petrol engine indicator diagram. Diesel figures are approximately double: 1, introduction; 2, compression; 3, power; 4, exhaust

The combustion process creates energy within the cylinder in the form of heat from the burning fuel/air mixture. Owing to the enclosed nature of the cylinder, this heat energy creates a pressure rise in the cylinder above the piston. This pressure, applied over the piston area, in turn, creates a force pushing down on the piston and turning the crankshaft via the connecting rod, thus producing torque at the crankshaft. The pressure in the cylinder is shown plotted against cylinder volume in Figure 2.22. This is known as an indicator diagram.

The torque at the crankshaft is a function of the cylinder pressure and crankshaft angle; the maximum torque is produced when the connecting rod and crankshaft main/big end bearings are at right angles (i.e. 90° crank rotation from TDC position). Note that at TDC, any pressure on the piston produces no work as there is no turning moment (torque), just a force pushing down on the bearings.

The ignition and fuel settings of an engine are set by the manufacturer at the optimum position to achieve the best compromise of performance, economy and minimal exhaust emissions. With respect to combustion, it is important that the maximum cylinder pressure and energy release occur at the correct angle. Damage to the engine can occur if this happens too early or late in the engine cycle (Fig. 2.23). An example is early or advanced ignition, which causes engine knock and damages the piston if allowed to occur for any significant period. This is a characteristic noise caused by pre-ignition or early ignition of the fuel/air mixture. Advanced or early ignition causes an early pressure rise that is applied to the piston at TDC. At this crank angle, no engine torque can be produced and this means that all the combustion

75

Figure 2.23 Ideal, retarded and advanced ignition point

energy is applied directly to the engine mechanical components (piston crown, bearings, etc.), causing them to generate the 'pinking' noise. Although the noise is quite subtle, the forces are massive and cause considerable damage to the engine.

When pinking occurs, the combustion energy precipitates through the engine components, causing damage. In addition, heat is generated that is not dissipated normally and this causes excessive temperature of engine components (e.g. pistons, valves and valve seats) and consequent heat-related damage.

Over-retarded ignition causes incorrect timing of the energy release from the fuel that, in turn, means less energy to do work and therefore more energy to dissipate via the cylinder boundaries. This causes an increase in engine temperatures, damages components and reduces overall engine efficiency. This excess energy also has to be rejected via the exhaust and this causes increased exhaust gas temperatures that can damage exhaust valves and seats, as well as exhaust gas components (catalytic converter).

Key fact
Overadvanced (early) or retarded (late) ignition can cause serious damage.

2.1.2.2 Compression ignition (CI)

In a diesel engine, the compression process must create sufficient energy to cause the temperature of the compressed gas (air) to rise above the self-ignition temperate of the fuel that is injected into the cylinder at the end of the compression stroke.

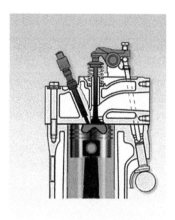

Figure 2.24 Direct diesel injection

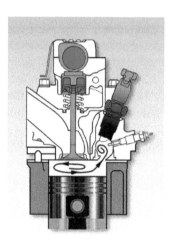

Figure 2.25 Indirect diesel injection

Inlet charge movement is particularly important in a diesel engine, to ensure that the fuel droplets have sufficient oxygen for complete combustion. The required air flows during the induction and compression processes are created by the design of the inlet tract and combustion space. There are two designs of combustion chamber in common use and these are named after the position in the chamber where the fuel is introduced. They are known as direct and indirect injection (Figs 2.24 and 2.25).

Key fact
A direct injection combustion chamber has a 'bowl' formed in the piston crown.

A direct injection combustion chamber has a 'bowl' formed in the piston crown. This is designed to promote a tumble movement of the incoming air mass; this helps to ensure good distribution of the fuel in the cylinder and reduced soot emissions.

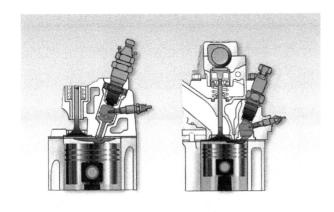

Figure 2.26 Precombustion chamber and swirl chamber (indirect injection)

The indirect type combustion chamber incorporates a precombustion chamber within the cylinder head (Fig. 2.26). The compressed inlet charge is forced into this chamber at high velocity and pressure. This creates a swirl movement that ensures complete mixing of fuel droplets with air for maximum combustion efficiency. During the combustion process, the burning gases are ejected from this chamber with high pressure and energy. This ensures sufficient turbulence in the main combustion chamber for efficient combustion.

The compression ratio of a direct injection engine is typically between 16:1 and 21:1. This is sufficient to raise the induced charge temperature for self-ignition of the fuel under all engine operating conditions without creating excessive combustion noise (or diesel knock). Indirect injection engines have higher compression ratios of 22:1 to 25:1. This is necessary to generate the extra heat energy required due to losses via the increased surface area of the cylinder head. Diesel knock is less apparent in indirectly injected engines as the energy release is more controlled and less spontaneous.

Key fact

The compression ratio of a direct injection engine is typically between 16:1 and 21:1.

The diesel engine is designed to produce compression pressures that generate sufficient heat in the cylinder to ignite the fuel as it is injected into the combustion chamber. This is known as compression ignition (CI). Petrol engines are generally known as spark ignition (or SI) engines. During the combustion stroke, the engine power output or work is generated, hence the name 'power' stroke in the four-stroke cycle of

induction, compression, power and exhaust. Engine combustion is a fundamental process in the operation of the engine. This process must be efficiently executed and controlled via the engine subsystems (fuel, air, ignition, etc.) to ensure best efficiency and performance, with minimum harmful exhaust emissions.

Combustion in a diesel engine begins very rapidly as the fuel is being injected into the combustion chamber and heated. This causes a rapid energy release that generates the characteristic 'diesel' engine noise. For this reason, a simple diesel engine is noisier than the equivalent petrol engine. The combustion process is most rapid in a direct injection diesel engine (Figs 2.27 and 2.28) and, because of this, combustion losses are minimal and these are the most fuel-efficient type of internal combustion engine seen in road vehicles. Combustion is not fully completed in the prechamber before the combustion gases are expelled into the cylinder and continue the combustion process in the

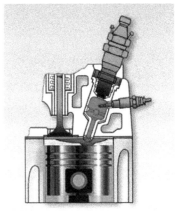

Precombustion Chamber

Figure 2.27 Indirect diesel injection

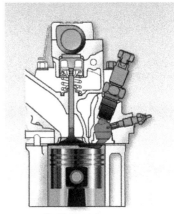

Swirl Chamber

Figure 2.28 Swirl chamber

main combustion chamber. The increased surface area means that more heat from combustion is lost to the cylinder boundaries (walls, head, etc.) than in the direct injection type. Indirect injection engines are, however, quieter in operation because of the longer, slower combustion process.

Diesel engines always operate with excess air (or weak mixture) and with high exhaust gas temperatures. This oxygen combines with nitrogen to form the pollutant NO_x (oxides of nitrogen). Exhaust gas recirculation (EGR) is often used to reduce cylinder temperatures and to reduce the amount of oxygen in the cylinder air charge, thus preventing the formation of NO_x (Figs 2.29 and 2.30).

For diesel engines, there are three main phases in the combustion process (Fig. 2.31). The first is the delay phase as the fuel absorbs heat from the cylinder air charge and vaporizes. The next phase occurs when the fuel has reached a sufficient temperature to self-ignite. This causes combustion and the flame front propagates rapidly out across the piston crown. This

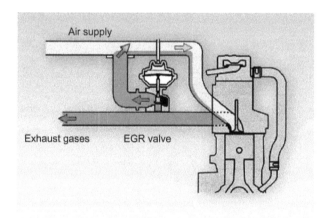

Figure 2.29 Exhaust gas recirculation (EGR)

Exhaust gas in air charge reduces oxygen content

Cylinder charge with exhaust gas recirculation

Figure 2.30 Reduced oxygen

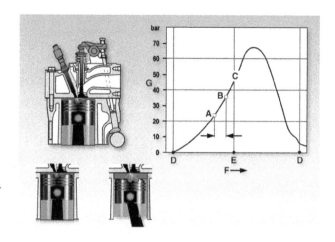

Figure 2.31 Phases of combustion, diesel engine: A, start of injection; B, start of combustion; C, controlled phase; D, bottom dead centre; E, top dead centre; F, piston travel; G, combustion chamber pressure (A–B is the delay phase)

is where the rapid energy release occurs and causes the characteristic diesel engine noise. Once initial burning of the fuel takes place, continued injection of fuel provides a controlled burning and energy release to provide sustained pressure on the piston and good torque generation at the crankshaft.

Key fact

For diesel engines, there are three main phases in the combustion process:
1 start of injection and delay
2 start of combustion and flame spread
3 controlled combustion.

Efficient and effective combustion promotes an engine with good power output and with minimal harmful emissions. This can only be achieved when the engine mechanical parts are in good condition and the engine control systems for fuel delivery, ignition and emission control are correctly optimized and set.

2.1.2.3 Engine terminology and methods

The following are some of the technical terms that are used to describe features of the engine:

▶ **Engine capacity** – the total, combined, displaced volume of all engine cylinders as a single value stated in units of cubic capacity. This is generally given in cubic centimetres (cm^3) or litres. In America, engine capacity is normally stated in cubic inches (in^3).

▶ **Swept volume** – the volume of a cylinder bore between the TDC and BDC piston positions, excluding the volume above the piston at TDC (Fig. 2.32).

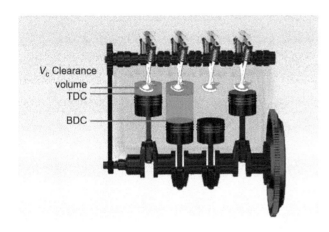

Figure 2.32 Clearance and swept volumes

▷ **Clearance volume** – the volume above the piston at TDC. Note that it is the volume of the combustion chamber itself (Fig. 2.32).
▷ **Bore** – the diameter of the engine cylinder.
▷ **Stroke** – the total linear distance travelled by the piston in the bore between TDC and BDC positions. Note that it is twice the crankshaft throw.
▷ **Compression ratio** – the total volume of the cylinder at BDC (swept + clearance volume), expressed as a ratio of the volume of the cylinder at TDC (clearance volume).

The information on cylinder dimension can generally be found in workshop or manufacturer manuals. In addition, these values can be measured directly or derived via calculations.

Swept volume can be calculated via the formula:

$$(l\pi d^2)/4$$

where d is cylinder bore and l is stroke. Note that units of bore and stroke must be consistent. Engine volume is mostly stated in litres by manufacturers, but remember that 1000 cc (cubic centimetres) equals 1 litre. The total engine displacement is the sum of all cylinders' individual displacements.

The formula used to calculate the compression ratio is:

$$CR = (V_s + V_c)/V_c$$

where CR is compression ratio, V_s is swept volume, and V_c is clearance volume.

Note the correct order of preference when carrying out this calculation (remember or Google BODMAS).

Two common terms used when expressing engine performance characteristics are 'torque' and 'power'. Torque is an expression relating to work and is a

measure of the turning force provided by the engine. Torque output can vary independently of engine speed and is a measure of the load on the engine. The SI units of torque are newton metres (Nm) and the imperial units are pounds/foot (lb-ft). Power is a derived unit and relates to the rate of work done, or the work done per unit of time. For an engine, the power is a product or torque and speed. Power output is given in kilowatts (kW) or horsepower (HP). Engine power is normally stated as measured at the flywheel, via a dynamometer or brake, hence the term 'brake horsepower' (Fig. 2.33).

Engine manufacturers often publish performance data in a graphical form showing torque and power curves against speed. Two examples are shown in Figures 2.34

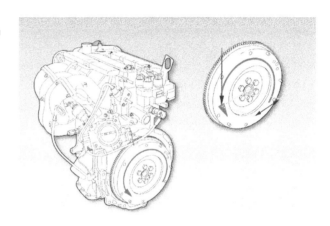

Figure 2.33 Torque and power measured at the flywheel

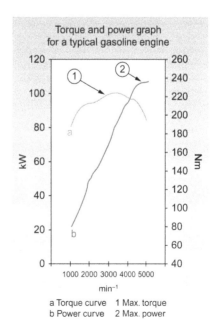

Torque and power graph for a typical gasoline engine

a Torque curve   1 Max. torque
b Power curve   2 Max. power

Figure 2.34 Spark ignition (SI) engine

79

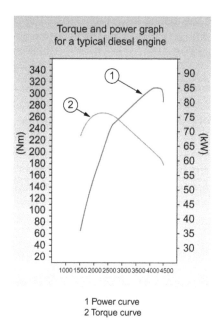

Torque and power graph for a typical diesel engine

1 Power curve
2 Torque curve

Figure 2.35 Compression ignition (CI) engine

and 2.35. Note that a petrol engine generally produces more power, at higher speed. A diesel engine produces more torque at lower speeds.

Definitions

Torque: The turning force produced by an engine. It is not affected by time.
Power: The rate at which energy is being converted. It is therefore related to time.
Or how about this version?: Power is how hard you hit a tree. Torque is how far you move it.

The optimum size of an individual engine cylinder is a compromise of a number of technical factors. The optimum displacement for a cylinder is generally between 250 and 600 cm³ for road vehicle applications. In this range, the combustion chamber size, surface area and individual components size (pistons, valves, etc.) produce an engine with optimum efficiency with respect to fuel consumption and emissions. Typically, engines with total displacements in the range of 1–2.5 litres have four cylinders (Fig. 2.36).

The number of power strokes per revolution can be found by dividing the number of engine cylinders by two (for a four-stroke engine). The greater the number of cylinders, the smoother the torque delivery owing to reduced peak torque firing pulses from each cylinder and the increased number of firing strokes per revolution. Over 2 litres, six-cylinder engines give smooth power delivery with optimum cylinder displacement sizes. An in-line six cylinder has a relatively long crankshaft that can be difficult to accommodate in a transverse engine installation layout; therefore by using two banks of three cylinders in a 'V' configuration, total length is reduced and torsional rigidity of the crankshaft is improved (Fig. 2.37).

The engine's flywheel acts as an energy buffer owing to its inertia. Energy stored in the flywheel maintains rotation between firing pulses and acts as a damper to smooth torque peaks as each cylinder fires. There are numerous engine configurations with respect to the arrangement of the engine cylinders, the number of cylinders, position and firing order

Figure 2.36 Displacement

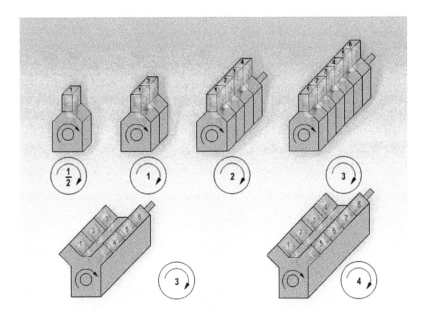

Figure 2.37 Cylinder layouts and power strokes per engine revolution

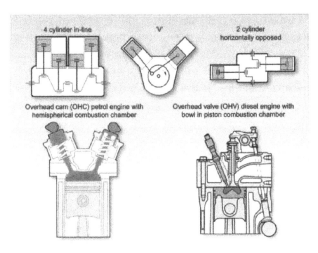

Figure 2.38 Engine designs

(Fig. 2.38). In addition, combustion chamber designs and valve train layout all dictate the basic properties of an engine. Engine installation and orientation is another important factor to be considered in a road vehicle. There are also two-stroke and rotary engine designs with their own particular characteristics, all of which are explained in this section.

2.1.2.4 Engine design variations

The following section outlines some of the many design variations that are or have been used for engines. The configuration of the engine depends on the number of cylinders, their relative position, the engine layout and the firing order, as well as combustion chamber design, fuel type, valve train design, engine location and mounting position. Further

parameters are engine type (reciprocating, rotary) and stroke (two or four).

The simplest engine design is the single cylinder, normally found in small engine applications (Fig. 2.39). If the engine is a single-cylinder four-stroke, then the engine fires only once every other engine revolution. This gives a large variation in the torque delivery at the engine crankshaft and hence torsional vibration is significant and the engine is not very smooth. Increasing engine capacity does not increase power output directly as there are other factors that contribute to the efficiency of the engine (heat and pressure losses); therefore, there is an optimum cylinder capacity that provides the best compromise of surface and valve area for maximum efficiency, which

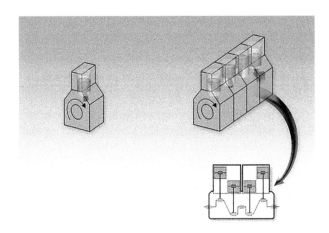

Figure 2.39 A single- (left) and four-cylinder (right) engine

81

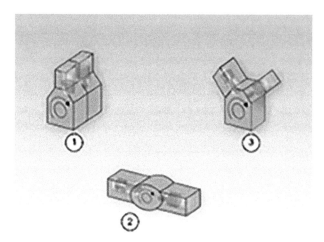

Figure 2.40 Twin-cylinder engines. 1, In-line; 2, horizontally opposed; 3, V type

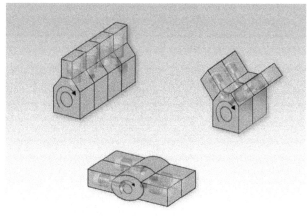

Figure 2.42 Four-cylinder in-line engines

is approximately 0.5 litre displacement volume. For this reason, increasing the engine capacity is normally done by adding extra cylinders.

Twin-cylinder engines can be configured as in-line, horizontally opposed or V types (Fig. 2.40). These engines have been used in car applications but are more commonly found in motor cycles. In-line engines have been built with both pistons operating in parallel and on alternate strokes; often, this depends on whether they are two- or four-stroke engines. Horizontally opposed cylinders have pistons that move out and return in opposing directions. They are well balanced as the forces generated by the reciprocating masses (pistons, con rod, etc.) cancel out exactly. Various V engine configurations have been used.

Three-cylinder in-line engines of around 1 litre displacement have been used by some manufacturers in car and motor cycle applications. The three-cylinder V engine is a possible design being considered as it has a very compact form (Fig. 2.41).

Four-cylinder in-line engines (Fig. 2.42) are commonly used as they provide a good compromise of performance, efficiency and smoothness with optimized individual cylinder displacements where the total engine displacement is in the range of 1–2.5 litres. This range is extremely common for passenger car applications. Another well-established design for this application is the opposed cylinder; less common is the V4 engine.

Where five-cylinder engines (Fig. 2.43) have been designed and used, they are generally of in-line construction only. There is an example of a V5 engine with a very narrow V-angle, two cylinders on one bank and three on the other. The main advantage of a five-cylinder engine is that the reciprocating forces are well balanced and this provides a smooth power delivery. The V5 cylinder engine has the added advantage that the overall length of the engine is reduced (compared with

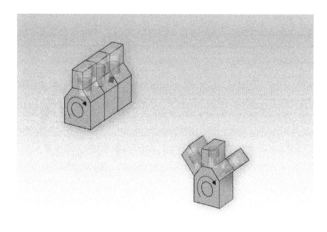

Figure 2.41 Three-cylinder engines

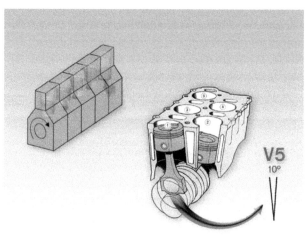

Figure 2.43 Five-cylinder in-line engines

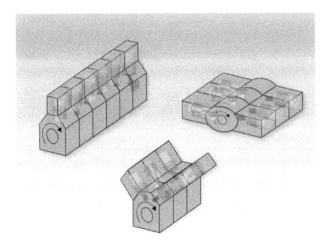

Figure 2.44 Six-cylinder engines

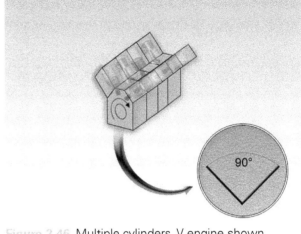

Figure 2.46 Multiple cylinders. V engine shown

the in-line design) and this allows for a transverse engine powertrain layout.

Six-cylinder engines (Fig. 2.44) are very common and have been built with in-line, horizontally opposed and V layouts. The construction and manufacturing costs of a six-cylinder engine are higher but they are well balanced and offer smoother power delivery than the equivalent four-cylinder engine. If greater engine capacity is necessary, then a six-cylinder engine is necessary to keep the individual cylinder displacements in the optimum range.

Horizontally opposed and V engines (Fig. 2.45) have shorter crankshafts and overall length than the equivalent in-line engines. This makes them appropriate for transverse or overhung installation in the powertrain. The optimum angle for a V6 cylinder engine is 60°.

Eight, ten, twelve and higher V configuration engines are manufactured but less common in vehicle

applications (Fig. 2.46). V8 engines are very common and used in most countries on larger vehicles. There are petrol and diesel engine designs that employ this layout with engine capacities greater than 3 litres. The optimum V-angle for cylinder banks in an eight-cylinder V engine is 90°.

For multicylinder engines, the firing order, the crankshaft big-end journal positions and the direction of rotation must be considered together in the engine design. In general, in-line four-cylinder engine crankshafts have cylinders numbered 1 and 4, and 2 and 3, paired and 180° apart. This gives two possible firing orders of 1-3-4-2 and 1-2-4-3 with alternate firing from each pair, giving a power stroke every 180° crank angle. In-line six-cylinder engine cylinders are paired; 1 and 6, 2 and 5, and 3 and 4. The big-end journals are positioned at 120° intervals and this gives the most common firing order of 1-5-3-6-2-4 (Fig. 2.47).

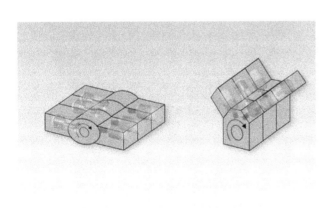

Figure 2.45 Flat and 'V' engines

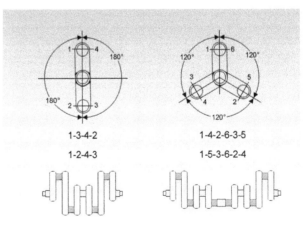

Figure 2.47 Firing order: 4-cylinder (left) and 6-cylinder (right) engines

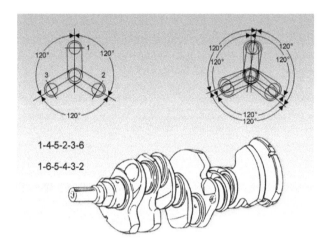

Figure 2.48 V6 crank

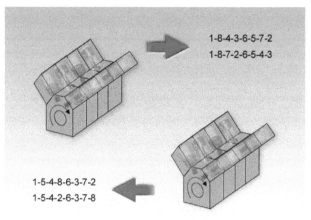

Figure 2.50 Typical firing orders – V8 engine

In a V6 engine (Fig. 2.48), the big-end journals carry two connecting rods each, one from each cylinder bank. The journals are positioned at 120° intervals and are either a single journal or offset journals with the two big-end crank pins offset to match the connecting rod angle to the journal.

V6 engine cylinder numbering is generally implemented in one of two ways. The first is that one bank has 1, 2 and 3 and the other bank has 4, 5 and 6. The alternative numbering sequence uses alternate banks with 1, 3 and 5 on one side and 2, 4 and 6 on the other. The firing orders are either one bank followed by the alternate bank or one cylinder from one bank followed by a cylinder from the other bank.

A V8 engine has four paired journals placed at 90° intervals to each other (Fig. 2.49). Each journal carries one connecting rod from each of the opposite banks. The cylinders can be numbered in a similar method to the six-cylinder engines with either odd numbers on

one side and even numbers on the other, or 1–4 on one cylinder bank and 5–8 on the other bank. Typical firing orders are shown in Figure 2.50.

2.1.3 Engine components

This section explores various different engine designs and components found on modern road vehicles (Fig. 2.51).

2.1.3.1 Engine block and liners

Engine cylinders, when cast in a single housing, are known as the engine block. Usually, the engine block

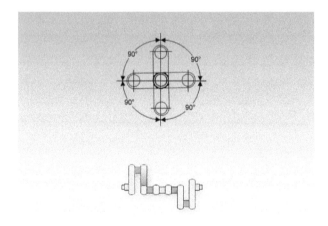

Figure 2.49 V8 crank

Figure 2.51 Engines vary, here is a four cylinder inline design. (Source: Ford Media)

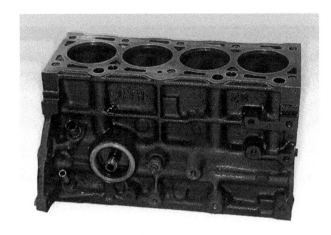

Figure 2.52 Cast iron engine block

is manufactured from cast iron (Fig. 2.52) or aluminium alloy. In the latter case, cast iron or steel liners form the cylinder bore. The engine block forms the major component of a 'short' motor.

The cylinder bores are formed via a machining process with a boring tool to give the correct form to the cylinder within closely specified tolerances. Cast iron is a mixture of iron with a small amount of carbon (2.5–4.5% of the total). The carbon added to the iron gives a crystalline structure that is very strong in compression. In addition, it is slightly porous and this helps to retain a film of lubricating oil on working

surfaces. This property makes cast iron particularly suitable for cylinder bores that can be machined directly into the casting. On many engines cylinder liners are used.

Cylinder liners fall into two main categories, wet and dry (Fig. 2.53). Wet liners are installed such that they are in direct contact with the coolant fluid. They are fitted into the block with seals at the top and bottom and are clamped into position by the cylinder head. Spacers are fitted at the bottom to adjust the protrusion of the liner to achieve the correct clamping force.

Definition
Cast iron: A mixture of iron with a small amount of carbon (2.5–4.5% of the total).

Key fact
Cylinder liners fall into two categories, wet or dry.

Dry liners are not in direct contact with the coolant. In general, they are fitted into the casting mould and retained by shrinkage of the casting via an interference fit. Alternatively, they can be pressed into place in a precast cylinder block. When repairing or reconditioning the engine, the former type can be rebored whereas the latter type is replaceable.

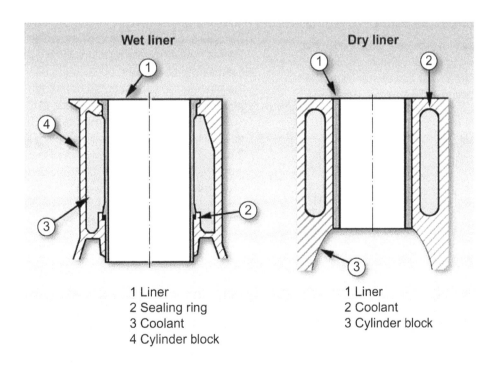

1 Liner
2 Sealing ring
3 Coolant
4 Cylinder block

1 Liner
2 Coolant
3 Cylinder block

Figure 2.53 Wet and dry liners

Figure 2.54 Modern engine block

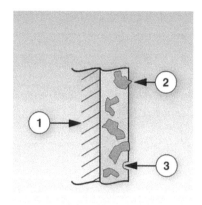

Figure 2.56 Cylinder coating: 1, engine block; 2, silicon carbonate; 3, nickel phosphate

Most modern engines (Fig. 2.54) have specific treatments applied to the cylinder bores and, as such, cannot be rebored or honed. Replaceable liners mean that the liner and piston assembly can be replaced without the need for specialist reboring equipment. Commercial vehicle engines often use replaceable liners to reduce repair times.

Cast iron has been used for cylinder block construction in the past as the cylinders can be bored directly into the material; in addition, these bores can be remanufactured or repaired by reboring oversize. Cast iron is porous and hence the cylinder bore is capable of retaining lubricating oil for lubrication of the contact surfaces. The disadvantage of cast iron is weight. Modern engines use aluminium and can achieve the same strength and stiffness as cast iron via advanced design techniques.

Aluminium alloy cannot provide a suitably durable surface for piston ring contact. Therefore, cylinder liners or sleeves, made from cast iron or steel, are normally fitted into an aluminium cylinder block (Fig. 2.55).

Key fact

Aluminium alloy cannot provide a suitably durable surface for piston ring contact, so liners made from cast iron or steel are used.

Recent developments in material technology have produced a coating of nickel phosphate and silicon carbonate which provides a suitably durable surface for the cylinder bores (Fig. 2.56). Note that these bores cannot be rebored, so if excessive wear occurs the block must be replaced.

A crankcase is usually integrated into the cylinder block and is machined in-line to form the crankshaft main bearings. This process is known as line boring. The main bearings are split in two halves; one half locates in the block, the other in the bearing cap. The bearing caps are secured before the machining process and thus each cap is matched in position with its opposite half. It is important to note this when disassembling and reassembling the bearings. The caps are located via dowels and fastened via high-tensile steel bolts (Fig. 2.57). It is important to follow manufacturer guidance if the bolts are removed and refitted; replacement of the bolts and tightening procedures must be followed if specified.

Between the cylinder walls and the outside surface of the cylinder block, voids and channels are formed during the casting process; this is known as the water jacket (Fig. 2.58) and is used for engine cooling purposes. A sand former creates this space during casting and when the cast block has cooled, the sand is evacuated via holes in the side of the block. These holes are then sealed using core plugs (Fig. 2.59).

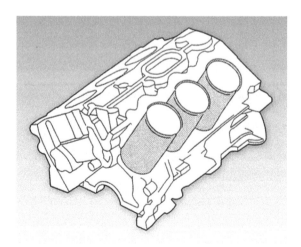

Figure 2.55 Liners in an aluminium block

Figure 2.57 Main bearing bolts

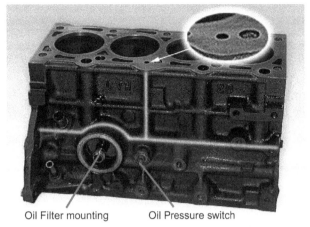

Oil Filter mounting   Oil Pressure switch

Figure 2.60 Oilways

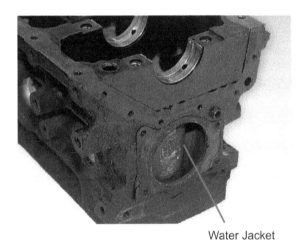

Water Jacket

Figure 2.58 Water jacket

Figure 2.59 Core plugs

crankshaft and cylinder head. Additional drillings connect the oil pump and pressure control valve to complete the oil supply system. The block is prepared, drilled and threaded to attach additional components such as the oil sump pan and oil pump assembly.

> **Key fact**
> Pistons are generally manufactured from an aluminium alloy, which reduces weight and increases heat dissipation.

2.1.3.2 Pistons and connecting rod

Pistons are generally manufactured from an aluminium alloy, which reduces weight and increases heat dissipation. There are numerous designs to accommodate thermal expansion according to engine type and application (Fig. 2.61).

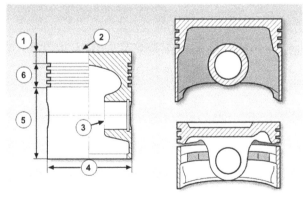

Figure 2.61 Piston features: 1, fire land; 2, crown; 3, piston pin bore; 4, skirt diameter; 5, skirt; 6, piston rings

In order to supply pressurized oil to the engine moving surfaces, an oil gallery is formed along the length of the cylinder block (Fig. 2.60). This has drillings to supply oil directly to the bearings in the block,

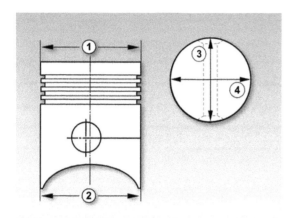

Figure 2.62 Piston dimensions: 1, crown diameter; 2, skirt diameter; 3, diameter in piston pin direction; 4, diameter at right angles to the pin

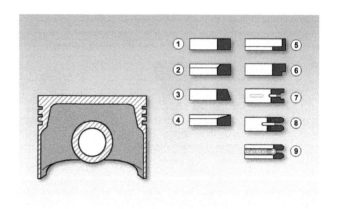

Figure 2.64 Piston rings: 1, rectangular; 2, internally chamfered; 3, taper faced; 4, trapezoidal; 5, L shaped; 6, stepped; 7, slotted oil control; 8, oil ring with expander; 9, oil ring with spiral expander

Aluminium has greater thermal expansion than cast iron used for the block and cylinder liners. This means that the piston expands more than the block as the engine temperature increases. When the engine is cold, the working tolerances are greater to allow for expansion. The piston has design features to allow for expansion and correct tolerances at running temperatures; for example, a cold piston is slightly oval and tapered inwards towards the crown (Fig. 2.62).

The piston pin or gudgeon pin has an offset by a small amount toward the thrust face of the cylinder bore, which allows the thrust forces at the piston crown to maintain the piston against the cylinder wall (Fig. 2.63). This has an effect when the engine is cold by reducing piston movement due to excessive clearance, which creates a noise known as 'piston slap'. Note that pistons are marked so that they can be installed correctly and this should be carefully observed.

Around the upper portion of the piston, grooves are cut to accommodate sealing rings, known as piston rings (Fig. 2.64). In general, there are three or four grooves and rings, the lowest of which is known as the oil control ring and is used to control the amount of lubricant remaining on the cylinder bore surface to lubricate the piston. The upper rings are known as compression rings and these provide the gas-tight seal, maintaining the cylinder pressures that create force to move the piston.

Piston rings seal the combustion chamber to prevent the escape of combustion gases and loss of cylinder pressure; these are known as 'compression rings' (Fig. 2.65). In addition, the piston rings must control the oil film on the cylinder bore surface, and these are known as 'oil control rings'. Combustion pressure is allowed to act on the back of the cylinder-sealing compression rings to help maintain a gas-tight seal of the piston assembly.

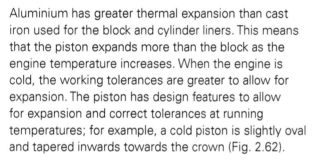

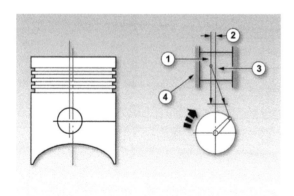

Figure 2.63 Piston design features: 1, piston; 2, offset; 3, centre line; 4, block thrust face

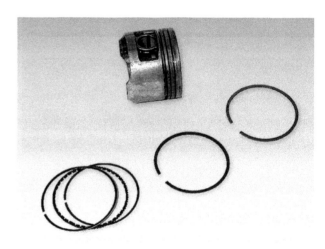

Figure 2.65 Compression and oil control rings

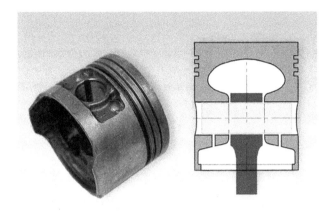

Figure 2.66 Cross-section of a piston

Key fact

Piston rings seal the combustion chamber to prevent the escape of combustion gases.

Compression rings are manufactured from cast iron, with a surface coating to promote fast bedding in. This means that the rings quickly wear in to give a gas-tight seal against the cylinder pressures. It is important not to damage this coating during fitting. Note that rings have different cross-sections according to their mounting position on the piston (Fig. 2.66).

Oil control rings can be one of two designs. A multipart ring consists of two thin alloy rings used in conjunction with an expander between them. A cast iron ring has a groove and slot arrangement to allow oil flow back to the sump via the ring and piston.

The piston pin or gudgeon pin bore is machined into the piston to accept the piston pin, also known as the gudgeon pin (Fig. 2.67). The fixing mechanism of the piston pin to the piston and the connecting rod

Figure 2.68 A selction of con rods

can vary (Fig. 2.68). It can be an interference fit in the connecting rod, or a push fit in both the piston and connecting rod end. If the piston pin is clamped in the connecting rod, the piston pin bore is smooth (Fig. 2.69). Circlip grooves are formed in the piston pin bore when a push fit piston pin is used.

The piston crown forms part of the combustion chamber and experiences the full cylinder pressure applied by the expanding gases. Many different

Figure 2.67 Piston pin (gudgeon pin)

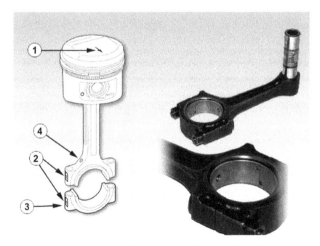

Figure 2.69 Con-rod features: 1, front of engine; 2, identification marks; 3, big end cap; 4, oil spray hole for cylinder wall lubrication

89

Figure 2.70 Piston pin in position

designs are available depending on engine type. Complex shapes can be formed in the piston crown to allow for valve movements and to create an effective combustion chamber space, promoting the correct charge motion for efficient combustion.

The piston or gudgeon pin provides the mechanical link between the piston and connecting rod. The pin locates in the piston body and the little end of the connecting rod (Fig. 2.70). The pin can be a clearance fit into the little-end bearing or bush, and hence a corresponding interference fit, or located via circlips, in the piston.

An alternative to circlips is that the pin is an interference fit in the little end, or is clamped by the connecting rod. In this case, the piston pin bore is the bearing surface and there are appropriate drillings in the piston to allow for lubrication.

The main purpose of the connecting rod is to transfer the linear force from the piston and apply it to the rotating crankshaft (Fig. 2.71). It is generally manufactured from carbon steel in a process known as drop forging to form the required shape and profile.

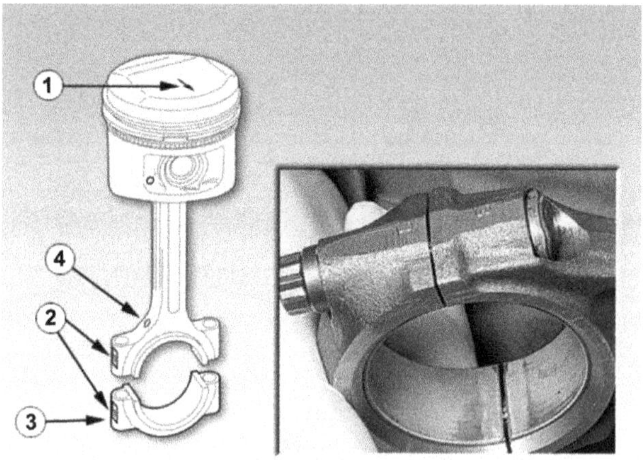

Figure 2.72 Big end bearing: 1, front of engine; 2, identification marks; 3, big end cap; 4, oil spray hole for cylinder wall lubrication

The connecting rod is designed specifically with a high resistance to bending, compressive and tensile forces via an I-section profile. The piston end, known as the little end, has an appropriate bush, bearing or clamping arrangement for the piston pin. The crankshaft end of the connecting rod is known as the big end (Fig. 2.72). This consists of a split bearing with a removable bearing cap. The bearing cap is attached to the connecting rod via bolts or nuts.

It is important to note that the connecting rod and bearing cap are machined as one unit, and hence, the parts are matched. Therefore, they must always be reassembled as a pair and fitted correctly oriented.

The crankshaft receives the linear force of the pistons, via the connecting rods, and converts this force into a rotating torque (Fig. 2.73). The crankshaft is generally manufactured from cast iron or steel alloy via a forging or casting process.

Key fact

The crankshaft of a four-cylinder engine generally has five main bearings.

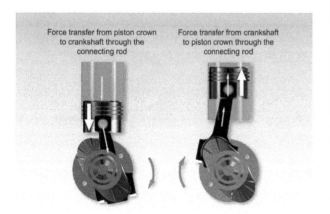

Figure 2.71 Force transfer

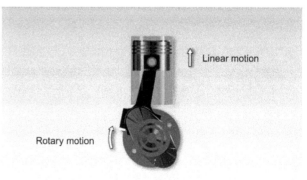

Figure 2.73 Conversion of linear to rotary motion

2.1.3.3 Crankshaft

The crankshaft of a four-cylinder engine usually has five main bearings. At the front of the crankshaft provision is made to locate and drive the crankshaft pulley and timing gear via keyways and securing bolts (Fig. 2.74). The oil pump drive is located behind this, and then the first or front main bearing.

The big-end bearing for the first cylinder is fitted in between the crankshaft webs radiating from the main bearing journals (Figs 2.75 and 2.76). These webs form

counter-balance weights to the big-end journal. One of the main bearings is usually fitted with a thrust washer to control axial movement of the crankshaft.

At the rear of the main bearing journal, at the back of the engine, a machined face is formed on the crankshaft as a mating surface for a sealing ring. This is the main oil seal at the back of the engine (Fig. 2.77). In addition, there is a machined, threaded flange surface to accommodate mounting of the flywheel. For a four-cylinder engine, the big-end journals are paired and set at 180°. For most four-cylinder engines the firing order is 1, 3, 4, 2.

The crankshaft bearings are split-type, steel-backed shells with an alloy or coated bearing surface (Fig. 2.78). Correct bearing types to engine manufacturers' original specification must always be fitted. Bearing shell halves, when correctly fitted and tensioned in the bearing caps, form a perfectly round profile with equidistant clearance around the bearing journal. The bearings are 'nipped' and held in position

Figure 2.74 Crankshaft pulleys

Figure 2.75 Crankshaft: 1–5 main journals

Figure 2.77 Radial lip oil seal

Figure 2.76 Journals in position in the engine block

Figure 2.78 Bearings

91

Figure 2.79 Signs of 'nip'

Figure 2.81 Dual-mass flywheel

Figure 2.80 Details of the bearing: 1, locating lug; 2, oil groove; 3, steel support shell; 4, support coating; 5, nickel coating; 6, surface coating; 7, oil bore; 8, thrust flange

when fitted into the tightened bearing caps (Fig. 2.79). The bearing shell is also fitted with a locating lug on the back that mates with a slot in the bearing locating 'half-bores' (Fig. 2.80). This ensures that the bearing cannot rotate. Oil supply holes and slots are machined in the bearing surface to supply appropriate lubrication.

Key fact

Bearing shell halves, when correctly fitted and tensioned in the bearing caps, form a perfectly round profile.

Axial displacement of the crankshaft is controlled by thrust bearings to limit the axial movement. These are fitted at a main bearing journal either as two semicircular rings or as part of a main bearing shell.

A flywheel is fitted on the end of the crankshaft to reduce vibration by maintaining the engine rotation

between power strokes. The clutch on a manual transmission vehicle is mounted on the flywheel. On automatic transmission vehicles the torque converter also works as a flywheel. Particularly on modern small but high-torque diesel engines, a dual mass flywheel can be used as shown in Figure 2.81. Vibration on these systems is reduced considerably by splitting the mass of the flywheel into two parts and connecting them by damping springs.

Key fact

A dual-mass flywheel splits the mass of the flywheel into two parts and reconnects them by damping springs to reduce vibration.

2.1.3.4 Cylinder heads

The cylinder head gasket has to form a gas-tight seal at the interface between the cylinder head and the cylinder bores. In addition, it must seal and separate the cooling water supply jacket and the oil supply and return drillings. Traditional head gaskets were constructed from copper and asbestos. Modern material technologies allow head gaskets to be made from composite materials which have superior sealing and heat transfer performance (Fig. 2.82).

Key fact

Modern head gaskets are made from composite materials.

Head gaskets must always be replaced when the cylinder head is removed and refitted. In addition, when refitting a cylinder head, it is important that the manufacturer's information is sought and applied with respect to replacement of cylinder head bolts where

Figure 2.84 Valve seats

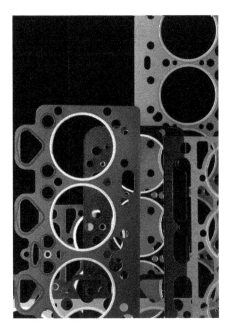

Figure 2.82 Head gaskets

Figure 2.85 Inside the combustion chamber

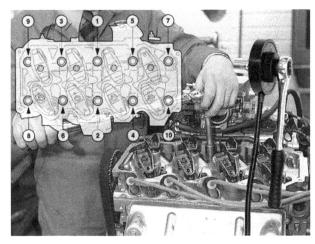

Figure 2.83 Tightening sequence

Figure 2.86 Overhead valve (OHV) head

necessary. Correct torque and a tightening sequence of cylinder head bolts must be followed (Fig. 2.83).

Cylinder heads are cast from aluminium alloy or iron. Aluminium alloy is lighter but cast iron or steel valve seats and guides must be installed in the head (Fig. 2.84). Cast iron heads generally have valve seats and guides formed directly in the head material.

The combustion chamber (Fig. 2.85) is formed in the cylinder head such that, on assembly, it is located directly over the cylinder bore in the engine block. There are numerous designs in use depending on engine type, optimization parameters and application. The evolution of petrol engines can be seen in the design of combustion chambers that has been developed over the years to improve efficiency.

The next major development was overhead valve (OHV) engines (Fig. 2.86). These used in-line valves and bath-tub combustion chambers over the piston. Improved combustion and flame propagation could be achieved with a wedge-shaped chamber (Fig. 2.87). This had the valves offset from the vertical position. A problem associated with the wedge design is combustion knock, also known as pinking (Fig. 2.88). This is caused by uncontrolled ignition of the end gases prior to ignition from the advancing flame front. It occurs as a result of compression of the end gases

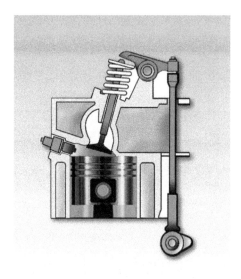

Figure 2.87 Combustion chamber wedge design

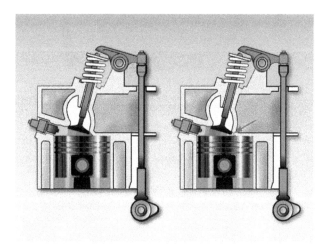

Figure 2.88 Engine knock can cause damage

in the thin end of the wedge. This generates pockets of combustion with high pressures that damage the piston crown and area above the compression ring.

Key fact

A problem associated with the wedge design is combustion knock, also known as pinking.

It is possible for inlet and exhaust ports to be on the same or opposite sides of the cylinder head, known as precross-flow or cross-flow. In addition, two cylinders can share a common inlet port and this is known as a siamese port. When a single inlet for each port is used, this is known as parallel ports.

Key fact

Most modern petrol engines use a hemispherical or pent roof combustion chamber design.

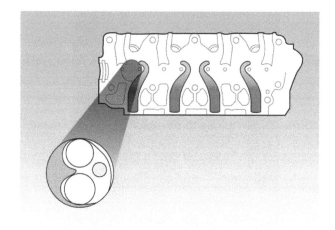

Figure 2.89 The need for a good mixture is why port and valve design is important

Most modern petrol engines use a hemispherical or pent roof combustion chamber design. This shape provides the best compromise of surface to volume ratio; this reduces heat energy loss during the expansion stroke, which in turn improves the thermal efficiency of the engine. The combustion chamber design allows single- or dual-inlet and exhaust ports with cross-flow engine breathing. This design easily accommodates two to five valves per cylinder.

The correct design of the inlet tract, including manifold, head and valves, is essential to provide the charge motion as the gases enter the cylinder during the induction stroke (Fig. 2.89). Charge motion is important to speed up the combustion process sufficiently in order to prevent excessive exhaust emissions.

One arrangement of combustion chamber and valves uses a hemispherical design with two valves per cylinder positioned opposite each other for cross-flow movement of the intake and exhaust gases. It is known as compound valve hemispherical (CVH). The valves are inclined such that they sit in the curved profile of the combustion chamber space. The spark plug is mounted as close to centric as possible in the combustion chamber via an appropriate drilling (Fig. 2.90). It is sealed by a compressible washer or conical sealing face.

The combustion chambers are surrounded by cooling water passages that are connected to the water jacket in the cylinder block (Fig. 2.91). The water jacket casting holes are sealed by core plugs in a similar way to the cylinder block. On the upper surface of the cylinder head, bearing journal surfaces are formed to locate the valve operating camshafts and mechanism. Oil supply drillings ensure adequate lubrication for the camshaft bearings and valve train components.

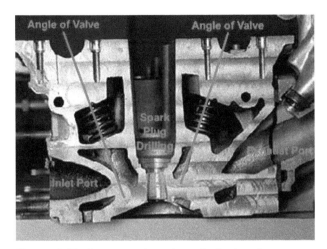

Figure 2.90 Cross-section of a cylinder head

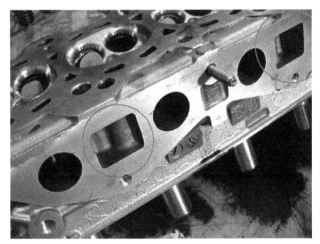

Figure 2.91 Engine coolant passages

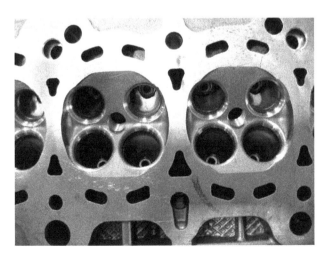

Figure 2.92 Hardened seats in an aluminium head

Hardened valve seat inserts are required in aluminium heads (Fig. 2.92). Certain cast iron heads will also use these. They are necessary to increase the durability of the head such that it can resist the heat of the exhaust

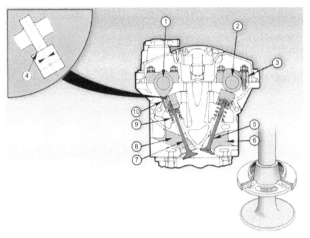

Figure 2.93 OHC value rotator features: 1, exhaust cam; 2, inlet cam; 3, cam bearings; 4, rotator; 5, inlet valve; 6, inlet port; 7, exhaust valve; 8, exhaust port; 9, spring; 10, follower

gases. In older engines, which ran on leaded fuel, the lead fuel additive provided an element of protection.

> **Key fact**
> The combustion chambers are surrounded by cooling water passages that are connected to the water jacket in the cylinder block.

> **Key fact**
> Hardened valve seat inserts are required in aluminium heads.

For some engines the cam followers and push rods are encouraged to rotate and this helps to extend their life and reduce wear (Fig. 2.93). In addition, in many engine designs, the valves rotate for the same reason. Rotation is promoted by a slight offset or taper on the tappet or rocker face in contact with the cam lobe. Some engines may be fitted with valve rotating mechanisms which are integral with the spring retainer. These are in two parts with opposing angle faces and rollers to provide a rotational drive as the valve is operated.

2.1.3.5 Valves

Inlet and exhaust valves are poppet-type valves with a circular sealing face recessed in the cylinder head (Fig. 2.94). The valves are located via the stem and slide inside valve guides mounted in the cylinder head. Valve heads are exposed to full, combustion chamber temperatures and pressures; the temperature of the exhaust valve can be as high as 800°C. The incoming

Figure 2.94 Valves in position on a cutaway engine. (Source: Ford Media)

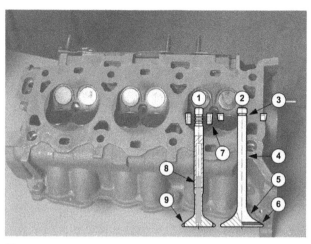

Figure 2.96 Valve details: 1, bimetallic valve; 2, single metal valve; 3, collet groove; 4, stem; 5, head; 6, face; 7, collets; 8, sodium filling; 9, armouring

gas charge has a cooling effect on the intake valve but, generally, heat dissipation from the valves is via the stem and guides to the cylinder head. Combustion and fuel deposits can cause problems on the valve; this can be avoided by the use of good quality fuels and oils.

The total valve opening area is always greater for the inlet valves; this is to increase the volumetric efficiency of the engine since the pressure difference across the inlet valve, when charging the cylinder, is much lower than the pressure difference across the exhaust valve, when evacuating the cylinder. Hence, a larger valve is needed to reduce restrictions to gas flow during the inlet stroke.

Valves seats and the valve sealing face are cut at slightly different angles (Fig. 2.95). This is to

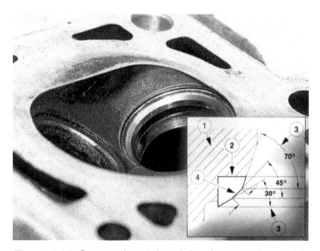

Figure 2.95 Seat angles: 1, head; 2, valve seat; 3, possible angles; 4, valve insert angle where contact is made

ensure that a complete seal is made under working conditions as, when the valve is installed and at running temperatures, the valve head will deform slightly, causing the sealing faces to meet correctly and seal efficiently. The angle of the sealing face is approximately 45°. The valves open via the force applied from the cam and valve gear and are held in the closed position by spring force. The springs are connected to the valve via a retainer and split collets as this allows removal and refitting. In operation, the valve head rotates and this helps to maintain the sealing face.

Some valve stems are hollow and contain sodium, which melts when the valve is hot and assists in transferring heat from the valve face to the valve stem for dissipation via the valve guides (Fig. 2.96). Cooling of the valve head can be improved significantly via a temperature reduction of up to 100°C. These valves are also known as bimetal type. Note that a sodium-filled valve of this type must be handled with care as sodium exposed to air is flammable and will self-ignite.

2.1.3.6 Camshaft

The camshaft, on most modern engines, is mounted in bearings formed into the cylinder head via an in-line boring process (Fig. 2.97). The camshaft is forged from steel or cast iron and the bearings and cam surfaces are a smooth, machined finish. The camshaft has cam lobes for each valve and to ensure the correct sequence of valve timing, the camshaft is timed and synchronized with the crankshaft position.

Figure 2.97 Head removed from an Alfa Romeo

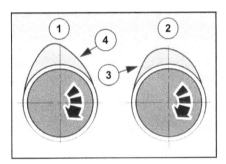

Figure 2.98 Cam profiles: 1, pointed cam; 2, cam with steep profile; 3, trailing face; 4, leading face

Key fact

The camshaft has cam lobes for each valve. To ensure the correct sequence of valve timing, the camshaft is timed and synchronized with the crankshaft position.

The cam lobes have a specific profile that consists of a base circle and lobe to provide the correct valve opening and closing characteristics (Fig. 2.98). The cam profile is not necessarily symmetrical and the profile may allow progressive opening of the valve but with a sharp closing action depending on the characteristics and optimization parameters of the engine.

Overhead valve (OHV) is used to describe an engine where the valves and operating mechanism are located in the cylinder head. The valve gear transfers reciprocating motion from the cam followers and camshaft. It is then passed to the valves via push rods and a rocker assembly, which acts on the valve stems (Fig. 2.99).

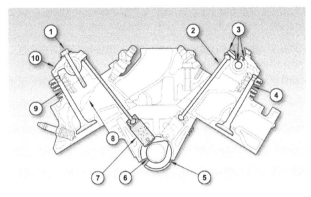

Figure 2.99 Overhead valves (OHV): 1, rocker shaft bolt; 2, push rod; 3, lubrication passages; 4, exhaust valve; 5, cam bearing journal; 6, cam lobe; 7, hydraulic tappet; 8, turbulence ramp; 9, inlet valve; 10, rocker

Overhead cam (OHC) refers to the position of the camshaft in the engine; that is, it is positioned in the cylinder head. There are various designs using direct or indirect mechanisms to convert the rotating cam motion into a reciprocating motion, and then transfer this motion to the valve stems. These designs are proposed to facilitate a close tolerance in operating clearances.

Key fact

Overhead cam (OHC) refers to the position of the camshaft in the engine.

A further development, now employed extensively, is the use of twin or double overhead camshafts (DOHC) (Fig. 2.100). These can use direct or indirect valve

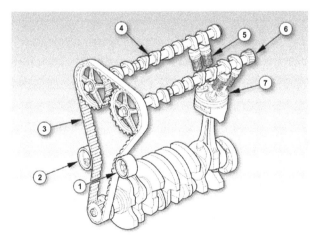

Figure 2.100 Double overhead camshaft engine (DOHC): 1 and 2: tensioner wheels; 3, timing belt; 4, inlet cam; 5, exhaust cam; 6, distributor drive; 7, valve

actuation and are well suited to multivalve engine designs, including those with variable valve timing.

Indirect, rocker arm-type valve actuators incorporate close tolerance adjusters. Two systems are commonly seen, a rocker shaft and pivot stud, or a rocker arm supported on a pedestal at one end and the valve stem at the other, the cam acting between these two points (Fig. 2.101). A hydraulic pedestal can be used for self-adjustment of the mechanism.

Figure 2.102 shows a typical engine oil lubrication circuit that feeds the self-adjusting followers with pressurized oil to maintain the correct valve clearances. Always refer to manufacturers' data for the service requirements of the valve train system. Often, special procedures are required when replacing and recommissioning

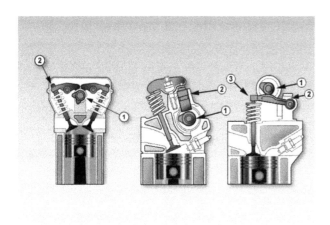

Figure 2.101 Valve operating mechanisms. Left, 1, cam; 2, adjusting screw in direct acting rocker. Centre, 1, cam; 2, hydraulic follower. Right, 1, cam; 2, pivot and adjuster; 3, finger follower

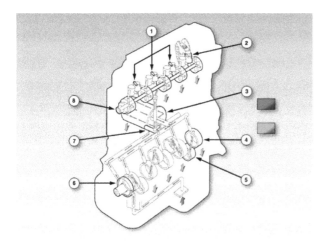

Figure 2.102 Hydraulic tappets and components: 1, oil to rocker arms; 2, hydraulic tappets; 3, filter; 4, crank main bearings; 5, big end bearings; 6, crank driven oil pump; 7, oil under pressure; 8, camshaft

Figure 2.103 Four camshafts on a V engine each with a variator to control the timing. The timing chain layout is complex to say the least (Source: Phil Miller)

self-adjusting valve mechanisms, and these must be followed to prevent engine damage.

2.1.4 Variable valve timing

Control of engine performance means using sensors to 'see' what is happening, making a decision based on this and stored information, and then operating some actuators to achieve the ideal outcome. Timing control and fuel control used to be the key areas but there are now many other options. It is all done to improve performance, reduce emissions and lower the fuel consumption. Controlling valve timing is now an integral part of the strategy. This can result in some complex mechanical systems such as the V engine shown in Figure 2.103.

Engines that always run at a relatively high speed, such as racing car engines, will have considerable overlap in their valve timings for maximum volumetric efficiency. Road car engines are different because they are also required to idle at less than 1000 rpm. Excessive valve overlap would make smooth idling impossible because of the mixing of fresh and exhaust gases. Variable valve timing can give both maximum power at high rpm and smooth idling at low rpm by making small changes to the relative angular position of the camshafts and thereby varying the valve overlap.

Figures 2.104–2.107 show the four main operating conditions and typical examples of how the valves are timed. This is achieved by rotating the camshaft drive relative to the camshaft position. In this example, and all engines are different, the only time overlap is used is for the EGR setting. The timing varies in between the examples shown, for example, when under light load conditions.

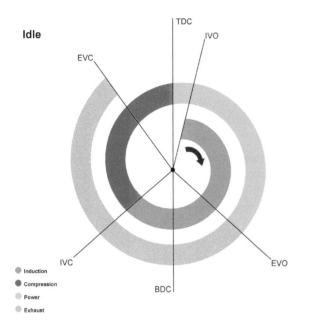

Figure 2.104 Idle

At idle speed, the camshafts are positioned so that the inlet opens late and closes late. The exhaust camshaft closes well before TDC. Because there is only a minimal gas residue at this speed, these settings result in a smooth idle.

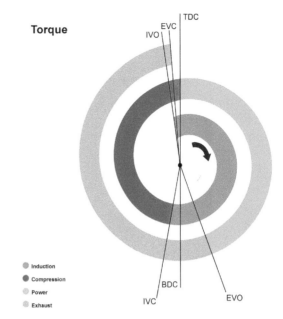

Figure 2.106 Torque

High volumetric efficiency is needed to provide maximum torque. The inlets are therefore opened early, and as a result they also close early (the actual open time is a mechanical function of the cams), this does, however, avoid fresh gases being expelled.

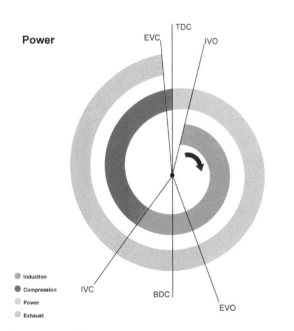

Figure 2.105 Power

At high engine speeds the exhaust valves are made to open later so that the expanding gases act against the piston for longer. Opening the inlet valves after TDC and closing them a long way after BDC, means the self-charging effect of the entering air is used to increase power.

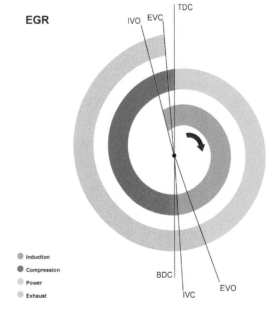

Figure 2.107 EGR

Internal exhaust gas recirculation (EGR) occurs during valve overlap (from IVO to EVC when both valves are open). During this time gas flows from the exhaust back into the inlet. The amount of overlap determines the amount of EGR. Internal EGR is much more efficient than external EGR as the system reacts faster.

Some engines now employ variable valve timing to optimize the inlet valve timing with respect to engine speed and load conditions. This is because it improves engine performance and efficiency. Most VVT engines can vary the inlet and exhaust camshafts.

As air enters the engine through the inlet manifold it forms a column of moving air that possesses kinetic energy. The pulsating nature of the engine's air consumption creates pressure waves in this air column. The energy in these pressure waves can be harnessed to assist in charging the cylinder, which increases volumetric efficiency. To do this, the valve opening point must be optimized according to the engine condition, and with variable valve timing (VVT) this can be achieved to increase engine torque and power at various points in the operating speed range.

There are several technologies used to provide the required phase angle between the drive and the camshaft for variable valve timing. The angle is usually achieved by using a hydraulic mechanism in the camshaft wheel, which is controlled via a valve assembly from the engine ECU. Camshaft wheel actuators like the one in Figure 2.108 use a helix that is moved by oil pressure. There are three main control methods (Table 2.1).

> **Key fact**
> Variable valve timing (VVT) optimizes the valve opening point to increase engine torque and power at various points in the operating speed range.

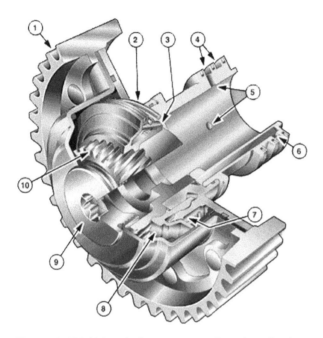

Figure 2.108 Valve timing actuator, often described as a variator: 1, camshaft gear; 2, spring; 3, outer helical teeth; 4, adapter with ring grooves; 5, oil supply to front chamber via hollow bolt (not shown); 6, oil supply to rear chamber; 7, rear chamber; 8, front chamber; 9, blanking plug; 10, inner helical teeth

2.1.5 Variable compression (VC) engine

Nissan has developed a variable-compression-ratio gasoline engine. The VC-Turbo, a 2.0 L inline four-cylinder house, features an ingenious crank train and control system that enables the effective compression ratio to be varied between 8:1 and 14:1, depending on load. Typically, VCR engines alter the compression ratio by raising or lowering the height of the piston at TDC, but the Infiniti engine achieves this in a different way.

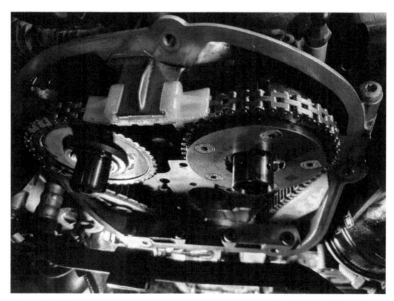

Figure 2.109 Variator fitted to a Volkswagen engine (Source: Clayton Millar)

Figure 2.110 VVT variators – the vanes are moved by oil pressure (Source: Clayton Mille)

Figure 2.112 VC-T operation at high and low compression ratios (Source: Nissan)

Figure 2.111 Components of the Nissan (Infiniti) variable compression turbo (VC-T) engine (Source: Nissan)

An electric motor drives a reduction gear, which moves an angled actuator arm. The arm in turn rotates a control shaft with four aligned eccentric cams, one for each cylinder. An intermediate link with bearings at each end connects the eccentric cam at the bottom end to the multi-link at the top end. The centre of the multi-link runs in a bearing around the crankshaft journal.

A second bearing on the multi-link, positioned 180° degrees from that connecting the intermediate link effectively serves as the piston connecting rod big end bearing. This arrangement produces a 17° offset of

Table 2.1 There are three main VVT control methods

Cam switching	This method uses two cam profiles, with an actuator to swap between the profiles (usually at a specific engine speed). Cam switching can also provide variable valve lift and variable duration; however, the adjustment is fixed by mechanical design rather than continuous.
Oscillating cam	These designs use a rocking motion in a cam lobe, which acts on a follower. This follower then opens and closes the valve. Some oscillating cam systems use a conventional cam lobe, while others use an eccentric cam lobe and a connecting rod. The advantage of this design is that adjustment of lift and duration is continuous. However, lift and duration cannot be separately adjusted.
Cam phasing	Most current VVT systems are the cam phasing type, using a device known as a variator. This is moved by oil pressure, which in turn is controlled by a valve. This allows continuous adjustment of the cam timing; however, the duration and lift cannot be adjusted. The valve is controlled by an ECU. On some systems, the entire variable valve timing system is adaptive. This compensates for component and assembly tolerances as well as wear occurring during engine use.

Figure 2.113 Hydraulic lifters

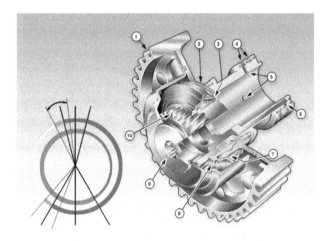

Figure 2.114 Timing is changed under electronic control: 1, camshaft gear; 2, spring; 3, outer helical teeth; 4, adapter with ring grooves; 5, oil supply to front chamber via hollow bolt (not shown); 6, oil supply to rear chamber; 7, rear chamber; 8, front chamber; 9, blanking plug; 10, inner helical teeth

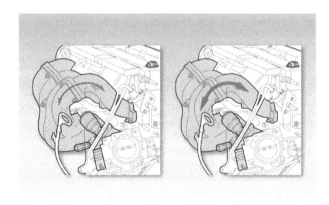

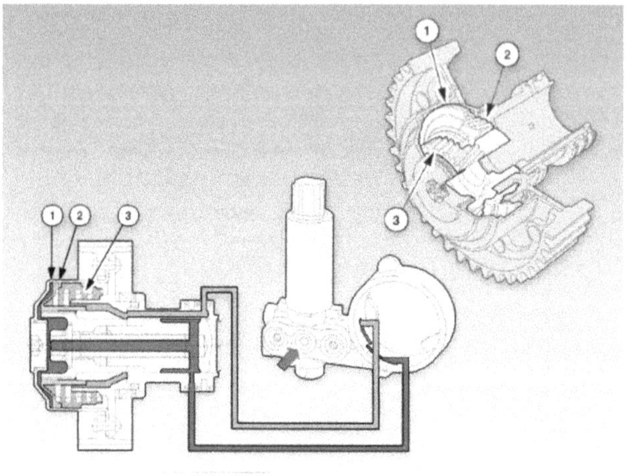

Figure 2.116 Electrohydraulic variable cam control method: 1, adjusting piston with inner and outer helical teeth; 2, outer helical teeth connected to camshaft pulley; 3, inner helical teeth connected to the camshaft

Figure 2.117 Variable cam timing. (Source: Ford Media)

the con rod from the crankshaft journal centre point. The harmonic drive is controlled by a dedicated ECU which gathers data from engine sensors to determine the compression ratio required for the current load. Since a low compression ratio is desirable when power is required and a high compression ratio when efficiency is preferable, the piston height at TDC can be continuously varied as required by rotating the harmonic drive, which will determine the position of the multi-link and hence the height of the piston in the cylinder bore.

As with other VCR engines, complexity, mass and cost are greater than in a conventional four-cylinder gasoline engine. On this system, there are three con rods for example. The motion described by the piston con-rod big end is not circular as in a conventional engine's big

Figure 2.115 Inlet manifolds

Figure 2.118 Cam drive belt in position on the Duratec engine; not to be confused with the serpentine alternator drive belt. (Source: Ford Media)

Figure 2.119 Two types of cam belt

Figure 2.120 Belt drive components: 1, cam gear; 2, belt; 3, crank gear; 4, water pump; 5, tensioner

Figure 2.121 Cam drive gears

end bearing, it is more elliptical with the con rod not passing through the vertical axis between the big- and small-end bearings. During the power stroke, the con rod remains almost vertical. This reduces the side force on the piston and helps to reduce the vibration.

The development team addressed the fuelling needs generated by the range of compression ratios by fitting both multipoint injection (MPI) for low compression and direct injection (GDI) for high-compression operation. Since GDI engines inherently generate high particulate emissions, the continual phasing between GDI and MPI helps to contain particulate emissions. Both sets of injectors are brought into use under high load and engine speed conditions.

Both the inlet and exhaust camshafts are fitted with variable valve timing, electronically controlled on the inlet side and hydraulically actuated on the exhaust

side. Forced induction is provided by a single scroll turbocharger, equipped with an electronic wastegate actuator.

2.1.6 Belt-in-oil (BIO) or wet cambelts

Manufacturers started using timing belts instead of chains or gears in the 1970s. Oil contamination of the belt was often a cause of failure because oil attached to the rubber content. The belt could lose its teeth or snap. There are now some belts that are designed to run in oil and are known helpfully as 'belt-in-oil' (BIO)!

These BIO systems are used for oil pump drives as well as for camshaft timing drives. Belts in general have three key advantages:

▷ absorb and isolate crankshaft harmonics from the valve train
▷ quieter in use
▷ reduced power consumption (less friction and noise)

103

Figure 2.122 Cam drive chain in position on the 2012 Focus engine. (Source: Ford Media)

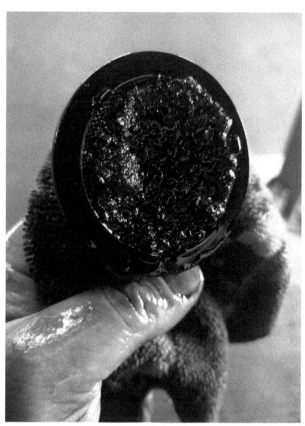

Figure 2.124 Consequences of not servicing can be very serious (Source: Donald Anthony)

Figure 2.123 Belt-in-oil (Source: Robert Chalmers)

Figure 2.125 Electrohydraulic valve actuator (Source: Freevalve.com)

temperature-resistant, less prone to stretching than conventional dry belts and have a life expectancy of up to 150,000 miles. The first automotive BIO system was introduced in 2008, in a 1.8L Ford diesel. Other manufacturers such as Volkswagen now use this technology.

There is nothing new in this advice: regular servicing with the correct materials is essential. When these belts do break or shred, other areas of the engine are seriously affected – the oil pump for example. Always refer to the manufacturer's recommendations.

BIO technology can offer up to 30% reduction in friction compared to chains or dry belts. The result is lower emissions and improved fuel economy by about 1%. Wet belts use improved materials, that are more

2.1.7 Camless engines

Not yet a mainstream technology but camless engines have been under development for many years. This engine not only allows complete control of valve timing, but also the valve lift profile.

Camless engines rely on the use of actuators to open and close valves and these are commonly electro-magnetic or electro-hydraulic. In an interesting system developed by Koenigsegg, called Freevalve, the actuators are electro-hydraulic-pneumatic.

The system allows a greater degree of independent control for each cylinder valve compared to normal VVT systems. The valves can be opened at any crankshaft angle, which allows an engine to respond more efficiently to demands for high performance, or high efficiency.

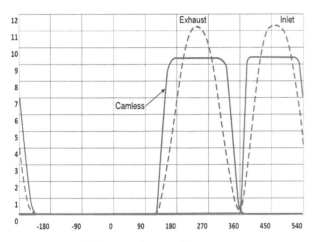

Figure 2.126 Valve opening profiles

Figure 2.127 On this diagram yellow represents the oil sump and other areas, green coolant passages, red exhaust ports and blue inlet air. (Source: Ford Media)

Fig. 2.125 shows how an actuator acts differently from a traditional cam operated valve, which causes valves to open pretty much in the same way that the cam is profiled. This is shown by the smooth peaks on the graph. Under actuator control the opening and closing time is almost instant, therefore the duration can be closely controlled.

Key fact

The advantages of timing belts are that they are cheap and run quietly.

2.2 Engine lubrication

2.2.1 Friction and lubrication

Lubrication is the introduction of a substance, called a lubricant (e.g. oil to create an oil film), between two moving contact surfaces, to reduce friction. This reduction of friction greatly reduces the wear of the surfaces and thus lengthens their service life. It also reduces the energy required for the movement. Lubrication is important in all moving parts of the vehicle but the engine has the greatest need.

Under a microscope, even the smoothest engine components have a surface that looks very rough. If these surfaces made contact they would rub together, overheat and destroy themselves. To prevent this happening, engines have a lubrication system that pumps or drips a constant supply of oil on all the moving metal components.

Although the basic function of a lubricant is to reduce friction and wear, it may also perform a number of other functions. It carries off generated heat and it helps to form a gas-tight seal between piston rings and cylinders. It also carries away harmful combustion waste products. Lubrication helps to control corrosion by coating parts with a protective film. A detergent added to the lubricant helps to removes sludge deposits.

Key fact

Under a microscope, even the smoothest engine components have a surface that looks very rough.

Some engine oils are refined from crude oil to which are added viscosity index enhancers, reduced friction enhancers, antioxygenates, sludge, lacquer and corrosion inhibitors, and cleaning agents for carbon, acids and water. Synthetic and semi-synthetic oils have improved performance for environmental or

Figure 2.128 Engine oil (Source: Motul)

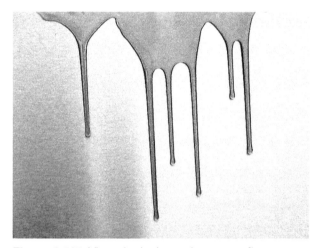

Figure 2.129 Viscosity is the resistance to flow

special purposes. Multigrade oils have been developed to modify the viscosity index and give thin oils at low temperatures that do not become excessively thin at higher temperatures (Fig. 2.128).

Viscosity is a measure of an oil's resistance to flow (Fig. 2.129). Thin oil will flow more easily than thicker oil. A viscosity index is the measure of a change in an oil's flow rate with a rise in temperature. The higher the viscosity index, the smaller the change in viscosity. Manufacturers' recommended viscosity ratings generally reflect the lowest temperature at which the vehicle is being used and may be different for summer and winter use. The viscosity rating is an indicator not of oil quality but of oil flow under particular conditions. Some low-grade oils carry recommendations that limit the use of the vehicle, particularly for high engine speeds, heavy loads and long journeys.

Definition

Viscosity: A measure of resistance to flow.

Engine oils are not normally biodegradable and should not be allowed to enter the environment

as either vapour or liquid. Total loss lubrication systems used on small two-stroke engines, such as those on motorbikes and outboard motors, use a 'petroil' mixture of petrol and a specially formulated biodegradable oil. Other types of oil should not be used.

Several international standards (described in a later section) describe the properties of an oil, but some original equipment manufacturers (OEMs) are now specifying their own. The safest advice on what oil to use is always to follow manufacturers' recommendations.

If a film of lubricant is used between the surfaces, the process is called boundary lubrication (Fig. 2.130). This type of lubrication is introduced between the moving surfaces by splash feed or an oil mist. The best example is on the side of the pistons but all sliding components use this method.

Boundary lubrication also commonly occurs on some bearings during starting of the engine. Pressure is generated once the engine is moving as a result of

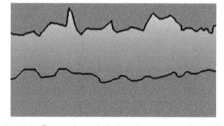

Figure 2.130 Boundary lubrication uses the oil to prevent two rough surfaces from touching

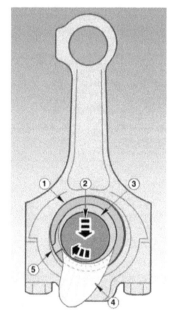

Figure 2.131 Con-rod big-end lubrication: 1, bearing shell; 2, load; 3, shaft; 4, lubrication pressure; 5, oil supply

the shape and motion of the surfaces. This is known as hydrodynamic lubrication. For example, the rotation of the crankshaft in a bearing forces lubricant into the wedge-shaped space between the shaft and the bearing (Fig. 2.131). The clearance between the bearing and the shaft, the load on the shaft, the speed of rotation and the viscosity of the lubricant have a marked effect on this process.

2.2.2 Methods of lubrication

Bath lubrication or splash lubrication (used in gearboxes and rear axles) may be used for gears, chains, bearings and other moving parts that can be partly submerged in an oil reservoir. In the bath system the gear simply picks up oil as it dips into the reservoir and sprays or carries it to other parts along its path. The splash system increases the efficiency by attaching a special splash ring to a moving part so that the oil is splashed against other parts that need to be lubricated. This is similar to oil-mist lubrication, created by the oil escaping from the engine's rotating crankshaft, in which the oil is atomized in a stream of air.

Force-feed lubrication uses an oil pump to force the oil under pressure to the parts to be lubricated, normally the engine crankshaft and camshaft. On some high-performance vehicles the mainshaft in the gearbox is pressure fed. Some parts are self-lubricating and require no external lubrication; the lubricant may be sealed in against loss as in sealed ball bearings, or a porous material such as porous bronze can be used so that oil impregnated in the material can penetrate to the point of contact of the moving parts through pores in the material. In small two-stroke gasoline engines the oil is mixed in with the fuel to bring it to the moving parts inside the engine.

> **Key fact**
> Force-feed lubrication uses an oil pump to force the oil under pressure to the parts to be lubricated, normally the engine crankshaft and camshaft.

Although lubricating oil is used elsewhere in a car, the lubrication of the engine is of greatest importance because it reduces the friction and wear between moving metal parts and also removes heat from the engine. A supply of oil is kept in the engine crankcase. An oil pump, which is powered by the engine, forces oil from the crankcase under pressure to the cylinder block main oil gallery. Passages in the engine block channel the oil to various moving parts, such as the crankshaft and camshaft, and the oil eventually drains back down in the crankcase. An oil filter is fitted in the oil circuit to filter out metal shavings, carbon

deposits and dirt. Because the filter is not completely effective, and because of prolonged exposure to high temperatures, the oil eventually becomes contaminated, decomposes and loses its lubricating qualities. This is why routine maintenance calls for changing the oil and oil filter at regular intervals.

2.2.3 Lubrication system

From the sump reservoir under the crankshaft oil is drawn through a strainer into the pump (Fig. 2.132).

Oil pumps have an output of tens of litres per minute and operating pressures of over 5 kg/cm² at high speeds. A pressure relief valve limits the pressure of the lubrication system to between 2.5 and 4 kg/cm². The pressure relief valve is a spring-loaded conical, or ball, valve that opens when the pressure in the oil exceeds the spring force acting on the valve seat (Fig. 2.133). When the valve opens, a return drilling is uncovered and the excess oil flows through this to return to the sump. This control is needed because the pump would produce excessive pressure at high

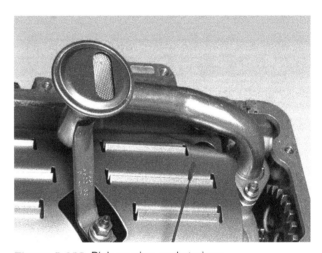

Figure 2.132 Pick-up pipe and strainer

Figure 2.133 Plunger and spring

107

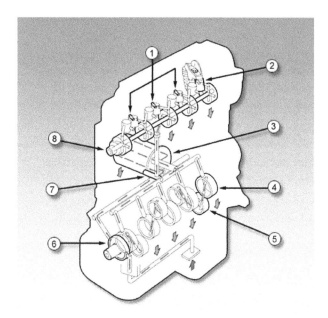

Figure 2.134 Oil flow: 1, oil to rocker arms; 2, hydraulic tappets; 3, filter; 4, crank main bearings; 5, big end bearings; 6, crank driven oil pump; 7, oil under pressure; 8, camshaft

Figure 2.135 Drillings in the block and crank

speeds. After leaving the pump, oil passes into a filter and then into a main oil gallery in the engine block or crankcase.

Drillings connect the gallery to the crankshaft bearing housings, and when the engine is running, oil is forced under pressure between the rotating crank journals and the main bearings. The crankshaft is drilled so that the oil supply from the main bearings is also to the big end bearing bases of the connecting rods.

The connecting rods are often drilled near the base so that a jet of oil sprays the cylinder walls and the underside of the pistons (Fig. 2.135). In some cases the connecting rod may be drilled along its entire

Figure 2.136 Oil drillings for the valve gear

length so that oil from the big end bearing is taken directly to the gudgeon pin (small end). The surplus then splashes out to cool the underside of the piston and cylinder.

The camshaft operates at half crankshaft speed, but it still needs good lubrication because of the high-pressure loads on the cams (Fig. 2.136). It is usual to supply pressurized oil to the camshaft bearings and splash or spray oil on the cam lobes. On overhead camshaft engines, two systems are used. In the simplest system the rotating cam lobes dip into a trough of oil. Another method is to spray the cam lobes with oil. This is usually done by an oil pipe with small holes in it alongside the camshaft. The small holes in the side of the pipe aim a jet of oil at each rotating cam lobe. The surplus splashes over the valve assembly and then falls back into the sump.

On cars where a chain drives the cam, a small tapping from the main oil gallery sprays oil on the chain as it moves past, or the chain may simply dip in the sump oil.

Some specialized vehicles use an oil cooler (Fig. 2.137). The oil cooler commonly used is an air radiator similar to an engine-cooling radiator, with tubes and fins to transfer heat from the oil to the passing air stream. This cooler is fitted next to the cooling-system radiator at the front of the vehicle. Pipes from the filter housing carry oil to and from the oil-cooler radiator.

A key component of the lubrication system is the dipstick. No matter how clever the system is it will not work if the oil level is low. The dipstick is marked to show the maximum and minimum acceptable levels.

Many modern engines are now also fitted with an electronic sensor that supplies information to the driver on the level of oil in the engine (low oil pressure indicating low oil level). A warning light, or a gauge in the instrument panel, indicates whether the oil level is

Figure 2.137 Oil cooler for a racing car. (Source: www.prcracing.com Media)

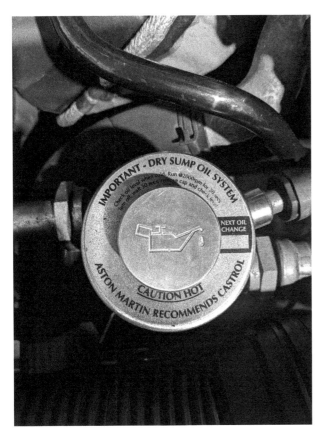

Figure 2.138 Dry sump filler cap

within acceptable levels. The sensor is fitted into the sump or the engine block. Some engines now have oil quality sensors to indicate when the oil should be changed.

Key fact

Typically engines have a dipstick – but there are some that rely on sensors.

To warn the driver about low oil pressure, a pressure-sensitive switch is fitted into the main gallery. It makes an electrical contact when the pressure is below about 0.5 bar (7 psi). The switch may be fitted in the same circuit as the oil level warning lamp. When the switch contacts make a connection, the lamp lights, and this should occur before the engine is started. Once the engine is running, oil pressure builds up and the switch contacts separate and the warning lamp will go out. This indicates that a minimum oil pressure is being maintained in the system. Oil pressure gauges are also used and employ a piezoelectric pressure sensor fitted into the main gallery and a gauge unit.

2.2.4 Oil filters

Even new engines can contain very small particles of metal left over from the manufacturing process or grains of sand that have not been removed from the crankcase after casting. Old engines continually deposit tiny bits of metal worn from highly loaded components such as the piston rings. To prevent any of these lodging in bearings or blocking oil ways, the oil is filtered (Fig. 2.139).

The primary filter is a wire mesh strainer that stops particles of dirt or swarf from entering the oil pump. This is normally on the end of the oil pick-up pipe. An extra filter is also used that stops very fine particles. The most common type has a folded, resin-impregnated paper element. Pumping oil through it removes all but smallest solids from the oil.

Key fact

Most engines use a full-flow system to filter all of the oil after it leaves the pump.

Figure 2.139 Oil filters

109

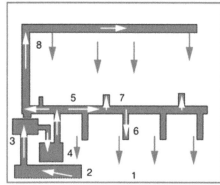

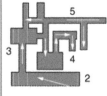

Figure 2.140 Oil circuits: 1, full flow (bottom left bypass flow); 2, sump; 3, pump; 4, filter; 5, main gallery; 6, main bearings; 7, big ends; 8, camshaft

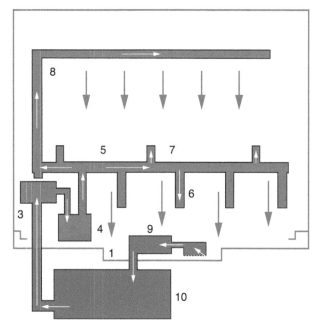

Figure 2.141 Dry sump: 1, small collection area; 2, sump (not shown); 3, pump; 4, filter; 5, main gallery; 6, main bearings; 7, big ends; 8, camshaft; 9, return pump; 10, remote tank

Most engines use a full-flow system to filter all of the oil after it leaves the pump (Fig. 2.140). The most popular method is to pump the oil into a canister containing a cylindrical filter. From the inner walls of the canister, the oil flows through the filter and out from the centre to the main oil gallery. Full-flow filtration works well provided the filter is renewed at

regular intervals. If it is left in service too long it may become blocked. When this happens the build-up of pressure inside the filter forces open a spring-loaded relief valve in the housing and the oil bypasses the filter. This valve prevents engine failure, but the engine will be lubricated with dirty oil until the filter is renewed. This is better than no oil!

A bypass filtration system was used on some vehicles (Fig. 2.140, bottom left). This system only filters a proportion of the oil pump output. The remainder is fed directly to the oil gallery. At first view this seems a strange idea but all of the oil does eventually get filtered. The smaller amount through the filter allows a higher degree of filtration.

For many high-performance applications, a larger oil supply is needed so that engine heat can be removed by the engine oil as well as by the engine-cooling system (Fig. 2.141). A separate reservoir of oil is held in a remote tank and drawn into the main oil pump for distribution throughout the engine in the same way as a wet-sump system. The oil returns to a small sump below the engine. A scavenge pump, with a pick-up pipe in the sump, draws oil out of the sump and delivers it back to the reservoir. An oil cooler is usually fitted in this return circuit.

2.2.5 Oil pumps

The oil pump is the heart of the system. It pumps oil from the sump into the engine. The main types of oil pump are gear, rotor, gerotor, vane and crescent. The gear type uses two gears in mesh with each other (Fig. 2.142). Drive is made to one gear which, in turn, drives the other. The housing has a figure-of-eight internal shape, with one gear in each end. Ports are machined in the housing and align with the

Figure 2.142 Gear pump

areas where the teeth move into, and out of, mesh. As the teeth separate, the volume in the inlet side of the housing increases and atmospheric pressure in the sump is able to force oil into the pump. The oil is carried around inside the pump in between the teeth and the side of the housing. When the teeth move back into mesh, the volume in the outlet side of the housing is reduced, the pressure rises and this forces the oil out into the engine.

Key fact

The main types of oil pump are gear, rotor, gerotor, vane, and crescent.

The rotor-type pump uses the same principle of meshing but with an inner rotor with externally formed lobes that mesh with corresponding internal profiles on the inside of an external rotor (Fig. 2.143). The inner rotor is offset from the centre of the pump and the outer rotor is circular and concentric with the pump body. As the rotors rotate, the lobes mesh to give the outlet pressure of the oil supply, or move out of mesh for the intake of oil from the sump.

The gerotor (gear rotor pump) is a variation on the smaller rotor pump (Fig. 2.144). The gerotor pump is usually fitted around, and driven by, the crankshaft. There are inner and outer rotors, with the inner rotor externally lobed and offset from the internally lobed outer rotor. During rotation, the pumping and carrying chambers are formed by the relative positions of the lobes.

The crescent pump is named after the solid block in the gear body. This pump is a variation on the gear pump, and also uses gear teeth to create the pumping chambers and to carry oil from the inlet port

Figure 2.144 Gerotor pump driven by the crankshaft

Figure 2.145 Vane pump

to the outlet port of the pump. The operation of this pump is based on the meshing of the gear teeth, the positioning of the ports in the housing and alignment at each end of the crescent where the teeth move in and out of mesh. Oil is carried from the inlet port to the outlet port in the spaces between the teeth and the crescent. This type of pump is used for engine lubrication and for automatic transmissions.

The vane-type pump uses an eccentric rotor with vane plates set at right angles to the axis of the rotor and sitting in slots in the rotor (Fig. 2.145). As the rotor rotates, the vanes sweep around inside the pump housing. The pump chambers increase in volume as the vanes move away from the housing walls, and reduce in volume as the vanes approach the walls. Oil is carried between the vanes and the pump housing from the inlet port to the outlet port.

Most modern engines now use the crankshaft to give a direct drive to the oil pump. These pumps are of the

Figure 2.143 Rotor pump

111

gerotor or crescent design, and are fitted around the front of the crankshaft. This arrangement is used on many overhead camshaft engines because it provides a low position for the pump.

Key fact

Most modern engines now use the crankshaft to give a direct drive to the oil pump.

2.2.6 Oil viscosity and grades

Oil viscosity refers to how easily oil pours at a specified temperature. Thin oils have lower viscosity and pour more easily than thicker oils that have a higher viscosity. Viscosity is defined as resistance to shear.

There are two basic categories for motor oil: mineral or synthetic. Mineral oils are easy to produce but they also offer less performance. Modern passenger car engines would not work properly with mineral oil, so its use has been reducing for decades. Synthetic oils are more expensive and more powerful. A blend of mineral and synthetic oil is called semi-synthetic. Because of the limited performance of the mineral oils, they do not play an important role in the market and their share will continue to diminish.

It is therefore the synthetics that really 'make the grade'. Poly alpha olefin PAO oils are the traditional synthetic oils which were first developed in the 1970s. They offered great performance, but were also very costly. The industry worked on more cost-effective alternatives and developed HC oils. In this context HC stands for hydro-cracked which describes how these oils are generated. They offer the same performance level as PAO oils but for a better price. This is why new oils for passenger cars are formulated with modern HC oils and not with the traditional type.

One of the newest fully synthetic 'thin' oils is graded as SAE 16. Society of automobile engineers 16 oils are only suitable for use in vehicles where the engine is designed to operate with such a low viscosity oil. The number 16 is the name of the viscosity grade, it does not mean anything else. For an oil to be described as SAE 16, its viscosity on 100°C must be between 6.1 and 8.2 mm²/s and its high temperature high shear (HTHS) viscosity must be at least 2.3 mPa*s (millipascal second).

The official SAE viscosity grades are now: 0W, 5W, 10W, 15W, 20W, 25W, 16, 20, 30, 40, 50 and 60. SAE will probably introduce 12, 8 and 4 viscosity grades when there is a demand by manufacturers.

Figure 2.146 SAE ratings. 0W-16 oils address emerging engine technology as vehicle manufacturers strive to optimize fuel economy. Honda, Lexus and Toyota recommend 0W-16 oils for some petrol and hybrid engines. (Source: LIQUI MOLY)

Figure 2.147 Low viscosity oil (Source: LIQUI MOLY)

LIQUI MOLY recently launched the lowest viscosity oil that they have produced. It is suitable for selected models from Honda, Lexus, Suzuki and Toyota. Up until now, this 0W-16 oil was very uncommon outside of Japan. However, this is changing and technologically these low-viscosity oils are very interesting.

The lower the oil viscosity, the less power the engine has to use to pump the oil. This lowers fuel consumption and emission levels accordingly. A low-viscosity oil is one of the many tricks car manufacturers use to reduce emissions. A thicker viscosity 10W-40 used to be the most widely spread viscosity, today it is 5W-30 and increasingly more 0W-20. And now 0W-16 has been launched.

The challenge is to develop an oil that has a low viscosity, on the one hand, but also ensures the lubrication of the engine under extreme loads. The actual base oil used in the motor oil cannot manage this because its viscosity is too low. Motor oil is made from base oil (the oily substance) and additive packages which bring the actual performance into the motor oil. About one third of a modern motor oil is additives.

2.2.7 Mineral, synthetic or fully synthetic?

Motor oil has developed from a simple lubricant to a high-tech liquid. The number of oil grades and specifications has increased significantly alongside these developments.

Motor oils can be roughly divided into two categories: mineral-based and synthetic oils. Mineral oils are becoming less and less important for cars because they do not offer such high performance. Modern engines demand much more from the oil than old engines. Only synthetic oils can meet these requirements. You can also mix a mineral oil with a synthetic oil. However, such semi-synthetic oils are playing an increasingly smaller role in the market, because their performance is not high enough, due to their mineral content.

There are two different ways in which synthetic oils can be produced. By one of these methods, a so-called PAO oil is obtained at the end. This is the classic synthetic oil as it came on the market in the 1970s. It is chemically very pure and therefore very effective, but also very expensive to produce. With the other approach, the oil is produced by hydro-cracking, which is why these oils are also called HC oils in technical jargon. HC oils are more modern and came up in the 1990s. Today, they offer the best possible performance for all modern engines.

Almost all oil development nowadays takes place using a base of HC oils. Many oil specifications can only be met with HC oils. The decisive factor is that the oil meets the specifications which the car manufacturer has set for that model.

Motor oils are becoming less viscous, and this is down to the car manufacturers. Their cars are designed to consume less and less fuel and emit fewer and fewer emissions. Oil does lubricate the engine; however, it also restricts it at the same time. A thick, viscous oil restricts the motor more than a thin oil. Thanks to a low-viscosity oil the consumption can be reduced by 2–3%. The industry is already working on 0W-12 grade viscosities. These viscosities are moving ever more towards water.

Figure 2.148 Classic oil (Source: Motul)

Figure 2.149 Society of Automotive Engineers (SAE) logo

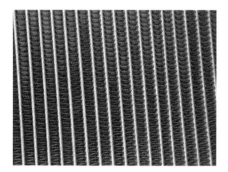

Figure 2.150 Elements of a radiator; a key component in the cooling system

There's a simple bottom line here. Oil is not really something that is bought in bulk and used on lots of different cars any more. It is a specific part for a specific vehicle – always use the manufacturers' recommended oil and use a well-known, high-quality brand.

2.3 Engine cooling

2.3.1 Introduction

Heat is a form of energy that can be sensed by a change in temperature. The engine uses chemical energy in the fuel and converts it into heat and then into movement. The energy conversion process in an engine is not very efficient and only about 30% is converted into movement energy. Of the remaining heat, up to 50% goes out of the exhaust and the rest heats the engine. Excessive heating of the engine must be controlled to prevent damage. Components expand with heat and, at high temperatures, this expansion can cause seizure, and burning of pistons and valve seats. High temperature would also produce rapid deterioration of the engine oil.

Cooling systems are designed to maintain engines at an optimum temperature. This allows the design of components that expand on heating to form very tight fits and running tolerances. The adjustment of ignition and fuel settings is equated to the optimum temperature required for the clean and efficient combustion of fuel. Because a cold engine produces

high levels of unwanted exhaust emissions, a rapid warm-up is needed to keep emissions to a minimum. The 'normal' running engine coolant temperature is maintained at about the boiling point of water, which enables efficient combustion. A further reduction in harmful exhaust emissions is achieved by keeping the warm-up time to a minimum. There are two types of cooling system.

> **Key fact**
> Cooling systems are designed to maintain engines at an optimum temperature.

Air-cooled systems have the air stream passing directly over the cylinder heads and cylinders to remove heat from the source (Fig. 2.151). Fins are cast into the cylinder heads and cylinders to increase the surface area of the components, thus ensuring that sufficient heat is lost.

Liquid-cooling systems use a coolant to carry heat out of the engine and dissipate the heat into the passing air stream (Fig. 2.152). The liquid coolant is contained in a closed system and is made to circulate almost continuously by the impeller on the water pump. Heat is collected in the engine and dissipated from the radiator into the passing air stream. Almost all

Figure 2.151 Air-cooled system

Figure 2.152 Bespoke water-cooled system

modern cars and light vehicles use liquid cooling systems.

2.3.2 System operation

The coolant is a mixture of water, antifreeze and inhibitors. The antifreeze is usually ethylene glycol, which needs inhibitors to prevent corrosion and foaming. These inhibitors have a lifespan of about two years, which means that the coolant should be changed at these intervals. Selection of the correct coolant mixture must be made to meet the manufacturer's specifications. Aluminium alloy engines are more prone to corrosion than cast-iron engines.

Antifreeze is mixed to a specified ratio with water. Many manufacturers specify a 50/50 mixture of water and antifreeze, which allows higher engine temperatures before the coolant boils and prevents freezing. An ethylene glycol antifreeze solution has an added advantage. It forms a semi-solid wax solution prior to solidification and this enables any expanding ice crystals to move within the water passages.

A 50/50 coolant mixture will increase the boiling point to about 105°C (222°F) and provide protection down to −34°C (−30°F). For colder temperatures down to −65°C (−90°F), a maximum mixture of 65% ethylene glycol can be used. Higher concentrations begin to freeze at higher temperatures and therefore no more than 65% ethylene glycol should be used.

Many areas have 'hard' water that contains calcium or chalk. This separates from the water when it is heated. Deposits can form inside the water jacket or radiator where they can block small water passages. Frequent topping up with tap water in hard-water areas should be avoided. In some areas, it may be recommended to use distilled water, or water from outside the area.

All three forms of heat transfer are used in the cooling system:

▷ Convection occurs in the water jacket, creating flows of internal coolant from the cylinder block to the cylinder head (Fig. 2.153).
▷ Conduction occurs through the cylinder and combustion chamber surfaces as heat passes to the coolant (Fig. 2.154).
▷ Radiation of heat occurs from the radiator and cooling fins when heat is dissipated to the atmosphere (Fig. 2.155).

The amount of heat transfer is dependent on four main factors:

▷ the temperature difference between the engine and coolant
▷ the temperature difference between the coolant and the air stream passing through the radiator
▷ the surface area of the radiator tubes and fins
▷ the rate of flow of air and coolant through the radiator.

Modern engines use water cooling because this is capable of giving the precise engine temperature control needed for exhaust emission regulations. Warming up to the optimum temperature as quickly as possible is important, because it helps not only

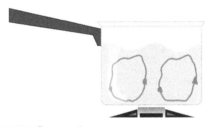

Figure 2.153 Convection

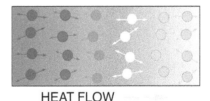

HEAT FLOW

Figure 2.154 Conduction

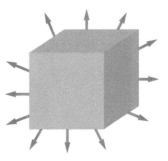

Figure 2.155 Radiation

115

to reduce exhaust emissions, but also to prevent the formation of water particles in the combustion chamber and exhaust when the engine is cold. Any water that does not evaporate can enter the engine and contaminate the engine oil, or remain in the exhaust system and cause premature corrosive damage.

Key fact

Modern engines use water cooling because this is capable of giving the precise engine temperature control needed for exhaust emission regulations.

The water jacket is cast into the cylinder block and cylinder head (Fig. 2.156). Casting sand is used to shape the inside, or core, of the casting for the water passages. The sand is removed after casting through a series of holes in the sides, ends and mating faces of the cylinder block and head.

The holes in the sides and ends of the block and head are machined to provide accurate location for core plugs. The holes in the mating faces are aligned to allow coolant flow from the cylinder block to the cylinder head. These components are also machined for the fitting of the water pump and a water outlet to the radiator.

The internal designs of the head and block vary to give different coolant flow patterns (Fig. 2.157). An even flow to all areas of the engine is very important.

The main areas where cooling is needed are around the combustion chambers and the upper cylinder walls (Fig. 2.158). The need for inlet ports, exhaust ports and valves makes cooling of these regions difficult. These areas are prone to cracking and other deterioration from overheating, freezing and the use of incorrect, or old, antifreeze solutions.

Figure 2.156 Engine block water jacket and core plugs

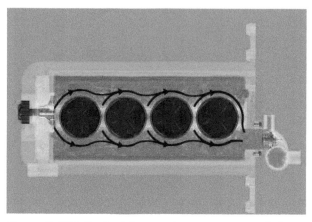
Figure 2.157 Coolant flow and temperature variations across the block

Figure 2.158 Cylinder head coolant passages

Key fact

The internal designs of the head and block vary to give different coolant flow patterns.

Developments in coolant circulation give improved control of engine temperature (Figs 2.159 and 2.160). Mixing cold and hot water as it enters the engine achieves this, as opposed to the cold fill of earlier systems.

Many engines use a heated inlet manifold that has a coolant flow from the engine water jacket running continuously through it (Fig. 2.161). As soon as an engine is started, some heat is produced and this rises into the inlet manifold very quickly. The heat vaporizes the fuel in the air stream into the engine. This improves atomization and fuel distribution in the new air and fuel charge.

Liquid cooling systems traditionally used a thermostat in the outlet to the top hose to control engine temperature. A thermostat is a temperature-sensing

Figure 2.159 Engine coolant ports with a thermostat in position

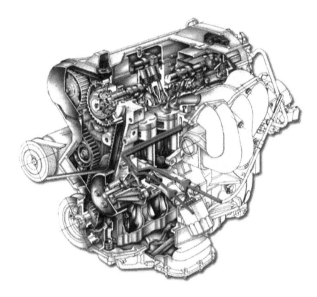

Figure 2.160 Engine coolant circuit (red is hot water, blue is cooler water)

Figure 2.161 Inlet manifold

valve that opens when the coolant is hot and closes as the coolant cools down.

This allows hot coolant to flow from the engine to the radiator, where it cools down and returns to the engine. The cooled coolant in the engine acts on the thermostat and it closes.

The coolant reheats in the engine, the thermostat opens, and the cycle of hot coolant flow to the radiator and cool coolant returning to the engine repeats itself (Fig. 2.162). Although this system provides a reasonably effective method of engine temperature control, it produces a fluctuating temperature. However, a steady temperature is required for very clean and efficient combustion.

Many engine designers are moving towards a system with the thermostat in the radiator bypass channel (see Figs 2.159 and 2.160). When the thermostat opens, it allows cold water from the radiator to mix with the hot-water flow in the bypass as it enters the water pump. This system provides a steady engine temperature and prevents the fluctuating temperature cycle of the earlier system. The modern system is shown in Figure 2.160 with arrows indicating the coolant flow.

The heat distribution within the engine needs to be controlled. The temperature around all cylinders and combustion chambers should be identical. The heat removed by the cooling system has, therefore, to be consistent for all areas of the engine. All modern engines have a fairly rapid coolant circulation within the engine so that an even temperature distribution is achieved.

The water (or coolant) pump draws the coolant through a radiator bypass channel when the engine is cold and from the radiator when the engine is hot (Fig. 2.163). The impeller on the water pump drives the coolant into the engine coolant passages or water jacket. Water-jacket passages are carefully designed to direct the coolant around the cylinders and upwards over and around the combustion chambers.

The density of coolant falls as it heats up and, as the temperature approaches boiling point, bubbles begin to form. These bubbles can create areas in the water jacket where the coolant is at a lower density and the actual mass of coolant in those areas is reduced. The reduced mass of coolant therefore cannot effectively absorb heat efficiently to cool the engine.

Key fact

The density of coolant falls as it heats up and, as the temperature approaches boiling point, bubbles begin to form.

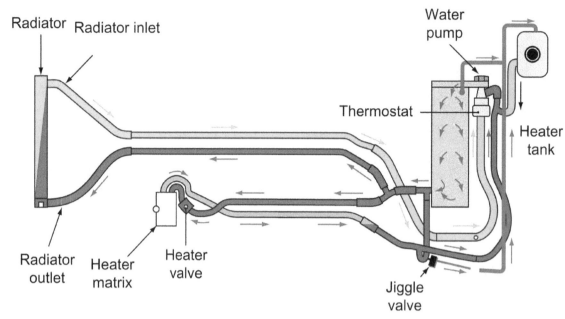

Radiator Radiator inlet

Water pump

Thermostat

Heater tank

Radiator outlet Heater matrix Heater valve

Jiggle valve

Figure 2.162 Basic cooling system

Figure 2.163 Water pump and its position in the engine block

To overcome this problem, all liquid cooling systems are pressurized. When hot, most modern systems have an operating pressure equivalent to about one atmosphere (1 bar, or 100 kPa). The pressure is obtained by restricting the loss of air above the coolant in a radiator header tank or an expansion tank. As coolant heats up it expands. If the air above the coolant has less space to occupy, and it cannot immediately escape, it increases in pressure. A pressure-sensing valve in the radiator or header tank cap allows excess pressure to escape but retains the operating pressure.

The pressure cap was traditionally called the radiator cap because it was fitted to the radiator. On most vehicles, the cap is fitted to the expansion tank.

There are many different designs and operating pressures. Vehicles are often fitted with a plastic, or nylon, cap that is specific to one manufacturer.

The main parts of all pressure caps are the sealing ring, pressure valve, vacuum valve and a bayonet, or screw, fitting (Fig. 2.164). The pressure valve consists of a spring-loaded seal that rests on a seat, either in the filler neck or in the cap. The vacuum valve allows air to return to the system as it cools to prevent a pressure lower than atmospheric. It is fitted in the centre of the pressure valve. Both the pressure valve and the vacuum valve are one-way valves and operate in opposite directions. The pressure valve allows air out and the vacuum valve allows air in. The advantages of a pressurized system are more efficient cooling with a

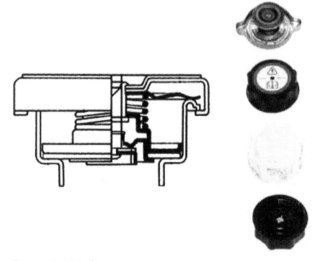

Figure 2.164 Pressure caps

higher safe operating temperature. It can also be used at high altitudes without the need for modification.

Key fact

The main parts of all pressure caps are the sealing ring, pressure valve, vacuum valve and a bayonet, or screw, fitting.

In a liquid-cooling system, the coolant carries heat from the engine to the radiator. Air flow through the radiator dissipates the heat into the atmosphere. Air is forced through the radiator by the forward movement of the vehicle, or is assisted by a fan fitted behind the radiator.

The fan can be driven by an electric motor (Fig. 2.165) or by a belt from the crankshaft. Early engines had the fan mounted on the front of the water pump with a 'V' belt driving the fan and pump. Fan designs that have been used include variable-pitch (to reduce noise) and viscous-hub types.

The thermostat uses a wax pellet in an enclosed cup (Figs 2.166 and 2.167). Inside the wax is a rubber sleeve enclosing a pin. The pin is connected to a plate that acts as the valve. All these components are held in the thermostat body, together with a spring to hold the valve closed when the coolant is not hot.

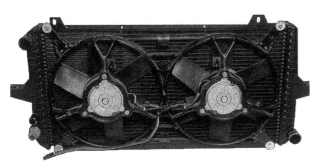

Figure 2.165 Twin electric fans

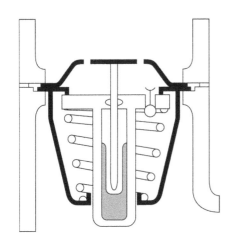

Figure 2.166 Thermostat closed

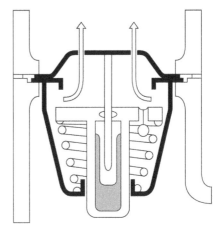

Figure 2.167 Thermostat open

The thermostat body includes a flange that fits into a housing in the coolant outlet from the cylinder head, or a radiator-bypass channel.

Key fact

The thermostat uses a wax pellet that expands with temperature.

When the temperature of the coolant acting on the wax pellet reaches the operating temperature, there is sufficient heat to cause the wax to expand, press on the pin and force it out of the cup to open the valve. Coolant is then free to flow through the valve.

The wax pellet must always be fitted so that it sits on the hot side of the coolant flow through the thermostat (Fig. 2.168).

Some thermostat flanges are fitted with a small subvalve to allow air to flow through the thermostat as the system is filled with coolant (Fig. 2.169). This small valve must be fitted towards the top if the thermostat

Figure 2.168 Thermostat with its wax pellet in the hot-coolant area

119

Figure 2.169 Thermostats

Figure 2.171 Radiator tubes

Figure 2.172 Traditional radiator

is fitted on its side. Some manufacturers fit the thermostat in a radiator hose. The thermostat may also be fitted directly into its own housing and, if so, has to be replaced as a complete assembly.

The various designs and manufacturing materials used for radiators (Fig. 2.170) all consist of a series of small tubes through which the coolant flows (Fig. 2.171). Very thin sheets of metal are used to form a large surface area surrounding the small tubes. This large surface area makes radiators efficient heat exchangers for engine-cooling purposes.

The radiator tubes are fitted to tanks at each end, and these tanks are fitted with connections for the top and bottom or cross-flow hoses. The traditional radiator had the core tubes set vertically and the coolant flowing downward from the header tank to the bottom tank (Fig. 2.172). The air space required for expansion of the coolant could be either in the header tank or in a separate expansion tank.

However, because of the lower frontal area of most modern cars and light vehicles, a different radiator

layout is needed. The cross-flow radiator has tubes and thin sheet fins forming the core (Fig. 2.173). The core tubes run across the vehicle and the coolant flows from one side to the other. The tanks at each end of the radiator are joined to the core and have connections for the hoses.

Key fact

Cross-flow radiators usually have a remote expansion tank to which the pressure cap is fitted.

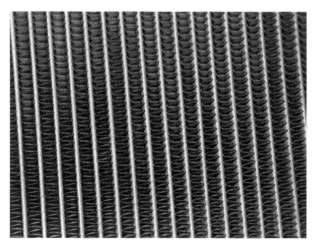

Figure 2.170 Radiator core construction

Figure 2.173 Cross-flow radiator

Cross-flow radiators usually have a remote expansion tank to which the pressure cap is fitted (Fig. 2.174).

Modern radiators are constructed from an aluminium core with nylon, or plastic end-tanks that are cinched together (Figs 2.175 and 2.176). This is a method of folding the edges of the radiator core ends over a sealing ring and a lip on the end-tank. Aluminium radiators are lighter and cheaper to produce than traditional copper/brass radiators.

The water pump is usually fitted into the water jacket of the cylinder block (Fig. 2.177). However, there have been some engines where it has been fitted into the cylinder head. An external water pump is used on some engines and connected to the water jacket by pipes or hoses (Fig. 2.178). The water pump is driven from the engine crankshaft by a belt.

Figure 2.176 The heater matrix (radiator) is similar to the main radiator but smaller

Figure 2.177 Water pump in water jacket

Figure 2.174 Remote expansion tank during coolant fill

Figure 2.178 External water pump

Figure 2.175 Aluminium and plastic radiator

Running through the centre of the water pump is a spindle mounted on a bearing. The bearing is prepacked with grease, and fitted with seals for retaining the grease and keeping the coolant in the engine. The drive pulley is fitted to the spindle on the outside of the pump. The movement of the impeller assists coolant flow through the water jacket. Water pumps are supplied as a replacement part fully assembled in a housing holding the bearing, spindle, impeller and drive flange for the pulley (Figs 2.179 and 2.180).

121

The drive components for the water pump on earlier cars consisted of a 'V' belt that also drove the alternator (Fig. 2.181). On many other vehicles multi-V belts are also in common use (Fig. 2.187). An adjuster for the belt is provided on the alternator mounting, or as a separate tensioner. The toothed camshaft drive belt is used to drive the pump on some vehicles (Fig. 2.179).

A fan is used to ensure an adequate air flow through the radiator when this is not provided by the forward speed of the vehicle. The fan was traditionally fitted to the front of the water pump and attached with the same bolts as the drivebelt pulley. Some longitudinal engines still use this system, but the fan, formerly a pressed-steel component, now incorporates a thermostatic viscous hub and nylon fan blades (Figs 2.188 and 2.189). The viscous hub is a fluid clutch using silicon oil. The operation of the clutch is temperature controlled with a bimetallic valve. When the air flow temperature over the viscous hub is cool, the valve remains closed and the clutch is inoperative. When the air flow temperature over the viscous hub increases, the valve in the hub opens and the viscous fluid is driven outwards by centrifugal force. The increased force in the fluid locks the plates in the hub together to engage the clutch drive to the fan.

An improved temperature-sensing arrangement is for the fan to be driven by an electric motor mounted on a cowl frame attached to the radiator (Fig. 2.190). A plastic fan is fitted to the motor spindle and operates when a temperature-sensitive switch closes. The electrical supply to the motor is connected through a relay.

Key fact

Many vehicles, particularly those fitted with air conditioning, have two-speed fan circuits.

Many vehicles, particularly those fitted with air conditioning, have two-speed fan circuits. These have a control circuit to switch the motor (or motors) to half speed at 95°C, and full speed at 100°C. This arrangement can also be operated by the engine management system.

Cooling system hoses are manufactured from fabric-reinforced rubber, and are moulded to suit the vehicle application. Connectors are cast, or formed, with a raised lip on the pipes leading into, and out of, other components. The hoses are held with round clips that can be drawn tight to give a watertight seal (e.g. jubilee clips).

2.3.3 Interior heater

The vehicle interior heater is made from an air box with a heat exchanger inside. The heat exchanger, called a heater matrix or a heater radiator, is very similar to the cooling radiator in that it consists of a similar series of tubes and fins. Hot coolant from the engine flows through the matrix, thus heating the tubes and fins. Air flows across the outside and collects some of the heat for distribution inside the vehicle.

Air is drawn into the heater through ducts on the vehicle exterior. The design of the ducts provides a dust and water trap, and usually the ducts have an outlet hose for water drainage. Many vehicles have a pollen filter fitted in the air-intake ducts (Fig. 2.198). The filter is a microporous paper element that traps pollen and dust particles.

The distribution of air inside the vehicle is provided by a series of ducts and outlets. These are positioned on the underside of the fascia, at fascia level, and adjacent to the front and side screens (and in the doors on some cars). The outlets can be selected by operating the controls to the required positions (Fig. 2.199). The controls are connected to flaps in the heater air box by a cable, a vacuum system or an electrical actuator. The flap position directs air to the appropriate outlet.

On older vehicles, temperature selection was achieved by regulating the coolant flow through the heater, by means of a valve. More often now it is by a flap in the heater air box that directs how much air flows through the heater matrix (Fig. 2.200). The flap is controlled in the same way as the air direction flaps. Thermostatic devices are used to control air temperature on some vehicles.

A blower motor in the air-intake duct boosts air flow through the heater system (Fig. 2.202). The motor may be fitted with a series of resistors or be electronically controlled to provide a range of speeds.

Most modern heaters are linked with an air conditioning system. This is covered in Chapter 3.

Key fact

Efficient cooling systems help reduce emissions, which is better for the environment.

2.3.4 Thermal management system (TMS)

Thermal management is about controlling air and coolant flows intelligently. However, it goes beyond cooling system technology because it considers all the

Figure 2.179 Water pump drive gear (cam belt)

heat flow systems on the vehicle. With this approach it is possible to improve efficiency, reduce emissions and improve passenger comfort.

Giving intelligence to the cooling system of a vehicle is the first step. This can be locally or in microprocessor-controlled systems. This intelligence will, for example, operate shutters (air control), coolant thermostats, bypass or mixing valves, and electrically driven coolant pumps (water/glycol control). Warm up times are faster and aerodynamic drag can be reduced by using radiator shutters. There is a reduction in energy consumed, passenger compartment warm-up time is reduced, and temperature control of the engine is far more precise. Taking further steps, it is possible to warm up and control the temperature of transmission fluids using engine coolant. Service life and efficiency of the transmission are therefore improved.

Thermal management systems need to be integrated with powertrain control systems, and a holistic view of the whole system should be taken. The range of potential applications appropriate for thermal management is considerable; here are some example systems where temperature can be controlled:

▷ coolant pump(s)
▷ thermostat
▷ radiator air flow using shutters
▷ cooling fan viscous clutch
▷ coolant-cooled intercooler
▷ exhaust gas cooling (EGR system)
▷ transmission oil temperature
▷ HVAC
▷ turbocharger
▷ throttle body.

In passenger vehicles the potential fuel saving could be as much as 7.5%, along with the consequential reduction in CO_2 emissions. Just as the use of electronic components has increased in the engine,

Figure 2.180 Water pump

the same is true of the TMS. Modern engines demand more extensive cooling, but in a controlled manner. This is delivered in part by adding electronic functionality to the TMS, and in part by redesigning TMS components.

2.4 Air supply, exhaust and emissions

2.4.1 Air pollution and engine combustion

Atmospheric pollution has become a serious problem to the health of people and to the environment. Many urban areas are heavily polluted, with people suffering medically from the effects of vehicle exhaust pollution. There have been many changes in climatic conditions in the world. Many of these have occurred over a long period and animals and plants have adapted to the changes naturally. However, the rapid burning of fossil fuels during the past century has increased carbon dioxide (CO_2) levels in the atmosphere.

Carbon dioxide allows the sun's heat in, but reduces the ability of the heat to radiate outward, causing the Earth to warm up. Many studies of the warming process indicate that the rate of Earth warming is increasing too quickly and preventing animals and plants from adapting. Vehicle exhaust gas also includes other toxic components (Fig. 2.206). Environmental regulations are now in place to find safer alternatives, or to reduce the production and use of the most harmful pollutants. Other regulations and agreements are seeking to reduce the production of carbon dioxide by improving the efficiency of fossil fuel burners.

Lead was at one time used as an additive in petrol to slow down the combustion process. This was to eliminate knocking or pinking in the engine. It made engines more efficient but the lead did not burn and passed into the atmosphere from the exhaust, producing airborne concentrations that were capable of causing many physical disabilities, including brain damage. Lead additives are no longer used and modern engines are now designed to run on lead-free fuel.

Another naturally occurring substance in fossil fuels, particularly diesel, is sulphur. This does not burn but, during combustion, chemically reacts with oxygen in the air to form sulphur dioxide (SO_2). This passes from the engine exhaust into the atmosphere, where it combines with water to form sulphuric acid (H_2SO_4) and falls back to Earth as acid rain, which destroys trees, plants, other vegetation and aquatic life in streams, rivers and lakes. Fuel suppliers now remove, or significantly reduce, the amount of sulphur during the refining process.

Air consists of about 80% nitrogen which, under normal circumstances, is an inert gas. An inert substance is one that has very little chemical reactivity and does not burn, or mix easily with other chemicals. Nitrogen, however, will mix with oxygen at high temperatures to form nitrogen oxides (NO_x). These combine in exceptional geographical and meteorological conditions to form smog, acids and increases in low-level ozone. This serves to make a very unpleasant atmosphere in which to live. Many respiratory and asthmatic fatalities occur under these conditions.

The combustion of fuel inside the engine is a chemical process that combines the carbon and hydrogen in the fuel with oxygen to release energy. Slightly less than 20% of air is made up of oxygen. Complete combustion produces carbon dioxide (CO_2) and water (H_2O). Neither of these is directly harmful. Both are naturally occurring substances in the atmosphere (but note the comments in the previous section).

Definition

Inert substance: A substance with very low chemical reactivity.

Incomplete combustion leaves some of the carbon and oxygen not fully combined. The product of this is carbon monoxide (CO), which is toxic. Small quantities of carbon monoxide molecules are dangerous because they attach themselves to red blood cells. This reduces the oxygen that the cells normally carry around the body, resulting in oxygen deprivation, brain damage, and fatality.

Another product of incomplete combustion is particles of fuel that have not been burnt. These are carried,

with the exhaust gases, into the atmosphere and are called unburnt hydrocarbons. Very small amounts of hydrocarbons in the atmosphere can cause respiratory problems.

Engine oil drawn into the combustion chamber, either from the inlet valve stem or by bypassing the pistons, can also be a source of hydrocarbon pollution. Oil vapours form in the engine crankcase and can escape into the atmosphere. A positive crankcase ventilation system is now used to draw the vapours into the engine so that they are burnt to form water and carbon dioxide.

On old vehicles, vapour in the fuel tank was directly vented to the atmosphere. This is no longer the case, but the fuel tank must still be vented in some way to allow air to flow into the tank as fuel is used. A charcoal filter is now used to prevent the loss of fuel vapour and for the expansion of the fuel when the weather is hot. The fuel vapour in the charcoal canister is drawn into the engine and burnt.

Good fuel economy is obtained with a lean air-to-fuel mixture. However, this mixture produces higher combustion temperatures and greater risks of NO_x being formed. In order to prevent, or reduce to a minimum, the formation of NO_x, the combustion temperature has to be kept as cool as possible and the amount of oxygen limited to match the quantity of fuel delivered. To reduce the amount of oxygen in the air charge, a gas that is low in oxygen can be introduced. This maintains the total air-charge mass to give good compression pressures and efficient operation of the engine. The available gas is the exhaust gas that has already used up its oxygen content during combustion. The addition of a regulated charge of exhaust gas reduces the oxygen content of the new charge to suit the amount of fuel delivered. This, in turn, reduces the combustion temperature and limits the formation of NO_x. Catalytic conversion of any remaining harmful gases can result in a much cleaner exhaust gas.

Definition

The Evaporative Emission Control System (EVAP) is used to prevent fuel vapours from escaping into the atmosphere.

2.4.2 Worldwide harmonized light vehicles test procedure

The worldwide harmonized light vehicles test procedure (WLTP) is a global standard for determining levels of pollutants and CO_2 emissions, fuel or energy consumption, and electric range. It is used for passenger cars and light commercial vans. Experts from the EU, Japan and India, under guidelines of UNECE

Figure 2.181 V' belt and pulley

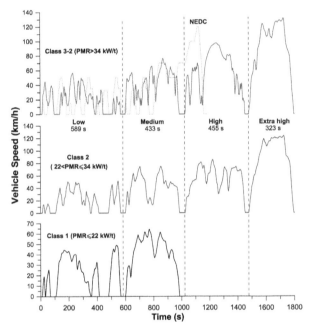

Figure 2.182 WLTP test cycles

World Forum for Harmonization of Vehicle Regulations, developed the standard. It was released in 2015.

Like all previous test cycles, it has its drawbacks, but it is a good attempt to make the test more realistic. The key thing is that a standardized test allows vehicles to be compared accurately, even if the figures differ from real-world driving. It includes strict guidance regarding conditions of dynamometer tests and road load (motion resistance), gear changing, total car weight (by including optional equipment, cargo and passengers), fuel quality, ambient temperature, tyres and their pressure. Three different cycles are used depending on vehicle class defined by power/weight ratio PWr in kW/tonne (rated engine power/kerb weight):

▷ class 1 – low power vehicles with PWr <= 22
▷ class 2 – vehicles with 22 < PWr <= 34
▷ class 3 – high-power vehicles with PWr > 34

Most modern cars, light vans, and buses have a power–weight ratio of 40–100 kW/t, so belong to class

3, but some can be in class 2. In each class, there are several driving tests designed to represent real-world vehicle operation on urban roads, extra-urban roads, motorways and freeways. The duration of each part of the tests is fixed between classes, the difference is that the acceleration and speed curves are shaped differently. The sequence of tests is also restricted by maximum vehicle speed (V_{max}).

Because there is a wide range of manual gearboxes with 4, 5, 6 and 7, or even 8 gears, it is impossible to specify fixed gear shift points. To overcome this, the WLTP uses an algorithm for calculating optimal shift points, which considers total vehicle weight and full load power curves within normal engine speeds. This covers the wide range of rpm and engine power available on current vehicles. To reflect normal use and a fuel-efficient driving style, gear changes are filtered out if they occur in less than 5 seconds.

WLTC (worldwide harmonized light vehicles test procedure) is an improvement over the New European Driving Cycle (NEDC), but transitions are still very slow. For example, the most rapid 0 to 50 km/h (0 to 30 mph) time is 15 seconds. Most drivers in Western Europe accelerate from rest to 50 km/h in 5 to 10 seconds. There is no hill climbing in the cycle. Perhaps there should be, as even modest gradients increase pollutant emissions because the engine load increases 2 to 3 times. Nonetheless, it is a step in the right direction.

The WLTC driving cycle for a class 3 vehicle is divided in four parts:

1 low
2 medium
3 high
4 extra high speed

If V_{max} < 135 km/h, the extra high-speed part is replaced with low speed part (the low speed part is completed twice).

2.4.3 Reducing pollution

Vehicle engine and component manufacturers have put a great deal of effort into reducing pollution. For example, lead is no longer needed in petrol (gasoline) because other less-damaging substitutes have been found. The changes in the fuel have necessitated the use of hardened valves and valve seats and changes to the ignition timing and fuel delivery systems.

Air-intake systems have been developed from a simple ducting to a complex air-flow design adapting to the changing speed and load conditions of the engine. Filtration is also an important aspect.

Electronic control of the combustion process has been the key development, and has achieved reductions in

125

Figure 2.183 Valves

Figure 2.184 Air filter

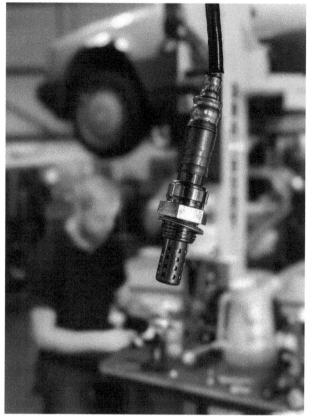

Figure 2.185 Lambda sensor

CO, NO_x and hydrocarbon emissions. Exhaust gases are monitored by the electronic engine control module by signals sent from a lambda, or oxygen, sensor in the exhaust. This monitoring allows fuel and air supplies to be accurately controlled for optimum combustion.

The remaining pollutants in the exhaust gases, which cannot be eliminated or reduced by the electronic systems, are converted into less harmful substances. This is achieved by using a catalytic converter.

Developments to improve the atomization and the mixing of the fuel in the incoming air stream include heating the inlet manifold, or heating the air as it enters the inlet manifold. This is achieved by preheating the air by ducting the air supply over the exhaust manifold.

Oil and fuel vapours are trapped and routed through the engine to be burnt. Positive crankcase ventilation and a charcoal filter in an evaporative canister are used for this purpose. Nitrogen oxide formation is reduced with EGR.

Engine performance has been increased, without an increase in weight, by the use of supercharging

Figure 2.186 Catalytic converters

Figure 2.187 Multi-V belt and pulley

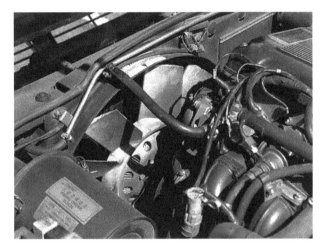

Figure 2.188 Viscous-fan hub on an earlier engine

Figure 2.189 Viscous-fan hub

Figure 2.190 Motor-driven fan

Figure 2.191 Air temperature control valve

Figure 2.192 Evaporative emission control system (EVAP) canister

Figure 2.193 Turbocharger (Source: Aeristech)

and turbocharging. Other emission-control devices that correct the ignition timing and fuel delivery are covered in the appropriate sections. These devices

127

improve the performance of those systems, as well as reducing harmful exhaust emissions.

2.4.4 Air supply system

The air supply system has to provide clean air in sufficient quantity to the engine and supply equal quantities of air to each cylinder. This will assist fuel vaporization and an even mixture distribution. Creating a swirl in the air flow as it enters the cylinders is also desirable. A system of warm air for cold starts, followed by temperature-controlled air for normal running, is essential. Finally, the system must silence the air flow and provide a flame trap in the event of fire in the inlet manifold.

> **Key fact**
> Creating a swirl in the air flow as it enters the cylinders improves efficiency.

The air supply systems for most vehicles are similar. They consist of an air-intake duct, an air-temperature

Figure 2.194 Air supply components with integrated intercooler

Figure 2.195 Throttle body (Source: Bosch Media)

Figure 2.196 Air flow meter

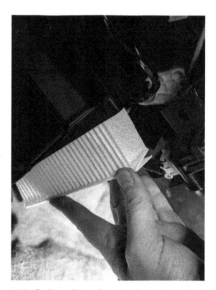

Figure 2.197 Pollen filter located next to heater matrix inside cabin

control mechanism, an air-cleaner housing and filter, an inlet manifold and inlet ports. A position for an EGR system may also be included. For fuel injection engines, the system will also include a throttle-body housing and an air-flow meter.

Clean air is required in the engine to prevent particles of dust and grit from damaging, or blocking, engine and fuel-supply components. Air is filtered through an element in the air cleaner. Most air-cleaner elements are made from microporous paper, which allows a good flow of air but traps airborne dust. Other elements have included oiled wire gauze and foam rubber. The air-cleaner housing and the filter elements are cleaned, or replaced, at scheduled service intervals.

Paper elements are folded to provide a large surface area and long service life. The element can be wrapped to form a circular element if required.

Figure 2.198 Heater intake with pollen filter

Figure 2.199 Air flow and heater controls

Figure 2.200 Heater control with flaps

Figure 2.201 Plastic inlet manifold

The outside edges are sealed with an integral, or separate, rubber sealing ring. Air-cleaner housings have internal ducting to distribute the air over the full surface of the filter. The air flow in some filter housings is made to swirl so that airborne dirt is thrown out and falls into a dust trap in the base of the filter. The air flow into flat filters is from the underside so that dirt falls out from below, rather than into the top, of the filter.

Key fact

Most air-cleaner elements are made from microporous paper, which allows a good flow of air but traps airborne dust.

The inlet manifolds on modern engines are usually of the same length and diameter to enable all cylinders to be supplied with the same volume of air and to provide the same air flow characteristics. Early engines, with manifolds using pipes of differing lengths, often produced slightly different combustion patterns in each of the cylinders.

At the entrance to the inlet manifold is the throttle body controlling the flow and quantity of air entering the engine.

On most engines, the temperature of the incoming air supply is controlled. Heating the air entering the inlet duct assists in atomization and fuel distribution in the air charge. To warm the air, it is passed over the exhaust manifold before being drawn into the air duct. This is only necessary when the air is cold. When the engine temperature increases, the air density, and therefore mass, would be reduced if heating of the air were continued. At an engine temperature of about 50°C, the full air supply is drawn from a cold position in the engine compartment, or from the front of the vehicle. Between a cold engine and 50°C, progressive mixing occurs.

Key fact

Heating the air entering the inlet duct assists in atomization and fuel distribution in the air charge.

The ducting of warm, or cool, air is controlled by a flap in the air-cleaner intake. This provides either a normal

129

Figure 2.202 Heater motor fan

air flow, or one from over the exhaust manifold. Several designs of thermostatically controlled air-cleaner operation are used. One type uses a vacuum motor and bimetallic vacuum valve, while another uses a wax-pellet actuator. On some vehicles an electrical actuator is used under the control of the engine management system.

The bimetallic valve responds to the temperature of the incoming air stream and opens or closes the vacuum supply from the inlet manifold to the vacuum motor. The motor reacts to the vacuum supply to move the flap and mix warm and cool air.

The air supply to each cylinder passes through equal length and diameter tubes of the inlet manifolds. Feeding the manifold tubes is a plenum chamber, which holds a large volume of air so that each intake tube receives an equal air supply. The air flow is made to swirl in the intake tubes, and careful design of the shape and direction of the tubes is required to make this happen. Another factor affecting the swirl is the volume and speed of the air flow.

Some engines have a dual-intake system that responds to low and high engine speeds. These systems have valves that open at higher engine speeds to balance the pressure in the two intake manifolds, or open to enable a secondary air supply to provide an adequate air flow for the higher engine speed. These systems have been developed to meet the changing air flow and swirl characteristics occurring with increases in air mass and speed.

2.4.5 Exhaust systems

The exhaust system has to carry the exhaust gases out of the engine to a safe position on the vehicle, silence the exhaust sound and cool the exhaust gases.

It also has to match the engine gas flow, resist internal corrosion from the exhaust gas and resist external corrosion from water and road salt.

The exhaust system consists of the exhaust manifold, silencers, mufflers, expansion boxes and resonators. It also has down or front pipes, intermediate and tail pipes, heat shields and mountings. Also included are one or more catalytic converters, one or two lambda sensors and a connection for the EGR system.

The exhaust gases are at a very high temperature when they leave the combustion chambers and pass through the exhaust ports. The exhaust manifold is made from cast iron to cope with the high temperatures. The remainder of the exhaust system is made from steel, which is alloyed and treated to resist corrosion. The downpipe, or front pipe, is attached to the manifold with a flat, or ball, flange. This joint is subject to bending stresses with the movement of the engine in the vehicle. To accommodate the movement and reduce stress fractures, some flange connections have a flexible coupling made from a ball flange joint and compression springs on the mounting studs.

Safety first

The exhaust gases are at a very high temperature when they leave the combustion chambers and pass through the exhaust ports.

Another system to accommodate movement is a flexible pipe constructed from interlocking stainless-steel coils or rings. Where a flexible joint is not required, the front pipe may be supported by a bracket welded to the pipe, which is bolted to a convenient position on the engine or gearbox. Where a catalytic converter is used, it is fitted to the front pipe so that the exhaust heat is used to

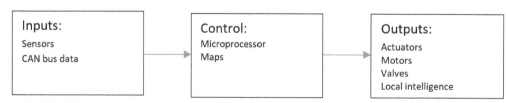

Figure 2.203 Simple representation of a thermal management control system

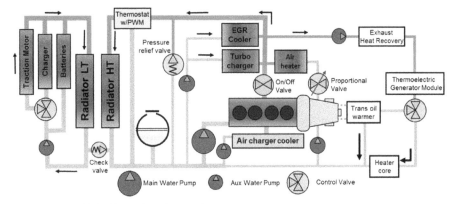

Figure 2.204 Hybrid engine thermal management system (Source: Solvay)

Figure 2.205 Audi coolant pump and control valves (Source: Solvay)

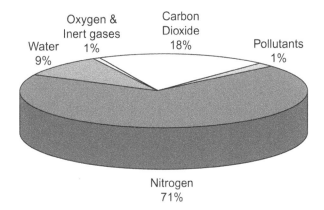

Oxygen & Inert gases 1%
Carbon Dioxide 18%
Water 9%
Pollutants 1%
Nitrogen 71%

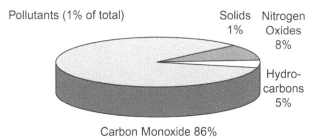

Pollutants (1% of total)
Solids 1%
Nitrogen Oxides 8%
Hydro-carbons 5%
Carbon Monoxide 86%

Figure 2.206 Approximate composition of vehicle exhaust gases

aid the chemical reactions taking place within the catalytic converter. The pipe continues and then connects to an expansion box or silencer (muffler). The exhaust gases are allowed to expand into this box and begin to cool. They contract on cooling and slow down.

Silencers or mufflers are constructed as single- or twin-skin boxes, and there are two main types: the absorption type, which uses glass fibre or steel wool to absorb the sound, and the baffle type, which uses a series of baffles to create chambers. In the baffle type, the exhaust gases are transferred from a perforated inlet pipe to a similarly perforated outlet pipe. These silencers have a large external surface area so that heat is radiated to the atmosphere. Additional pipes and silencers carry the exhaust gas to the rear.

Pipes are joined together by a flange, or clamp, fitting. Flange connections have a heat-resistant gasket and through-bolts to hold the flange together.

Clamp fittings are used where pipes fit into each other. The larger pipe is toward the front and the smaller pipe fits inside. A ring clamp, or 'U' bolt and saddle, are tightened around the pipes to give a gas-tight seal. An exhaust paste is usually applied to improve the seal of the joint. The exhaust system must be sealed to prevent toxic exhaust gases from entering the passenger compartment.

131

Figure 2.207 Flange connections

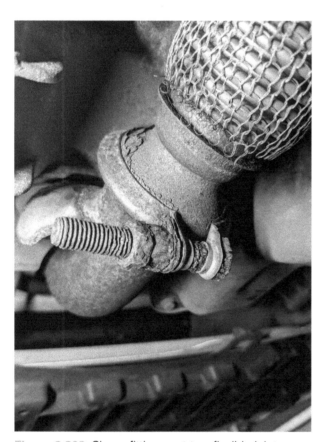
Figure 2.208 Clamp fitting next to a flexible joint

The exhaust is held underneath the vehicle body on flexible mountings. These are usually made from a rubber compound and many are formed as a large ring that fits on hooks on the vehicle and the exhaust-pipe brackets. Other mountings are bonded-rubber blocks on two steel plates.

Heat shields are fitted to the exhaust, or to the vehicle floor, to prevent the ignition of sound-deadening and anti-corrosion materials. Catalytic converters become

very hot during operation. It is important, therefore, that all heat shields are correctly fitted and positioned to insulate the vehicle from the high temperature of the catalytic converter.

Key fact

Heat shields are fitted to the exhaust, or to the vehicle floor, to prevent the ignition of sound-deadening and anti-corrosion materials.

Definition

Catalyst: A substance that will accelerate chemical changes in other substances without itself changing.

2.4.6 Catalyst systems

A catalyst is a substance that will accelerate (or, in some cases, slow down) chemical changes in other substances without itself changing. The purpose of the

Figure 2.209 Honeycomb substrate. (Source: Denso Media)

Figure 2.210 Steel substrate

catalytic converter on a vehicle is to convert potentially harmful chemicals in the exhaust gas into harmless water vapour, carbon dioxide, nitrogen and oxygen.

Several types of catalytic converters have been used on motor vehicles. However, almost all petrol (gasoline) vehicles now use a three-way catalyst. The main catalytic materials used consist of a mixture of platinum, palladium and rhodium, but less expensive materials are being investigated and developed. The catalytic material is applied as a thin coat to ceramic or stainless-steel 'honeycomb' or pellets. The exhaust gases flow freely through the honeycomb, or pellets, where the catalytic chemical reactions take place. The operating temperature of the catalyst is high, and the catalyst must be heated before it becomes effective. Exhaust heat is used for this.

Some catalysts require surplus oxygen in the exhaust gases for use in the conversion of hydrocarbons and carbon monoxide (CO) to water (H_2O) and carbon dioxide (CO_2). Oxidation catalysts are suitable for engines that run with a surplus of oxygen, such as diesel engines, and where additional air and, therefore, oxygen can be supplied.

Three-way oxidizing catalysts convert the hydrocarbon and CO to H_2O and CO_2 and additionally reduce the nitrogen oxides (NO_x). In these catalytic converters, the NO_x reacts with carbon monoxide to give nitrogen (N_2) and CO_2. The nitrogen oxides also react with hydrogen to give nitrogen and water vapour. The performance of catalytic converters relies on the correct exhaust gas constituents being produced. Modern engines do this by using electronic closed-loop control with an oxygen sensor in the exhaust manifold or downpipe.

> **Key fact**
> Three-way oxidizing catalysts convert the hydrocarbon and CO to H_2O and CO_2, and reduce the nitrogen oxides (NO_x).

The oxygen or lambda sensor is named after the Greek letter lambda, which is used as the symbol for a chemically correct air-to-fuel ratio, or stoichiometric ratio of 14.7 parts of air to 1 part of fuel. This sensor is known as an exhaust gas oxygen (EGO) sensor, or a heated exhaust gas oxygen (HEGO) sensor when it is preheated (Fig. 2.211). The sensor measures the presence of oxygen in the exhaust gas and sends a voltage signal to the engine electronic control module (ECM).

More fuel is delivered when an oxygen content is detected and less fuel when it is not. In this way, an

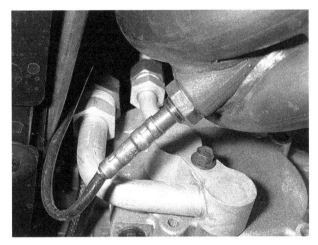

Figure 2.211 Heated exhaust gas oxygen sensor in the inlet manifold

accurate fuel mixture close to the stoichiometric ratio, or lambda, is maintained. This produces the correct exhaust gas constituents for chemical reactions in the catalytic converter.

2.4.7 Emission control systems

2.4.7.1 Crankcase ventilation

Oil vapour occurs in the engine crankcase because of heat, spray and the churning action of engine components as the engine is running. A fine mist of oil vapour is always present in a running engine. The engine crankcase pressure is never constant. Slight leakages into and from the combustion chambers, and the movement of the pistons, are responsible for most of the pressure variations.

A vent to atmosphere system was once used for ventilating pressure variations in the engine. This simple vent allowed a large quantity of oil vapour to escape. By fitting an oil separator the quantity of oil was reduced but still unacceptable quantities of oil vapour were emitted. Developments since that time have seen the introduction of a positive crankcase ventilation (PCV) system. This takes any escaping oil vapour into the engine for combustion.

One type of PCV system consists of a valve mounted in the crankcase vent oil separator (attached to the cylinder block) and two hoses. One PCV hose connects the PCV valve to the intake manifold; the other connects the valve cover to the air cleaner. Under idle and part throttle conditions, the crankcase vapour flows through the intake manifold into the combustion chambers where the vapour is burnt during combustion. Under full throttle conditions, the crankcase vapour flows from the valve cover into the air cleaner through the PCV hose.

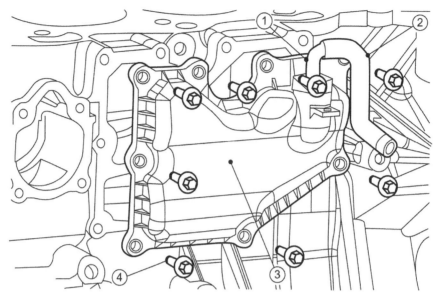

Figure 2.212 Positive crankcase ventilation (PCV) system oil separator on the side of a block: 1, PCV valve; 2, PCV hose; 3, crankcase vent oil separator; 4, crankcase vent oil separator retaining bolt. (Source: Ford Motor Company)

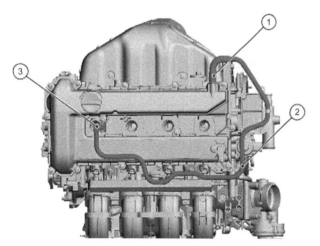

Figure 2.213 Positive crankcase ventilation (PCV) connection from rocker cover to the air cleaner box: 1, PCV hose to valve cover connector; 2, PCV hose; 3, PCV hose to air cleaner connector. (Source: Ford Motor Company)

Figure 2.214 Pick-up for hot air on an exhaust manifold

2.4.7.2 Evaporative emission control

Fuel vapour, and particularly petrol vapour, is harmful. It is given off from petrol at quite low ambient temperatures. Fuel is stored in underground tanks to reduce vapour formation. However, during filling up and when fuel is in the tank, vapour can escape into the atmosphere. Modern vehicles are fitted with fuel systems that prevent vapour loss from the vehicle.

Safety first

Fuel vapour, particularly petrol (gasoline) vapour, is harmful.

Figure 2.215 Fuel tank cap

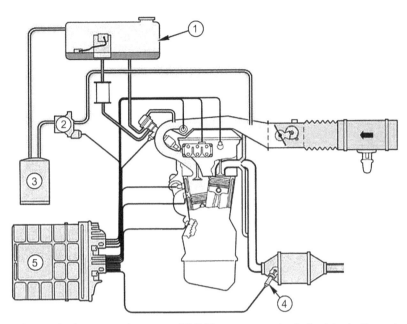

Figure 2.216 Evaporative emission control system (EVAP) components: 1, fuel tank; 2, control valve; 3, charcoal canister; 4, lambda sensor; 5, ECU

Figure 2.217 Persuasive tap applied to a failing EVAP solenoid

Figure 2.218 Warm-air control (vacuum type)

An evaporative emission control system (EVAP) has a sealed tank and fuel lines. It allows for expansion and reuse of the fuel through a charcoal canister. Air can pass through but the fuel vapour is trapped. To prevent the filter becoming saturated it is cleaned, or purged, by drawing air through the filter in the opposite direction. The air collects the deposited fuel vapour and carries it through pipes into the inlet manifold and engine, where it is burnt.

To prevent vapour loss through this route when the engine is not running, a canister purge control valve is fitted into the fuel vapour line. The valve is closed when the engine is stationary and during warm-up. When the engine is at normal running temperature the valve opens and the inlet manifold vacuum is able to cause air flow through the canister. This draws vapour out of the filter and into the inlet manifold.

The evaporative canister can be fitted almost anywhere on the vehicle. It may be near to the fuel tank, in the engine compartment or under a body panel. Fuel traps, to prevent fuel loss if the vehicle turns over in an accident, are also fitted.

2.4.7.3 Exhaust gas recirculation

Exhaust gas recirculation (EGR) has become a common feature on petrol and diesel engines (Fig. 2.219). The addition of exhaust gas to a fresh air and fuel charge lowers the combustion temperature and reduces the formation of NO_x. EGR operates during normal engine temperature and high vacuum conditions. Exhaust gases are piped from the exhaust manifold to the inlet manifold through a vacuum or electrically operated valve.

The amount of exhaust gas introduced into the air supply is usually less than 15% of the total charge. However, where closed-loop control is used, up to 50% can be used under some conditions on diesel engine systems. Some systems use a one-piece electrical solenoid valve, in place of the separate electronic vacuum regulator and valve. Some valves have a sensor fitted above the valve so that the ECM can monitor the opening performance.

> **Key fact**
>
> The addition of exhaust gas to a fuel/air charge lowers the combustion temperature and reduces the formation of nitrogen oxides.

2.4.8 Turbocharging and supercharging

Supercharging is a method of increasing the performance of internal combustion engines by boosting the air charge with an air pump. The most popular method is turbocharging, as this uses some of the lost energy in the exhaust gas flow (Fig. 2.220). Superchargers are driven from the engine crankshaft.

Figure 2.220 Turbocharger

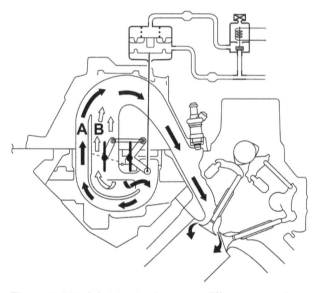

Figure 2.221 Inlet tracts change at different speeds: A: long tract, B: short tract

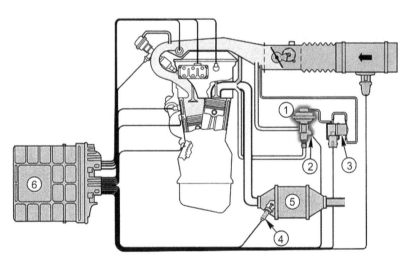

Figure 2.219 Exhaust gas recirculation (EGR) system: 1, vacuum actuator; 2, sensor; 3, vacuum control valve; 4, lambda sensor; 5, catalytic converter; 6, electronic control module (ECM)

Turbocharging (supercharging) of an engine is not strictly an emission control device but a method by which an increase in power and fuel efficiency can be obtained from a smaller engine. At the same time, there are improvements in exhaust emissions. Forced air induction has advantages over natural aspiration, because cylinder charging is more consistent over the full engine speed range. This helps to give high torque and power over a wider speed range, improved overall performance and improved fuel consumption.

> **Key fact**
>
> **Superchargers are driven from the engine crankshaft.**

Exhaust turbochargers use waste energy in the exhaust gas flow for power. This method of air boost charging is suitable for all types of engine. However, applications on small petrol engines are usually found only on high-performance vehicles.

Turbocharging of small high-speed diesel engines, used in cars and light vans, is now very popular. Diesel engine cylinder charging can be increased from about 60% for naturally aspirated engines to about 90% with exhaust turbocharging. The

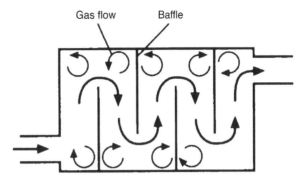

Figure 2.223 Baffle silencer

increased volume of air means that a corresponding increase in fuel can be delivered and more torque and power can be obtained per litre of engine capacity. Turbochargers use the energy in the exhaust gas to drive a turbine. The turbine is connected by means of a shaft to a compressor wheel in the engine air intake tract. The greater the flow of exhaust gas, the greater the speed of the turbine and compressor wheel and therefore the amount of additional charging.

> **Key fact**
>
> **Exhaust turbochargers use waste energy in the exhaust gas flow for power.**

The boosted air pressure is from 0.2 to 0.9 bar, depending on compressor speed. The maximum boost pressure is regulated by splitting the exhaust gas stream so that the excess gas flow and energy bypasses the turbine through a waste gate. The waste gate is a pressure-operated poppet or plate valve, which normally remains closed.

When the boost pressure in the inlet air stream rises, it is applied to the waste gate valve. The pressure acts on a diaphragm connected to the waste gate valve, and when it reaches the maximum operational pressure the valve opens. This allows exhaust gases to bypass the turbine. With the reduced gas flow, the turbine and compressor slow down, the pressure reduces and the waste gate closes. This opening and closing cycle maintains the boost pressure within operational limits. An intercooler cools the air and therefore it becomes denser, further increasing efficiency.

> **Key fact**
>
> **An intercooler cools the air and therefore it becomes denser, further increasing efficiency.**

The turbine and compressor fan wheels are radial flow types. The exhaust flows towards the centre and then

Figure 2.222 Exhaust system

137

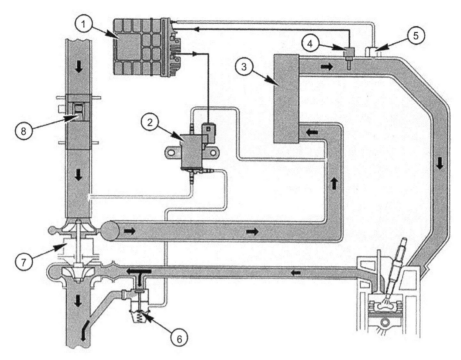

Figure 2.224 Boost control system: 1, electronic control unit (ECU); 2, boost control pressure solenoid; 3, intercooler; 4, temperature sensor; 5, boost pressure sensor; 6, boost pressure control valve; 7, turbocharger; 8, mass air flow (MAF) sensor

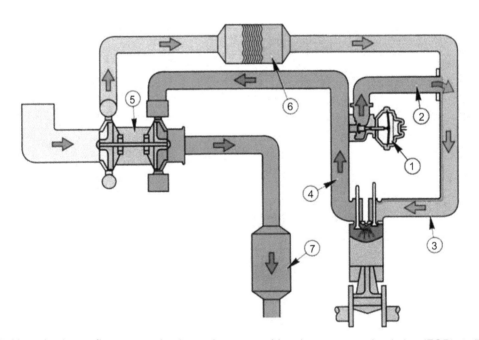

Figure 2.225 Air and exhaust flow on a turbocharged system with exhaust gas recirculation (EGR): 1, EGR valve; 2, tube; 3, inlet manifold; 4, exhaust manifold; 5, turbocharger; 6, intercooler; 7, catalytic converter

out. The inlet air flows in at the centre and outwards to the engine air intake duct. The air is forced out by the rotary centrifugal action of the compressor wheel. Air and gas flow is directed by spiral ducting in the turbocharger body. The spindle carrying the turbine and compressor wheel is mounted on special bearings with forced feed oil lubrication, which allows rotation with a minimum of metal-to-metal contact.

Figure 2.226 Turbine blades

The oil feed is made from the engine main oil gallery and returns to the oil sump/pan. The lubricating oil is also used for cooling in the turbocharger. Turbochargers must be allowed to slow down and to cool down before the engine is switched off. Usually about 30–60 seconds is required for this. The charged air increases in temperature through the turbocharger and becomes less dense and of lower mass. To overcome this loss, an intercooler is often fitted between the turbocharger and the inlet manifold. The intercooler is similar in construction to a coolant radiator but is an air-to-air heat exchanger.

Key fact

Lubrication is used for cooling in the turbocharger.

Some vehicles have a variable vane turbocharger instead of a waste gate. The turbocharger is designed to improve engine induction and engine performance.

A standard turbocharger has two key weaknesses (although these are much reduced on many systems):

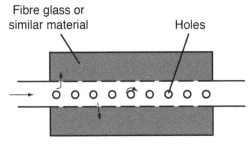

Figure 2.227 Absorption silencer

▷ High engine speed produces excessive turbine speed and therefore creates excessive turbocharger boost pressure.
▷ Low engine speed does not produce sufficient turbine speed and therefore not enough turbocharger boost pressure is achieved.

The variable vane turbocharger does not have a waste-gate control valve. Instead, it has variable turbocharger vanes which are located in the turbocharger turbine housing and can overcome the previous weaknesses. The turbocharger vanes act as the control for the turbocharger boost pressure.

The variable vane turbocharger produces its full turbocharger boost pressure over the entire engine speed range, not just at high engine speed. This is achieved through the adjustment of the vanes and the resulting change in the velocity of the exhaust gas. The speed of flow of the stream of exhaust gas is increased independently of engine speed by varying the intake cross-section in front of the turbocharger turbine. The variable vanes are controlled by the powertrain control module (PCM). A duty cycle signal from the PCM controls a vacuum supply to the turbocharger vacuum diaphragm unit using a solenoid valve.

Superchargers are mainly of the Roots blower or radial flow types. The radial flow types are similar to the compressor on the exhaust turbocharger. However, they are driven by belts and gears from the engine crankshaft. Vane radial superchargers have been used but are less popular.

The Roots blower uses two- or three-lobe intermeshed rotors to pump air. The rotors have helical rotor vanes to reduce noise and improve efficiency. The rotor vanes are driven and matched together with a pair of gears, so that they rotate in mesh with each other. They run on ball or needle roller bearings at each end of the rotor spindles. They must be lubricated with high-performance grease or synthetic oil in the bearing cases.

2.4.9 Electrical pressure chargers

UK company Aeristech has developed an eSupercharger. It claims that its proprietary electric motor technology delivers an engine's full load air requirements in under one second, without the cooling issues associated with conventional motors. This could solve a significant challenge facing the use of downsized engines in pursuit of low CO_2 emissions.

For serious engine downsizing, multi-stage or mechanical superchargers are the answer. This is because an engine using them would lack low-speed power. An electric supercharger with enough power to

139

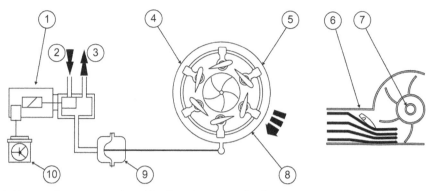

Figure 2.228 1, Vane adjustment solenoid valve; 2, atmospheric pressure; 3, vacuum; 4, adjusting ring; 5, vanes; 6, vanes; 7, turbine; 8, turbine; 9, vacuum diaphragm unit; 10, electronic control module (ECM). (Source: Ford Motor Company)

Figure 2.229 Twin Roots blowers

Figure 2.231 eSupercharger (Source: Aeristech)

Figure 2.230 Cutaway supercharger

supply all the low-speed boosting needs of the engine could be the answer. Multi-stage turbocharging, which combines large and small turbos, makes downsized engines easier to drive across a wider speed range. However, other issues such as cost, complexity, thermal management and catalyst performance become problematic. The catalyst issue is likely to be more of a problem as tailpipe emissions are further restricted. The thermal mass of a two-stage

turbocharger system located between the engine and the catalyst makes light-off (getting up to temperature) more difficult during light loads or low engine speeds.

An electric supercharger with sufficient power to meet all low-speed boosting needs could be achieved with the move to 48V systems. Aeristech uses permanent magnet motor technology, providing a faster response without the cooling challenges arising from the alternative switched reluctance (SR) motors. With a motor control strategy that separates commutation and power control, it means the electrical switching frequency need be no higher than running speed. The strategy would make permanent magnet technology cost-effective, meaning that an eSupercharger could run continuously at boost levels of 2.5 bar or more.

The company claims the eSupercharger motor will accelerate to 150,000 rpm with a transient response of idle-to-target speed in under 0.4s. The company successfully subjected its technology to independent evaluation by Ricardo and Mahle Powertrain UK.

Figure 2.232 Flexible rubber mounting supporting a rusty system

Figure 2.233 Rubber mounting

Audi's 2018 V6 diesel included a new electrically enhanced turbocharger system, as well as a new integrated NO$_x$ and PM after treatment suite. Many

Figure 2.234 Audi V6 Diesel with electric turbo (Source: Audi Media)

diesels could be electrically turbocharged due to the uptake of diesel engines in the USA.

It is claimed that using the e-turbo will easily gain you a lead of at least two car lengths in the first two seconds at the traffic lights! This illustrates the clear benefit of the system in that turbo lag is all but eliminated as the turbo compressor can be brought up to speed almost instantly. Other new features are to be incorporated:

▷ Piston rings have been optimized for minimal friction.
▷ Crankcase and new cylinder heads have separate coolant loops.
▷ A new thermal management system improves efficiency.
▷ The turbocharger and the fully variable-load oil pump have been updated.

Figure 2.235 Fitting a new system

141

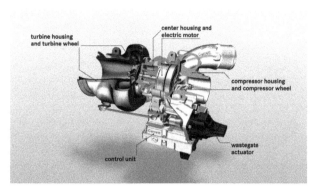

Figure 2.236 E-turbo technology (Source: Garrett media)

The engine will include a NO_x storage catalytic converter which has been combined with a diesel particulate filter and SCR injection in a single unit. The engine package satisfies the most stringent of emissions legislation, including Euro 6 and reduces CO_2 emissions by an average of 15 g/km. Emission and economy issues are what manufacturers are working towards with all their new developments.

Electrically driven turbo/superchargers are likely to become more common, particularly on vehicles with a 48 V system. This is because they offer an increase in performance, a reduction of emissions and reduced costs.

2.4.10 Diesel particulate filters (DPF) and selective catalytic reduction (SCR)

The main approach to lowering of diesel engine emissions involves internal engine improvements. This is because improved fuel combustion prevents, as far as possible, the formation of pollutants and reduces fuel consumption. In this respect, automobile manufacturers and their component suppliers have

Figure 2.237 Cordierite filter cores

Figure 2.238 Early catalytic system

already achieved a great deal. A diesel particulate filter (DPF) is a device designed to remove diesel particulate matter or soot from the exhaust gas of a diesel engine. Wall-flow diesel particulate filters usually remove 85% or more of the soot and under certain conditions can attain soot removal efficiencies of close to 100%.

The most common filter is made of cordierite (a ceramic material that is also used as catalytic converter cores). Cordierite filters provide excellent filtration efficiency, are (relatively) inexpensive and have thermal properties that make packaging them for installation in the vehicle simple. The major drawback is that cordierite has a relatively low melting point (about 1200°C) and cordierite substrates have been known to melt down during filter regeneration. This is mostly an issue if the filter has become more heavily loaded than usual and is more of an issue with passive systems than with active systems unless there is a system breakdown.

Cordierite filter cores look like catalytic converter cores that have had alternate channels plugged – the plugs force the exhaust gas flow through the wall and the particulate collects on the inlet face.

Silicon carbide, or SiC, has a higher (2700°C) melting point than cordierite; however it is not as stable thermally, making packaging an issue. Small SiC cores are made of single pieces, while larger cores are made in segments, which are separated by special cement so that heat expansion of the core will be taken up by the cement, and not the package. These cores are often more expensive than cordierite ones; however they are manufactured in similar sizes and one can often be used to replace the other.

Ceramic fibre filters are made from several different types of ceramic fibres that are mixed to form a porous medium. This medium can be formed into almost

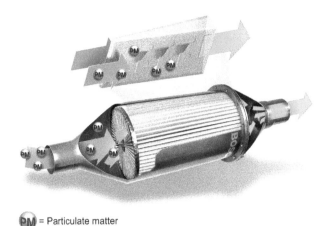

PM = Particulate matter

Figure 2.239 Diesel particulate filter (Source: Bosch Media)

any shape and can be customized to suit various applications.

The particulate filter shown here is made of sintered metal and lasts considerably longer than current ceramic models since its special structure offers a high storage capacity for oil and additive combustion residues. The filter is designed in such a way that the filtered particulates are very evenly deposited, allowing the condition of the filter to be identified more reliably and its regeneration controlled far better than with other solutions. This diesel particulate filter is designed to last as long as the vehicle itself.

The two main DPF systems are those with an additive and those without. To enable a vehicle to operate without an additive the particulate filter must be fitted close to the engine. Because the exhaust gases will not have travelled far from the engine, they will still be hot enough to burn off the carbon soot particles. In these systems, an oxidising catalytic converter will be integrated into the particulate filter. In other systems the particulate filter is fitted some distance

Figure 2.240 Different fitting methods for DPF

from the engine and as the exhaust gases travel along the exhaust they cool. The temperatures required for ignition of the exhaust gas can only be achieved using an additive.

Use of an additive lowers the ignition temperature of the soot particles and, if the engine management ECU raises the temperature of the exhaust gas, the filter can be regenerated. Regeneration is usually necessary after between 300 and 450 miles, depending on how the vehicle is driven. The process takes about 5–10 minutes and the driver shouldn't notice it is occurring, although sometimes there may be a puff of white smoke from the exhaust during regeneration. The additive is stored in a separate tank and is used at a rate of about 1 litre of additive to 3000 litres of fuel. It works by allowing the carbon particles trapped in the particulate filter to burn at a significantly lower temperature than would usually be required (250–450 °C rather than 600–650 °C).

On-board active filter management can use a variety of strategies:

- Engine management increases exhaust temperature through late fuel injection or injection during the exhaust stroke (the most common method).
- A fuel-borne catalyst (the additive) reduces soot burn-out temperature.
- A fuel burner after the turbo increases the exhaust temperature.
- A catalytic oxidizer increases the exhaust temperature, with after injection.
- Resistive heating coils increase the exhaust temperature.
- Microwave energy increases the particulate temperature.

Not running the regeneration cycle soon enough increases the risk of engine damage and/or uncontrolled regeneration (thermal runaway) and possible DPF failure. There are two types of regeneration – passive and active. Passive regeneration takes place, automatically, on motorway-type runs in which the exhaust temperature is high. If the exhaust is hot enough to ignite the soot particles, the regeneration process can carry on continuously and steadily across the platinum-coated catalytic converter.

Once the storage capacity of the particulate filter has been exhausted, the filter must be regenerated by passing hot exhaust gases through it which burn up the deposited particulates. To produce the necessary high exhaust gas temperatures, the EDC alters the amount of air fed to the engine as well as the amount of fuel injected and the timing of injection. In addition,

143

Figure 2.241 Sectioned view of a new filter

some unburnt fuel can be fed to the oxidizing catalytic converter by arranging for extra fuel to be injected during the expansion stroke. The fuel combusts in the oxidizing catalytic converter and raises the exhaust temperature even further.

A significant number of people don't use motorways, so passive regeneration will be possible only occasionally. In the case of a filter without additives when the soot loading reaches about 45% the ECU switches off the EGR and increases the fuel injection period so there is a small injection after the main injection. These measures help to raise the engine exhaust temperature to over 600°C which is high enough to burn off the soot particles.

A warning light is triggered at a 55% soot loading. In such circumstances the car needs to be driven hard in a lower gear so the temperature in the particulate filter will be sufficient to burn off the soot. If the driver ignores the warning and continues to use the car as normal, the soot will continue to build until it reaches 70%. Additional warnings will then be given using the malfunction indication lamp (MIL). Now it will not be

possible to clear the DPF by driving and it may need to be replaced. If loading reaches 95% then the DPF will need to be replaced.

If the EGR valve malfunctions, this may overload the DPF causing issues with regeneration. A malfunctioning EGR may present a light on the dash without immediate noticeable change to driving, a situation that can lead to a domino effect of component failure if left unaddressed.

Selective catalytic reduction (SCR) is a method of converting nitrogen oxides (NO_x) with the aid of a catalyst into nitrogen (N), and water (H_2O). A gaseous reductant, typically anhydrous ammonia, aqueous ammonia or urea, is added to a stream of exhaust gas and then into a catalyst. Carbon dioxide (CO_2) is a reaction product when urea is used as the reductant. A typical SCR fluid is known as AdBlue.

Emissions become an area of increasing concern as the acceptable limits become tighter to meet legislative requirements. This has developed areas of specific services for the maintenance and repair of these systems due to the expense of the components and possibility of MOT failure. Strategies to help protect emissions components include premium fuel, tank additives and specialist cleaning machines which have been developed for aggressively cleaning carbon that could cause component failure. The use of SCR is now essential to meet the EURO6 diesel emissions standards for heavy trucks, and for cars and light commercial vehicles. In many cases, emissions of NO_x and PM (particulate matter) have been reduced by

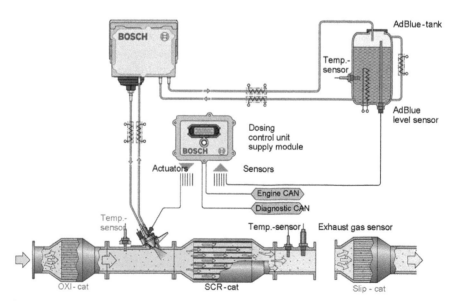

Figure 2.242 NO_x reduction Bosch Denoxtronic 2.2 (Source: Bosch Media)

upwards of 90% compared with vehicles of the early 1990s.

2.4.11 Cooled EGR

In petrol and diesel engines, exhaust gas recirculation (EGR) is a technique used to reduce nitrogen oxide (NO_x) emissions. It works by recirculating a small portion of an engine's exhaust gas back to the engine cylinders. This reduces the amount of oxygen in the incoming air stream and provides gases inert to combustion to act as absorbents of combustion heat. As NO_x is produced in a narrow band of high cylinder temperatures and pressures, it can be reduced by this method.

Cooling the gas during exhaust gas recirculation (EGR) can result in a further reduction of nitrogen oxides (NO_x). Cooler gas reduces its volume so it is possible the 'fit more in' to the combustion chamber. The system outlined here has been developed by a company known as Pierburg from Germany. When the combustion chamber temperatures are lower, it results in less NO_x production. In fact, this is the primary purpose of EGR, but this system takes it a step further by using a module with integrated control, bypass valves and a cooler.

Key fact

Cooling the gas during exhaust gas recirculation (EGR) can result in a further reduction of nitrogen oxides (NO_x).

In Figure 2.364 (on page 193), flap (6) allows exhaust gases to bypass the cooler during warm-up so that the engine and catalytic converter can reach their ideal operating temperature as quickly as possible. This also results in reduced HC emissions and less diesel knock. An additional advantage is that the cooler can be bypassed during the regeneration phase of a diesel particulate filter, when high temperatures are used for a short time.

2.5 Fuel systems

2.5.1 Introduction

Fuel systems have been extensively developed during the past fifty years. There were many developments in the traditional petrol supply and air mixing methods used in carburettors. However, the introduction of fuel injection systems has made carburettors almost obsolete. Fuel injection is now fitted to all petrol engine vehicles to meet the latest requirements for performance and the reduction of harmful exhaust gas emissions (Fig. 2.245). Diesel fuel pumps and injectors have seen similar developments with the introduction of electronic control and common rail systems.

Figure 2.244 Carburettor

Figure 2.243 Waste gate actuator under electronic control. (Source: Bosch Media)

Figure 2.245 New petrol/gasoline injection system

145

Figure 2.246 New diesel injection system

All fuel delivery systems have to supply a quantity of fuel that matches the amount of oxygen that is in the air entering the engine. For petrol engine vehicles, the quantities of hydrogen and carbon in the fuel and the oxygen content of the air should be chemically correct to allow a complete chemical change during combustion. The chemical formula for clean combustion is $CH_4 + 2O_2 = CO_2 + 2H_2O$, or hydrocarbon plus oxygen results in carbon dioxide and water.

For complete and clean combustion, the ratio of air to fuel should be as close as possible to the stoichiometric value. This is where λ (lambda) equals one. This is a ratio of 14.7:1 by mass of air to fuel. In petrol engines, the optimum for clean combustion is for these quantities to be delivered accurately to each cylinder in the engine.

Definition

Lambda (λ) value: A λ value of 1 is a ratio of 14.7:1 by mass of air to fuel.

Petrol is ignited in the combustion chamber by a spark arcing across the electrode gap of a spark plug. Diesel fuel ignites following injection into the high-temperature air charge. The high temperature is obtained by compression of the air charge. The air charge on petrol

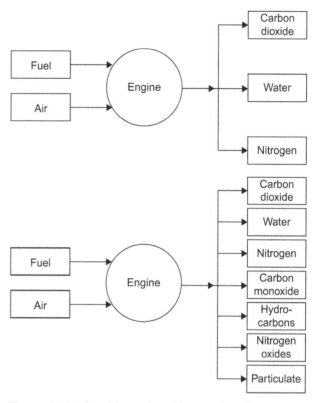

Figure 2.247 Ideal (upper) and incomplete (lower) combustion

engines is matched to the amount of fuel delivered. In diesel engines, a full air or gas charge is required to raise the temperature by compression.

The fuel on the vehicle is held in a tank fitted in a safe position. Recent construction legislation requires that the tank is unlikely to be ruptured in a vehicle collision. The positioning and protection of the tank are considered at the design stage of the vehicle and tested during development. The tank is fitted with a filler neck and pipework from the filler cap to the tank. Also fitted are the outlets to the atmospheric vent or evaporative canister and the fuel feed and return pipes to the engine. The fuel gauge is located in the fuel tank. Fuel supply and return lines are made from steel pipes, plastic pipes and flexible rubber joining hoses, depending on application and the type of fuel used.

A pump to supply fuel to the engine is fitted into or near the tank on petrol injection vehicles. On carburettor vehicles, a mechanical lift pump is fitted to the engine and is operated by a cam on the camshaft or crankshaft, or an electric pump is fitted in the engine compartment. Diesel-engined vehicles using a rotary fuel injection pump may use the injection pump to lift fuel from the tank. Alternatively, they may have a separate lift pump similar to the ones used on carburettor engines. A separate priming pump fitted in the fuel line may also be used.

Figure 2.248 Steel fuel tank in position

Figure 2.249 Mechanical fuel pump

The carburettor was the traditional method of mixing petrol with air as it enters the engine (Fig. 2.250). However, a simple carburettor is only capable of providing a correct air and fuel mixture ratio within a very small engine speed range. For road vehicles a wide engine speed range and a wide engine load are required. To respond to the speed and load variations, complex carburettors are used.

There are two basic carburettor designs, the fixed venturi and the variable venturi types. The term choke is often used to describe the venturi and this gives the alternative carburettor definitions of fixed choke and variable choke types. The usual meaning of the term choke is to describe the engine cold-start device fitted to the carburettor.

The function of the carburettor is to meter a quantity of petrol into the air stream entering the engine cylinders. As the pistons move down in the cylinders on the induction stroke the pressure in the space above the cylinders falls. On naturally aspirated engines, that is, those that are not fitted with pressure chargers, atmospheric pressure provides the force for the air flow into the cylinders.

The greater the difference in pressure, the greater will be the volume of air that enters the engine and the speed of the air flow through the carburettor and inlet manifold. A valve to meter the air flow is fitted at the base of the carburettor just in front of the inlet manifold. This valve is called the throttle and it consists of a round plate on a spindle. The spindle has a lever attached to one end and this is connected directly to the throttle pedal with a cable or rods. The throttle restricts the air flow in all positions except when wide open, and this gives a range of variable pressures in the carburettor and the inlet manifold.

Key fact

The function of the carburettor is to meter a quantity of petrol into the air stream entering the engine cylinders.

The basic carburettor consists of the venturi, through which the air flows, and the float chamber, which holds a supply of petrol at a constant level in relation to the supply beak in the venturi. The level of petrol in the float chamber is maintained by a needle valve that is lifted onto its seat by the float so that it stops the flow when the chamber is full. As petrol is used

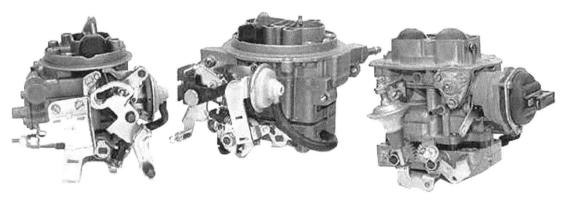

Figure 2.250 Single-choke (left) and twin-choke (middle and right) carburettors

147

Figure 2.251 Fixed-choke (left) and variable-choke (right) carburettors

Figure 2.252 Throttle butterfly

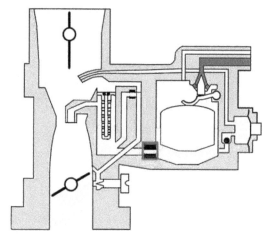

Figure 2.254 Main jet in the fuel feed to the venturi

should be checked and adjusted if necessary, if problems occur or if the carburettor is stripped for cleaning.

The main jet in the fuel feed to the venturi forms a restriction in the petrol flow and by virtue of its size acts as a metering device. The venturi is a tube with an inward curving restriction. Air flow through the venturi speeds up as it passes through the restriction. The effect of this is to reduce the air pressure at that point. Inside the float chamber, atmospheric pressure is applied to the top of the petrol held there. A vent in the top of the float chamber allows a free passage of air and atmospheric pressure.

A pressure differential exists at each end of the fuel supply tube between the float chamber and the venturi supply beak, when there is sufficient air flow to create a vacuum in the venturi. It is this pressure differential that is used to lift petrol up to the beak. From here, it passes into the air stream through the venturi and into the engine cylinders.

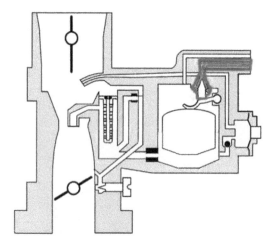

Figure 2.253 Fuel level is controlled by a float and needle valve

the level drops, the needle valve opens and the flow of petrol into the chamber resumes. In this way, a constant petrol level is maintained. The float level

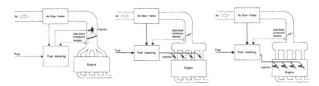

Figure 2.255 Injection methods (single point, multipoint port, multipoint direct)

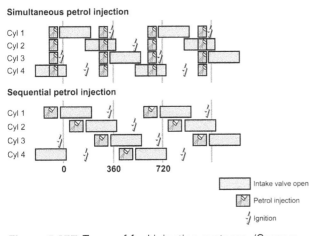

Figure 2.256 Electronic fuel injection system components. (Source: Bosch Media)

Simultaneous petrol injection

Cyl 1
Cyl 2
Cyl 3
Cyl 4

Sequential petrol injection

Cyl 1
Cyl 2
Cyl 3
Cyl 4

0   360   720

☐ Intake valve open

☒ Petrol injection

⚡ Ignition

Figure 2.257 Types of fuel injection systems. (Source: Bosch Media)

Although there is an increase in fuel delivery with an increase in air flow, these do not match sufficiently to maintain the correct air and fuel ratio over the full operating range. Other devices are needed to adjust the metering of petrol to the correct ratios. These are explained later in this section. The venturi can be positioned vertically with the air flow being downward or upward, or it can be positioned horizontally. This gives the expressions downdraft, updraft and sidedraft for descriptions of carburettors.

There are six clearly identifiable engine and vehicle use conditions, known as the stages of carburation.

Figure 2.258 Bosch gasoline direct injection (GDi). (Source: Bosch Media)

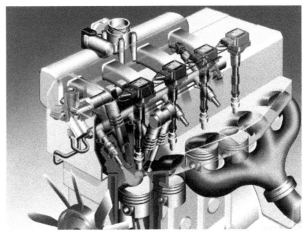

Figure 2.259 Ignition and fuel systems are now always combined

Figure 2.260 Fuel injection and ignition components from an earlier Motronic system. (Source: Bosch Media)

The development of twin-choke progressive carburettors was a way by which designers tackled the problems of maintaining correct mixture strengths, over the full range of engine operating conditions. The

149

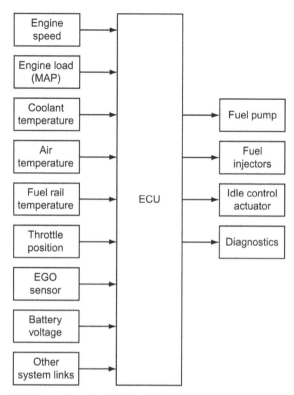

Figure 2.261 Fuel control electronic control unit (ECU) with some typical inputs and outputs

Figure 2.262 Inside an electronic control unit (ECU)

2.5.2 Electronic fuel injection

Petrol/gasoline fuel injection systems can be classified into two main categories:

▸ single point
▸ multipoint injection.

The multipoint systems (used by almost all cars now) can then be further divided into:

▸ manifold or port injection
▸ direct injection (into the combustion chamber).

The three methods are shown in Figure 2.255. Depending on the sophistication of the system, idle speed and idle mixture adjustment can be either mechanically or electronically controlled.

Electronic petrol injection systems have been in use for many years, first on expensive and sports vehicles and now standard equipment on all vehicles. The tougher

primary venturi works at the low throttle positions and a secondary venturi is added at the higher throttle positions. However, carburettors have had their chance, so fuel injection it is from now on!

> **Key fact**
>
> There are six clearly identifiable engine and vehicle use conditions, known as the stages of carburetion.

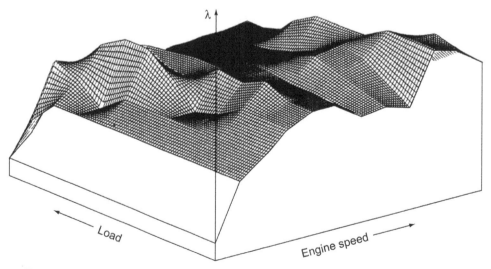

Figure 2.263 Fuel map

standards of exhaust emission regulations have made the use of microelectronic control systems for fuel delivery a virtual necessity. There are many different manufacturers of electronic fuel systems, and this programme covers the main points of the systems.

Some engines have been developed to include a gasoline direct injection (GDi) system, where the fuel is injected into the combustion chamber.

Modern petrol injection systems are linked to the ignition systems and are controlled by an engine control module (ECM). All modern fuel injection systems have closed loop electronic control using an EGO sensor.

The components for any electronic fuel injection system can be divided into three groups:

▷ electronic control system (ECU, sensors and actuators)
▷ air supply components
▷ fuel supply components.

2.5.2.1 Electronic control system

At the heart of EFi systems is the fuel control or ECU with a stored map of operating conditions (also referred to as the electronic control module or ECM). Electronic sensors provide data to the microprocessor in the ECM, which calculates and sends the output signals to the system actuators, which are the fuel pump, fuel injectors and idle air control units. The ECM will also switch some of the exhaust emission and auxiliary system components. ECU and ECM tend to be used interchangeably.

The ECU is an electronic microcomputer with a central processing unit (CPU) or microprocessor (Fig. 2.262). Inside the CPU are software programs that compare all sensor input data with a fixed map of operating conditions. It then calculates the required output signal values for the injection valves and other actuators.

The fixed map of operating conditions specific for each engine is held in a fixed value memory or read only memory (ROM). The operating data store of input values from the sensors is held in a random access memory (RAM). A 'keep alive' memory (KAM) of specific data such as adjustments, faults and deviations in component performance may also be used. The RAM data is erased when the ignition is switched off. The KAM data is erased when the battery is disconnected. New data is replaced in the RAM and KAM during engine start-up and operation.

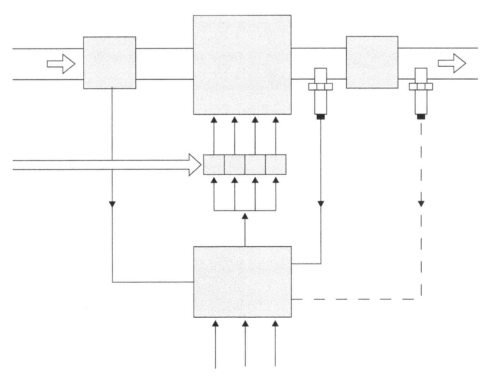

Figure 2.264 Closed loop control with lambda sensor (with a second sensor after the catalytic converter)

A computer program that demonstrates the operation of an engine management ECU and system is available from www.automotive-technology.org

During the cold-start and warm-up phases of engine operation the computer operates in an open loop mode based on the sensor data. Once the engine reaches a certain temperature and the signals from the EGO (or lambda) sensor are logical, the computer operates in a closed loop mode based on the data from this sensor.

Key fact

The fixed map of operating conditions specific for each engine is held in a fixed value memory or read only memory (ROM).

Other programs in the CPU monitor the system and sensor data. They provide fault diagnosis and limp-home or a limited operation strategy in the event of any defects being detected. Other components in the ECU provide signal amplification and pulse shaping. This includes analogue to digital (A/D) converters for direct current (d.c.) voltages, and pulse formers for

alternating current (a.c.) voltages. The CPU requires digital signals for all processing functions. On the output side, power transistors are used for switching the actuator supply voltages either to the components or to an earth or ground point.

The ECU also operates the emission control components at appropriate times depending on the engine operating conditions. Typical emission control actuators are the canister purge solenoid valve, the EGR valve and the secondary air solenoid valve.

The electrical harness for the engine management system is a complex set of cables and sockets. Cables have colour and/or numerical coding and the sockets are keyed so that they can be connected in one way only. Special low-resistance connectors are used for low-current sensor wiring. Follow manufacturer's data sheets for further technical detail.

Sensors provide data to the ECU. The engine speed and load conditions are used to calculate the base time value (in milliseconds) for the injector pulse width. A range of correction factors is added to or subtracted from the base time value to suit the engine operating conditions occurring at all instances of time.

In early electronic fuel injection systems, the engine speed was provided from signals obtained from the ignition low-tension primary circuit. In engine management systems, the engine speed and position are required for the ignition and fuel systems.

There are two methods of engine speed and position sensing. The older system is a conventionally geared distributor with an inductive or Hall effect generator. This provides an alternating signal current that is used by the ignition system. It is also used for engine speed sensing in the fuel ECM.

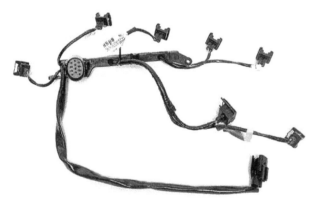

Figure 2.265 Injection wiring harness

Figure 2.266 An inductive speed sensor produces a sine-wave signal, the frequency being proportional to engine speed. (Source: Denso Media)

Figure 2.267 Distributor for engine speed sensing on early systems

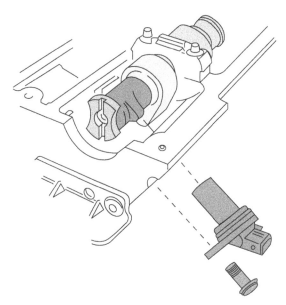

Figure 2.268 Camshaft position sensor. (Source: Ford Motor Company)

All of the latest systems have inductive pulse generators mounted close to, and responding to, a toothed wheel attached to the crankshaft pulley or flywheel. There is an air gap between the toothed wheel and the inductive generator, and as the teeth pass the inductive generator, an alternating electric current is produced. The waves of the alternating current are used to measure engine speed. For position sensing, a missing or different size of tooth or mask opening on the sensor ring is used. A distributor can also provide a reference for number one cylinder at TDC. When a sensor is fitted to determine the crankshaft position, this is suitable for continuous injection systems.

For sequential injection, a camshaft position sensor is used to recognize the position of number one cylinder. The ECU is then able to follow the engine firing order.

Key fact

For sequential injection, a camshaft position sensor is used to recognize the position of number one cylinder.

Inductive sensors produce an output pulse each time a lobe or tooth passes the inductive coil. The frequency and pattern of the pulses are used by the ECU to determine the engine speed and position.

The fuel requirement is calculated in the ECU from the engine speed and load conditions. An air flow meter is one method of measuring the engine load conditions. A variable voltage, corresponding to

the measured value at the air flow meter, is used by the ECU to calculate the amount of fuel needed to give a correct air/fuel ratio. Often now, engine load is determined from the inlet manifold absolute pressure (MAP). In these systems, an air flow meter is not used.

Key fact

Inductive sensors produce an output pulse each time a lobe or tooth passes the inductive coil.

Definition

MAP: Manifold absolute pressure

There are two main types of air flow meter. These are vane air flow (VAF) and the mass air flow (MAF). The vane type air flow meter consists of an air passage and damping chamber into which is fitted a fixed pair of flaps (or vanes), which rotate on a spring-loaded spindle. The spindle connects to and operates a potentiometer and switches.

Air flowing through the meter acts on the intake air flap and moves it in opposition to a spring force. The integral damper flap moves into the sealed damper chamber to smooth out the intake pulses. The degree of flap movement and spindle rotation is measurable at the potentiometer as a variable voltage dependent on position. The voltage signal, together with other signals, is used in the ECU to calculate the fuel requirement.

A bypass air duct is built into the housing. This provides for starting without opening the throttle, a smooth air flow during engine idle and a means to adjust the idle mixture.

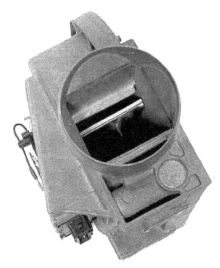

Figure 2.269 Vane type air flow meter

153

Mass air flow meters are fitted with two similar resistors inside an air tube. A measurement resistor is heated and often referred to as a hot wire The other resistor is not heated. It provides a reference value for use in the calculation of the air mass. The control circuit maintains the temperature differential between the two resistors. The signal sent to the ECU is proportional to the current required to heat the measurement resistor and maintain the temperature differential. The output signal from some mass air flow meters is similar to that of the air vane types. However, some produce a digital output signal.

On many systems MAP sensor signals are used by the ECU to calculate the fuel requirements. These systems do not have an air flow meter. The signals from manifold absolute pressure, engine speed, air charge temperature and throttle position sensors are compared in the ECU to calculate the injector pulse width.

The MAP sensor is a pressure-sensitive component consisting of a diaphragm and piezoelectric circuit. It can be a component fitted in the engine compartment or be integral with the ECU. It is connected by a rubber hose to the inlet manifold.

Two types of throttle position sensor are used: a throttle switch assembly and a throttle

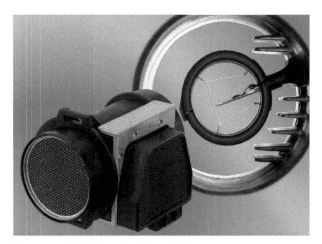

Figure 2.270 Hot wire/film air flow meter

Figure 2.271 Manifold absolute pressure (MAP) sensor. (Source: Denso Media)

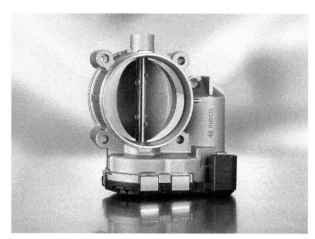

Figure 2.272 Throttle controller and position sensor. (Source: Bosch Media)

potentiometer. Both are fitted to the throttle body and operated by the throttle plate spindle. A throttle switch assembly has two switches, one to indicate the closed throttle or idle position and the other for the wide open throttle position. A throttle potentiometer is a variable resistor with a rotary sliding contact. The sliding contact is moved along the rotary resistance track to provide changes in voltage proportional to the position of the throttle.

The throttle potentiometer signals are used in the ECU for a number of functions. At the closed throttle position, idle speed and deceleration fuel cut-off are controlled. In the part open throttle position (about 5–70% open), there is normal operation with close control of fuel delivery and exhaust emissions. In the wide open throttle position (70–100%), full load enrichment and starting of a flooded engine are provided. During rapid movement of the throttle plate there is acceleration enrichment, depending on the rate of change of the throttle plate and signal voltages from the sensor.

In order for the ECU to correctly calculate the required fuel for a correct mixture ratio, an accurate figure for air mass is necessary. However, air volume and density are affected by changes in temperature. As the temperature rises, the air density falls. The air flow, or MAP measurement, therefore, must be corrected for temperature. The sensor is a temperature-dependent resistor with a negative temperature coefficient (NTC) (Fig. 2.273).

Key fact

In order for the ECU to correctly calculate the required fuel for a correct mixture ratio, an accurate figure for air mass is necessary.

Figure 2.273 Air temperature sensor

Figure 2.275 Exhaust gas oxygen (lambda) sensor

Figure 2.274 Coolant thermistor

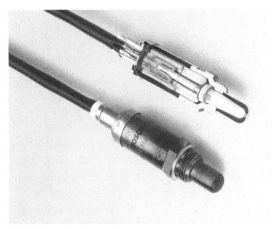

Figure 2.276 Construction of a zirconia type lambda sensor

The engine coolant temperature sensor is an NTC thermistor. It is of a similar type to the air temperature sensor. It is fitted into the water jacket close to the thermostat or bypass coolant circuit passages. The sensor measures the engine coolant temperature and provides a signal voltage to the ECU. This is used for cold-start and warm-up enrichment as well as fast idle speed control through the idle speed control valve.

The Greek letter (λ) lambda is used as the symbol for a chemically correct air to fuel ratio; hence the use of this letter for naming the sensor that is used to control the amount of fuel delivered, so that a very close tolerance to the stoichiometric ratio is maintained. The lambda sensor is often known as an EGO sensor (Figs 2.275 and 2.276). Some of these sensors are electrically heated. Preheating allows the sensor to be fitted lower down in the exhaust stream and prolongs the life of the active element. The sensor measures the presence of oxygen in the exhaust gas and sends a voltage signal to the engine ECU.

More fuel is delivered when oxygen content is detected and less fuel when it is not. In this way, an accurate fuel mixture close to the stoichiometric ratio is maintained. This produces the correct exhaust gas constituents for chemical reactions in the catalytic converter. Exhaust gases pass over the active element and when the oxygen concentration on each side is different an electric voltage is produced. Voltages of about 0.8 V for little or no exhaust oxygen and 0.2 V for higher content are typical outputs.

Key fact
The engine coolant temperature sensor is a negative temperature coefficient (NTC) thermistor.

Key fact
The lambda sensor is often known as an exhaust gas oxygen sensor.

The sensors for power steering and air conditioning are pressure or mechanically operated switches. They provide a voltage signal when the system is in operation. The ECU uses these signals to increase the engine idle speed to accept the increased engine load.

Switches are used in automatic transmission. They include the neutral drive switch, which is used for idle speed control, the kick-down switch for acceleration control, and the brake on/off switch, which is used to ensure that the torque converter lock-up clutch is released. This is to prevent the engine stalling as the vehicle comes to rest.

155

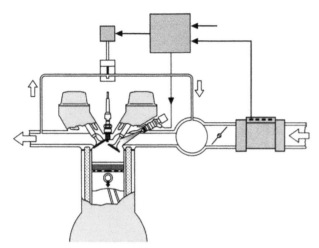

Figure 2.277 Exhaust gas recirculation (EGR) system

Figure 2.278 Throttle body assembly

Figure 2.279 Solenoid air valve

A sensor can be used to measure exhaust gas pressure. It uses a ceramic resistance transducer, which responds to the exhaust gas pressure applied through a pipe connection to the exhaust system. The signal voltage from the electronic pressure transducer is used to regulate the EGR valve. The valve may be operated directly from the ECU if electromechanical or by vacuum through a solenoid vacuum switch.

Service and on-board diagnostic (OBD) plugs are used for diagnostic and corrective actions with scan tools, dedicated test equipment and other test equipment. If faults are detected the system malfunction indicator lamp (MIL) will come on. Alternatively, it will fail to go out after the preset time duration after switching on the engine. All faults should be investigated as soon as possible. Many electronic systems have a limp-home or limited operation strategy (LOS) program, which allows the vehicle to be driven to a workshop for repair.

Key fact

Service and on-board diagnostic (OBD) plugs are used for diagnostic and corrective actions with scan tools, dedicated test equipment and other test equipment.

2.5.2.2 Air supply

The air supply components consist of ducting and silencing components between the air intake and the inlet manifolds. This will also include an air filter, a throttle body, throttle plate assembly and idle control components. The air supply components must provide sufficient clean air for all operating conditions. The air flow into the engine would be noisy and unbalanced

between cylinders without the use of resonators and plenum chambers. A plenum chamber is a large volume air chamber that can be fitted either in front of or behind the throttle plate housing.

Air filters on most modern petrol (gasoline)-engined vehicles consist of a plastic casing with a paper filter element. Air flow into the filter is upwards so that dust and dirt particles drop into the dust chamber, or

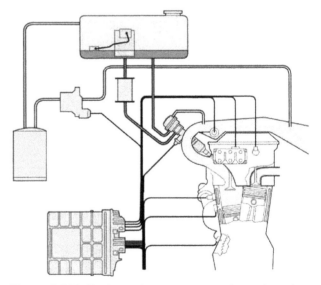

Figure 2.280 Fuel supply components shown in red

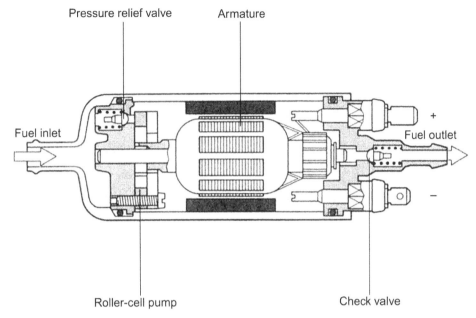

Pressure relief valve   Armature

Fuel inlet

Fuel outlet

+

−

Roller-cell pump   Check valve

Figure 2.281 Roller cell pump

rotary so that dust and dirt are thrown out before the air enters the engine. Crankcase ventilation and the air supply or pulse air exhaust emission systems are also connected to the filter assembly.

The throttle is a conventional circular plate in an air tube. For fast idle and warm up, an auxiliary air valve is fitted to bypass the throttle plate, or an electromechanical link is made to the throttle plate spindle. An auxiliary air valve, idle air control (IAC) or idle speed control (ISC) valve is operated from signals from the ECM.

The auxiliary air valve is often a rotary air valve. This has a special electric motor to move and hold the valve in position. The position is based on the electrical signals supplied by the ECU. Two electric windings in the motor work in opposition to each other so that the motor is variable over a 90° arc. Other designs have graduated opening values based on the signal supplied from the ECU. This type consists of a solenoid valve with a spring-loaded armature connected to the valve in the air channel. All idle control valves operate to hold the engine speed to the stored data specification for engine temperature and load conditions.

Stepper motors are also used to control idle speed and give graduated positions depending on the supply current to a number of electric windings. Sensors in the idle control mechanisms provide feedback signals to the ECU to provide data on operation and position.

2.5.2.3 Fuel supply

The fuel supply, from the fuel tank to the injector valves, for most systems except GDi, follows the

same basic layout. A basic layout of fuel supply components is shown in Figure 2.280. A fuel pump is fitted either in or close to the fuel tank. A fuel filter is fitted in the delivery fuel lines from the tank to a fuel rail. A fuel pressure regulator is located on either the housing for throttle body injector systems or the fuel rail for port fuel injection systems. Return fuel lines run from the pressure regulator to the fuel tank.

The fuel pumps on injection systems are usually roller cell types, driven by a permanent magnet electric

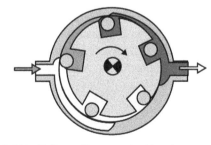

Figure 2.282 Roller cell pump: fuel intake

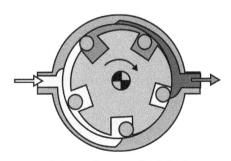

Figure 2.283 Roller cell pump: fuel discharge

157

motor. Fuel flows through the pump and motor, but there is no risk of fire as there is never an ignitable mixture in the motor. The pump delivery pressure is set by a pressure relief valve, which allows fuel to return to the inlet side of the pump when the operating pressure is reached. There is also a non-return valve in the pump outlet. Typical delivery pressures are between 300 and 400 kPa (3–4 bar).

The rollers in the roller cell pump are thrown out by centrifugal force when the motor armature and pump rotor spindle rotate. The rotor is fitted eccentrically to the pump body and as the rollers seal against the outer circumference, they create chambers that increase in volume to draw fuel in. They then carry the fuel around and finally discharge it as the chamber volume decreases.

The fuel pump electrical supply is live only when the engine is being cranked for starting or is running. The fuel pump electric feed is from a relay that is switched on with the ignition. Safety features are built into the electric control feed to the relay so that it operates only to initially prime the system or when the engine is running. The control functions of the fuel pump relay are usually provided by the fuel control module. A further safety feature is the use of an inertia switch in the feed from the relay to the fuel pump. This operates, in the event of an accident, to cut the electric feed to the fuel pump and to stop the fuel supply. It is an impact-operated switch with a weight that is thrown aside to break the switch contacts. Once the switch has been operated it has to be manually reset.

The fuel filter is an in-line paper element type that is replaced at scheduled service intervals. The filter uses microporous paper that is directional for filtration. Filters are marked for fuel flow with an arrow on the casing and correct fitting is essential. They are often matched specifically to a vehicle and fuel type, so, as always, refer to manufacturers' specifications.

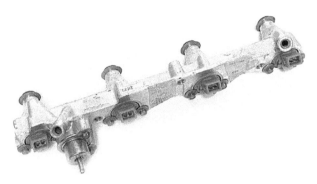

Figure 2.285 Fuel pressure regulator on rail

A fuel pressure regulator is fitted to maintain a precise pressure at the fuel injector valve nozzles. On port fuel injection systems, a fuel rail is used to hold the pressure regulator and the fuel feed to the injector valves. The injector valves usually fit directly onto or into the fuel rail. The fuel rail holds sufficient fuel to dampen fuel pressure fluctuations and keep the pressure applied at all injector nozzles at a similar level.

Key fact

A fuel pressure regulator is fitted to maintain a precise pressure at the fuel injector valve nozzles.

Fuel regulators are sealed units with a spring-loaded diaphragm and valve on the return outlet to the fuel tank. Fuel is pumped into the regulator and when the pressure is high enough, it acts against the diaphragm and compression spring to open the valve. Surplus pressure and fuel is allowed to return to the fuel tank. Once the pressure in the fuel regulator is reduced, the

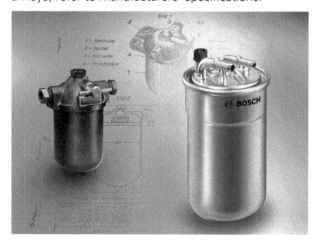

Figure 2.284 Fuel filters have been developed to match specific needs of system manufacturers

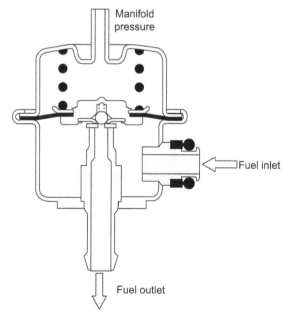

Figure 2.286 Fuel pressure regulator valve closed
Figure 2.260

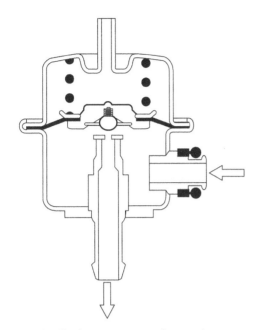

Figure 2.287 Fuel pressure regulator valve open

valve closes and the pressure builds up again. Throttle body injection systems operate in the region of 1 bar, and port fuel injection systems in the region of 2.5 bar.

> **Key fact**
>
> Throttle body injection systems operate in the region of 1 bar, and port fuel injection systems in the region of 2.5 bar.

In port fuel injection systems, the inlet manifold vacuum acts against the compression spring in the fuel pressure regulator. This is required to maintain a constant pressure differential between the fuel rail and the inlet manifold. With a constant pressure differential, the amount of fuel delivered during a set time will be the same irrespective of inlet manifold pressure.

For vehicles fitted with a turbocharger or supercharger, inlet manifold pressure is applied to the diaphragm and regulator valve. When the inlet manifold pressure rises above a certain value, the regulator valve is closed so that the full pump delivery pressure is applied to the injector valve nozzles. This raises the amount of fuel delivered to match the boosted air charge.

The injector valves spray finely atomized fuel into the throttle body or inlet ports, depending on the system. The electromagnetic injection valves are actuated by signals from the ECU. The signals are of a precise duration depending on operating conditions but within the range of about 1.5–10 milliseconds. This open phase of the injector valve is known as the 'injector pulse width'.

> **Key fact**
>
> Injector signals are of a precise duration depending on operating conditions but usually within the range of about 1.5–10 milliseconds.

There is a range of individual injector valve designs but all have the same common features. These are an electromagnetic solenoid, with a spring-loaded plunger, connected to a jet needle in the injector valve nozzle. The electrical supply to the solenoid is made from the system relay or ECU. Grounding the other connection energizes the solenoid. This lifts the plunger and jet needle so that fuel is injected for the duration that the electric current remains live. As soon as the electrical supply is switched off in the ECU, a compression spring in the injector valve acts on the solenoid plunger to close the nozzle.

One problem experienced with the arrangement of a top-feed fuel injector for port injection system is fuel vaporization and bubbles forming in the fuel rail. The bubbles can cause starting and running problems. To overcome this problem lateral or side or bottom-feed injectors are used (shown overleaf in Figure 2.289). When fitted in the fuel rail it can be seen that any bubbles that may form will be at the top of the rail. They will therefore be flushed out through the regulator as soon as the fuel pump is actuated.

The multipoint fuel injection system outlined above is now very common as it works well to meet stringent economy and emission requirements. However, as these requirements increase further, new ways of meeting them are being sought. One of these is direct injection, which is discussed in the next section.

Figure 2.288 Injector operating

159

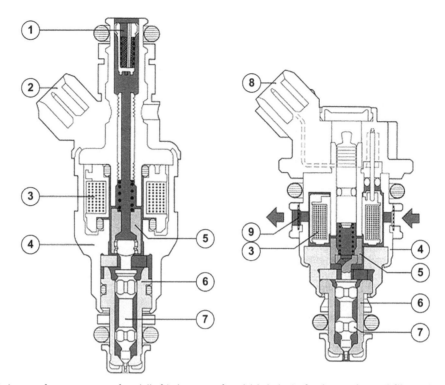

Figure 2.289 Injector features; top-feed (left), bottom-feed (right): 1, fuel supply and filter; 2, electrical connection; 3, solenoid (winding); 4, injector body; 5, armature (moving part); 6, valve body; 7, needle; 8, electrical connection; 9, filter

2.5.2.4 Gasoline direct injection

Bosch's high-pressure injection system for petrol engines is based on a pressure reservoir and a fuel rail, which a high-pressure pump charges to a regulated pressure of up to 120 bar. The fuel can therefore be injected directly into the combustion chamber via electromagnetic injectors.

The air mass drawn in is adjusted through an electronically controlled throttle valve and is measured

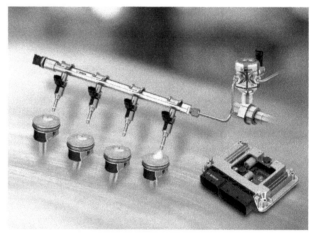

Figure 2.291 Components of gasoline direct injection. (Source: Bosch Media)

with the help of an air mass meter. For mixture control, a wide-band oxygen sensor is used in the exhaust, before the catalytic converters. This sensor can measure a range between $\lambda = 0.8$ and infinity. The electronic engine control unit regulates the operating modes of the engine with gasoline direct injection in three ways:

▶ stratified charge operation – with λ values greater than 1
▶ homogeneous operation – at $\lambda = 1$
▶ rich homogeneous operation – with $\lambda = 0.8$.

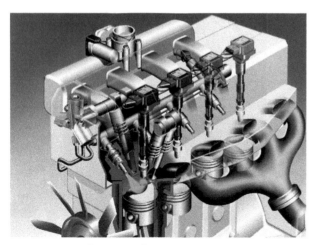

Figure 2.290 Gasoline direct injection on an engine. (Source: Bosch Media)

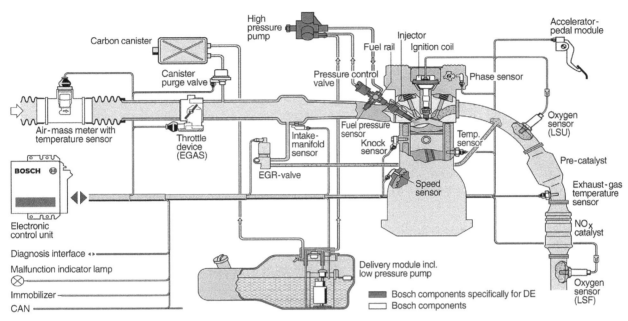

Figure 2.292 Bosch DI-Motronic. (Source: Bosch Media)

Key fact

This sensor on the Bosch GDi system can measure a range between lambda = 0.8 and infinity.

Compared to the traditional manifold injection system, the entire fuel amount must be injected in full-load operation in a quarter of the time. The available time is significantly shorter during stratified charge operation in part-load. Especially at idle, injection times of less than 0.5 milliseconds are required owing to the lower fuel consumption. This is only one-fifth of the available time for manifold injection.

Definition

Micrometre (µm): One micrometre (1 µm) is a millionth of a metre.

The fuel must be atomized very finely to create an optimal mixture in the brief moment between injection and ignition. The fuel droplets for direct injection are on average smaller than 20 µm (micrometres, i.e. a millionth of a metre). This is one-fifth of the droplet size reached with the traditional manifold injection and one-third of the diameter of a single human hair. It improves efficiency considerably. However, even more important than fine atomization is even fuel distribution in the injection beam. This is done to achieve fast and uniform combustion.

Key fact

The fuel droplets for direct injection are on average smaller than 20 µm.

Figure 2.293 Injector for direct injection under test. (Source: Bosch Media)

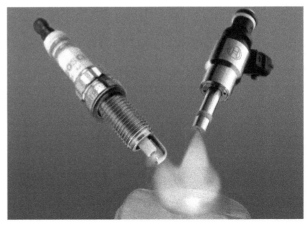

Figure 2.294 Fuel droplet size is important. (Source: Bosch Media)

161

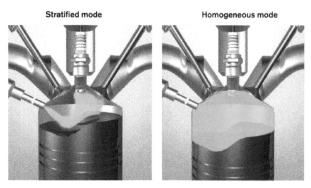

Figure 2.295 Operating modes. (Source: Bosch Media)

Conventional spark ignition engines have a homogeneous (well mixed up!) air/fuel mixture at a 14.7:1 ratio, corresponding to a value of $\lambda = 1$. Direct injection engines, however, operate according to the stratified charge concept in the part-load range and function with high excess air. In return, very low fuel consumption is achieved.

Fuel injection just before the ignition point and injection directly into the combustion chamber is to create a stratified (layered) mode. The result is a combustible air/fuel mixture cloud on the spark plug, cushioned in a thermally insulated layer, composed of air and residual gas. This raises the efficiency level because heat loss is avoided on the combustion chamber walls. The engine operates with an almost completely opened throttle valve, which avoids additional charge losses.

Definition

Stratified: Arranged in approximately horizontal layers.

With stratified charge operation, the lambda value in the combustion chamber is between about 1.5 and 3. In the part-load range, gasoline direct injection achieves the greatest fuel savings, with up to 40% at idle compared with conventional petrol injection processes.

With increasing engine load, and therefore increasing injection quantities, the stratified charge cloud becomes even richer and emission characteristics become worse. As in diesel engine combustion, soot may form. In order to prevent this, the DI-Motronic engine control converts to a homogeneous cylinder charge at a predefined engine load (Fig. 2.296). The system injects very early during the intake process to achieve a good mixture of fuel and air at a ratio of $\lambda = 1$.

Definition

Homogeneous: Of uniform composition throughout.

As is the case for conventional manifold injection systems, the amount of air drawn in for all operating modes is adjusted through the throttle valve according to the desired torque specified by the driver. The Motronic ECU calculates the amount of fuel to be injected from the drawn-in air mass and performs an additional correction via lambda control. In this mode of operation, a torque increase of up to 5% is possible. Both the thermodynamic cooling effect of the fuel vaporizing directly in the combustion chamber and the higher compression of the engine with gasoline direct injection play a role in this.

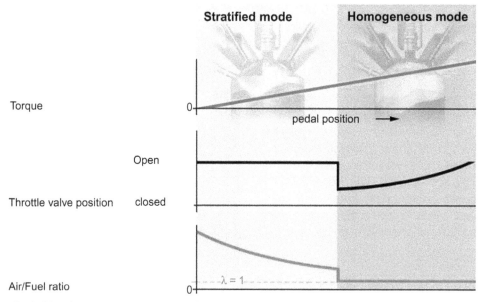

Figure 2.296 Switching between operating modes depending on engine load. (Source: Bosch Media)

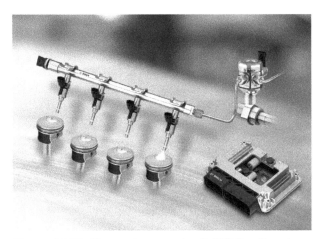

Figure 2.297 Electronic control unit (ECU), rail and injectors. (Source: Bosch Media)

For these different operating modes, two central demands are raised for engine control:

▸ The injection point must be adjustable between 'late' (during the compression phase) and 'early' (during the intake phase) depending on the operating point.

▸ The adjustment for the drawn-in air mass must be detached from the throttle pedal position to permit unthrottled engine operation in the lower load range. However, throttle control in the upper load range must also be permitted.

With optimal use of the advantages, the average fuel saving is up to 15%.

In stratified charge operation the nitrogen oxides (NO_x) segments in the very lean exhaust cannot be reduced by a conventional, three-way catalytic converter. The NO_x can be reduced by approximately 70% through exhaust returns before the catalytic converter. However,

this is not enough to fulfil the ambitious emission limits of the future. Therefore, emissions containing NO_x must undergo special treatment. Engine designers are using an additional NO_x accumulator catalytic converter in the exhaust system. The NO_x is deposited in the form of nitrates on the converter surface, with the oxygen still contained in the lean exhaust.

Key fact

In a GDi system, NO_x emissions are reduced by an accumulator catalytic converter in the exhaust system.

The capacity of the NO_x accumulator catalytic converter is limited. As soon as it is exhausted, the catalytic converter must be regenerated. To remove the deposited nitrates, the DI-Motronic briefly changes over to its third operating mode (rich homogeneous operation with lambda values of about 0.8). The nitrate together with the carbon monoxide is reduced in the exhaust to non-harmful nitrogen and oxygen. When the engine operates in this range, the engine torque is adjusted according to the accelerator pedal position via the throttle valve opening. Engine management has the difficult task of changing between the two different operating modes, in a fraction of a second, in a way not noticeable to the driver.

The continuing challenge, set by legislation, is to reduce vehicle emissions to very low levels. Bosch is a key player in the development of engine management systems. The DI-Motronic system, which is now used by many manufacturers, continues to reflect the good name of the company.

2.5.3 Diesel fuel injection systems

Diesel engines have the fuel injected into the combustion chamber where it is ignited by heat in the air charge. This is known as compression ignition (CI) because no spark is required. The high temperature needed to ignite the fuel is obtained by a high compression of the air charge. Diesel fuel is injected under high pressure from an injector nozzle, into the combustion chambers. The fuel is pressurized in a diesel injection pump. It is supplied and distributed to the injectors through high-pressure fuel pipes or directly from a rail and/or an injector. The high pressure is generated from a direct acting cam or a separate pump.

Key fact

The high temperature needed to ignite fuel in a diesel engine is obtained by high compression of the air charge.

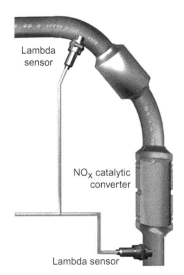

Figure 2.298 NO_x catalytic converter. (Source: Bosch Media)

Figure 2.299 Diesel fuel injection components. (Source: Bosch Media)

The air flow into a diesel engine is usually unobstructed by a throttle plate so a large air charge is always provided. Throttle plates may be used to provide control for emission devices. Engine speed is controlled by the amount of fuel injected. The engine is stopped by cutting off the fuel delivery. For all engine operating conditions a surplus amount of air is needed for complete combustion of the fuel.

Diesel engines used to be considered as indirect and direct injection. Now almost all are direct and a number of methods are used. The rotary pumped direct injection and common rail systems will be discussed further in this section.

Small high-speed diesel engine compression ratios are about 20:1 for direct injection systems. This compression ratio is capable of raising the air charge to temperatures of between 500°C and 800°C. Very rapid combustion of the fuel therefore occurs when it is injected into the hot air charge.

The combustion process in a diesel engine follows three phases: ignition delay, flame spread and controlled combustion. In addition, an injection lag occurs in the high-pressure pipes of earlier systems, as the pressure builds up just before injection.

The most important phase of controlled combustion is when fuel is being injected into a burning mixture. This must be at a rate that maintains an even combustion pressure onto the piston throughout the critical crankshaft rotational angles. This gives maximum torque and efficient fuel usage, because temperatures remain controlled and the heat lost to the exhaust is

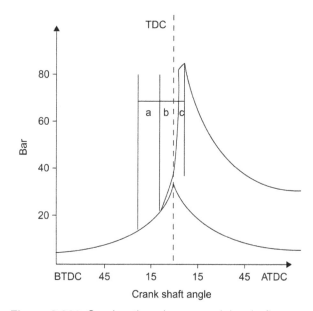

Figure 2.301 Combustion phases: a, delay; b, flame spread (rapid pressure rise); c, controlled combustion (afterburn); BTDC, before top dead centre; ATDC, after top dead centre

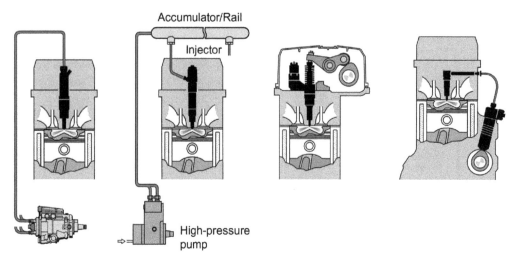

Figure 2.300 Types of diesel injection: direct using a rotary pump, common rail, unit injection and pumped unit injection. (Source: Bosch Media)

minimized. The low temperatures also help to keep NO_x emissions to a minimum.

The speed of flame spread in a diesel engine is affected by the air charge temperature and the atomization of the fuel. These characteristics are shared with the delay period. A sufficiently high air charge temperature, of at least 450°C, is a minimum requirement for optimum ignition and combustion.

The delay phase or ignition lag for diesel fuel combustion lasts a few milliseconds. It occurs immediately on injection as the fuel is heated up to the self-ignition temperature. The length of the delay is dependent on the compressed air charge temperature and the grade of fuel. The air charge temperature is also affected by the intake air temperature and the engine temperature.

Key fact

The delay phase or ignition lag for diesel fuel combustion lasts a few milliseconds.

A long delay period allows a high volume of fuel to be injected before ignition and flame spread occurs. In this situation diesel knock is at its most severe. When a diesel engine is cold, there may be insufficient heat in the air charge to bring the fuel up to the self-ignition temperature. When ignition is slow, heavy knocking occurs.

To aid starting and to reduce diesel knock, cold-start devices may be used. For indirect injection engines, starting at lower than normal operating temperatures requires additional combustion chamber heating. For direct injection engines, cold-start devices are only required in frosty weather.

An initial delay, known as injection lag, occurs in the high-pressure fuel lines of rotary pumped systems.

This occurs between the start of the pressure rise and the point when pressure is sufficient to overcome the compression spring force in the injectors. Diesel fuel pipes and injectors are shown in Figure 2.303.

Key fact

An initial delay, known as injection lag, occurs in the high-pressure fuel lines of rotary pumped systems.

Ignition of the fuel occurs in the combustion chamber at the time of injection into the heated air charge. The injection point and the ignition timing are therefore effectively the same thing. Diesel engine injection timing is equivalent to the ignition timing for petrol engines. Injection timing must fall within a narrow angle of crankshaft rotation. It is advanced and retarded for engine speed and load conditions. Injection timing is set by accurate positioning of the fuel injection pump. Incorrect timing leads to power loss. An increase in the production of NO_x when too far advanced, or an increase in the hydrocarbon emissions when too far retarded, can occur.

Particulate emissions result from incomplete combustion of fuel. Particulates are seen as black carbon smoke in the exhaust under heavy load or when fuel delivery and/or timing is incorrect. White smoke may also be visible at other times, such as when the injection pump timing is incorrect. It also occurs when compression pressures are low or when coolant has leaked into the combustion chambers.

Key fact

Diesel engines are particularly suitable for turbocharging.

Figure 2.302 Glow plugs. (Source: Bosch Media)

Figure 2.303 Pipes and injectors

Figure 2.304 Common rail injection. (Source: Bosch Media)

Recent developments in electronic diesel fuel injection control have made it possible to produce small direct injection engines. Diesel engines are built to withstand the internal stresses, which are greater than in other engines. Diesel engines are particularly suitable for turbocharging, which improves power and torque outputs.

Exhaust gas recirculation is used to reduce NO_x emissions from petrol engines; this is also true for diesel engines. In addition, a small quantity of hot exhaust gas in the air charge of a cold engine helps to reduce the delay period and the incidence of cold engine diesel knock.

Diesel-engined vehicles are fitted with oxidation catalytic converters that work in conjunction with other emission components to reduce hydrocarbon and particulate emissions.

The fuel systems for traditional direct and indirect injection are similar and vary only in injection pressures and injector types. Until more recently, all light high-speed diesel engines used rotary diesel fuel injection pumps, producing injection pressures of over 100 bar for indirect engines. However, these can rise up to 1000 bar at the pump outlet for turbocharged direct injection engines.

Figure 2.306 Bosch rotary injection pump. (Source: Bosch Media)

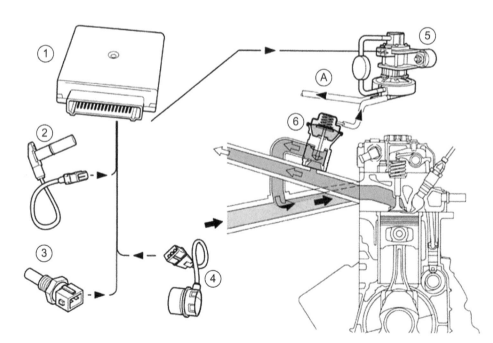

Figure 2.305 Diesel exhaust gas recirculation (EGR): 1, electronic control unit (ECU) and sensors; 2, EGR actuator and control valve; 3, coolant temperature sensor; 4, intake air temperature sensor; 5, ECU controlled valve; 6, vacuum actuator; A, connection to vacuum source

Safety first

Caution: pressure can be up to 1000 bar at the outlet of a rotary pump.

Injectors operate with a pulsing action at high pressure to break the fuel down into finely atomized parts. Atomization is critical to good fuel distribution in the compressed air charge. The air charge pressure may be in excess of 60 bar. The pressure differential between the fuel injection pressure and air charge pressure must be sufficient to overcome the resistance during injection. This will also give good fuel atomization and a shorter injection time.

Key fact

Injectors operate with a pulsing action at high pressure to break the fuel down into finely atomized parts

The main components of a diesel fuel system provide for either the low-pressure or the high-pressure functions. The low-pressure components are the fuel tank, the fuel feed and return pipes and hoses, a renewable fuel filter with a water trap and drain tap, and a priming or lift pump.

The high-pressure components are the fuel injector pump, the high-pressure pipes and the injectors. Other components provide for cold engine starting. Electronically controlled systems include sensors, an electronic diesel control (EDC) module and actuators in the injection pump.

Key fact

The internal components of the pump and injectors are manufactured to very fine tolerances.

All diesel fuel entering the injection pump and injectors must be fully filtered (Fig. 2.307). The internal components of the pump and injectors are manufactured to very fine tolerances. Even very small particles of dirt could be damaging to these components.

The most common rotary injection pumps are axial-piston designs having a roller ring and cam plate attached to an axial piston or plunger in the distributor head to generate the high pressure. The latest versions have full electronic control. The details are examined further in the next section.

The high-pressure pipes are of double-thickness steel construction and are all of the same length. This is so that the internal pressure rise characteristics are identical for all cylinders. The high-pressure connections are made by rolled flanges on the pipe ends and threaded unions securing the rolled flanges to convex, or occasionally concave, seats in the delivery valves and injectors.

The fuel injectors are fitted into the cylinder head with the nozzle tip projecting into the precombustion (indirect injection) or combustion chamber (direct injection). The injectors for indirect combustion are of a pintle or 'pintaux' design (similar to petrol injectors in many ways) and produce a conical spray pattern on injection. The injectors for direct injection (DI) are of a pencil-type multihole design that produces a broad distribution of fuel on injection (Fig. 2.308).

Fuel injectors are held closed by a compression spring. They are opened by hydraulic pressure when it is sufficient to overcome the spring force on the injector needle. The hydraulic pressure is applied to a face on the needle where it sits in a pressure chamber. The fuel pressure needed is in excess of 100 bar (1500 psi). This pressure lifts the needle and opens the nozzle, so that fuel is injected in a fine spray pattern into the combustion chamber.

Figure 2.307 Fuel filter

Figure 2.308 Direct injection (DI) injector

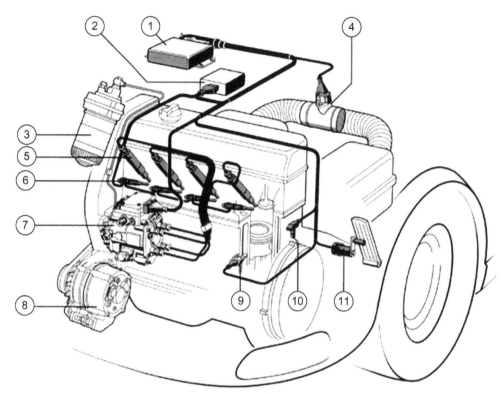

Figure 2.309 Electronic diesel control (EDC) rotary pump system: 1, engine ECU; 2, glow control unit; 3, filter; 4, air-mass meter; 5, injectors; 6, glow plugs; 7, rotary distributor pump with ECU; 8, alternator; 9, coolant temperature sensor; 10, crankshaft sensor; 11, throttle pedal sensor

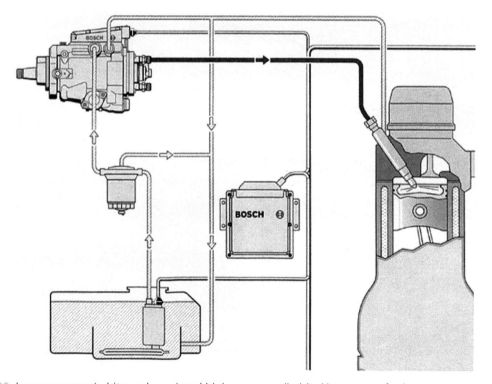

Figure 2.310 Low-pressure (white and grey) and high-pressure (in black) pressure fuel system

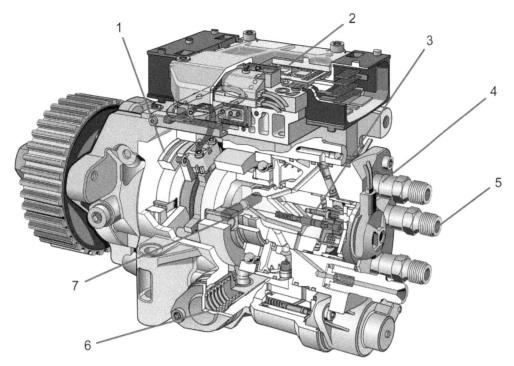

Figure 2.311 Solenoid valve-controlled radial-piston distributor pump: 1, sensor (position/timing); 2, ECU; 3, high-pressure solenoid valve needle; 4, solenoid; 5, outlets to injectors; 6, timing device (ignition advance mechanism); 7, radial-piston high-pressure pump. (Source: Bosch Media)

Key fact

The fuel pressure needed to open the fuel injector is in excess of 100 bar (1500 psi).

The next section examines the Bosch rotary pump electronic diesel control system, while the section after that looks at the system now in use on most modern cars: common rail direct injection.

2.5.3.1 Rotary pump system

Bosch rotary VR pumps are used on high-speed direct injection diesel engines for cars and light commercial vehicles (Fig. 2.309). They are radial-piston distributor injection pumps having opposing plungers that are forced inwards by cam lobes on the inside of a cam ring to produce high pressure, which can be up to 1400 bar in some applications. The cam is located in the pump body and the plungers are in the rotor driven by the pump spindle. Four-cylinder engines have two plungers and four cam lobes. Six-cylinder engines have three plungers and six cam lobes. The pump is driven from the engine at half crankshaft speed.

A low-pressure feed to the injection pump is provided by a submerged electrical pump in the fuel tank. This provides for priming and positive pressure in the injection pump. In common with all diesel fuel

systems, a fuel filter and water trap are used to ensure that only very clean fuel is delivered to the pump. Return pipes are used for excess fuel leakage, for purging the pump and for lubrication of the injectors.

Inside the distributor pump is a vane-type pump, which is used to produce the pump body pressure. Pump body pressure is used for charging the high-pressure chamber between the plungers and for injection advance. A pressure control valve is used to prevent excessive pressure. It is a spring-loaded plunger that is lifted by hydraulic pressure to expose ports in the valve bore. This will then allow fuel to flow back to the inlet side of the vane-type pump.

An overflow throttle valve, in the pump housing, is used to allow a defined quantity of fuel to flow back to the fuel tank at all times. This provides some cooling in the pump and venting of air during pump priming. A second, larger overflow bore in the valve opens at a given pressure to allow a flow of fuel from the distributor head.

The Bosch pump has full electronic control for fuel metering and for injection advance. The electronic diesel control unit consists of two ECMs to perform the control functions (Fig. 2.312). These two modules are the engine control ECU and the injection pump ECU. The pump ECU is fitted on top of the pump.

169

Key fact

Fuel metering is controlled by the ECU, which operates a high-pressure solenoid valve.

Fuel metering is controlled by the high-pressure solenoid valve. This is an electrically actuated valve set centrally inside the distributor rotor. There are connecting bores in the distributor rotor for filling of the high-pressure circuit, through the inlet port at pump body pressure, and for delivery at high pressure to the fuel injectors. These are either connected or separated by the position of the valve.

A high-pressure solenoid valve is closed by an electrical signal from the pump ECU. When the valve is closed, fuel under high pressure passes from the high-pressure pump chamber, through the bores in the rotor and distributor head, through the return-flow throw throttle valve (delivery valve) and out to the injectors. It is then injected into the engine combustion chambers. The few

Figure 2.314 Injector and common rail fuel pump (Source: Bosch)

microseconds of time during which the valve remains closed is referred to as the delivery or injection period.

The quantity of fuel that is metered for injection at any time is computed by the engine ECU, which sends signals to the injection pump ECU for control of the high-pressure solenoid valve. The electrical current for operating this valve is high, and the two ECUs are separated to avoid high current interference in the electronically more vulnerable engine ECU.

The electronic diesel control units are provided with data signals from sensors and switches attached to the engine, the pump and other vehicle systems. The sensors are used for comparisons to programmed operating parameters and for calculations for metering the amount of fuel delivered and for controlling the injection advance.

Injection advance is obtained by rotation of the cam ring by pump body pressure in the injection advance mechanism. The injection advance mechanism consists of a transverse timing device piston and control components and an electrical solenoid valve. The maximum advance is 40° of crankshaft rotation.

A needle motion sensor in the injector sends a signal to the engine ECU at the instant of opening of the injector. This point, relative to the crankshaft rotational angle before TDC, is used for load and speed injection timing calculations and for control of the EGR valve.

The Bosch VR electronic diesel control system uses a number of sensors and control actuators. This allows it to achieve optimum performance. However, even this sophisticated system has virtually been superseded, so read on to discover more about the common rail system!

2.5.3.2 Common rail system

The development of diesel fuel systems is continuing, with many new electronic changes to the control and

Figure 2.312 Electronic diesel control (EDC) unit

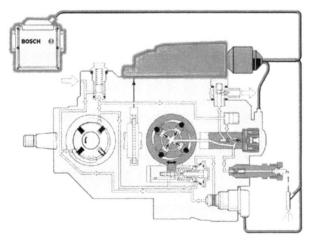

Figure 2.313 Electronic control units

injection processes. One of the latest developments is the common rail (CR) system, which operates at very high injection pressures. It also has piloted and phased injection to reduce noise and vibration.

The common rail system has made it easier for small high-speed diesel engines to have all the advantages of direct injection. These developments have resulted in significant improvements in fuel consumption and performance.

The combustion process, with common rail injection, is improved by a pilot injection of a very small quantity of fuel, at between 40° and 90° before top dead centre (BTDC). This pilot fuel ignites in the compressing air charge so that the cylinder temperature and pressure are higher than in a conventional diesel injection engine at the start of injection. The higher temperature and pressure reduce ignition lag to a minimum, so that the controlled

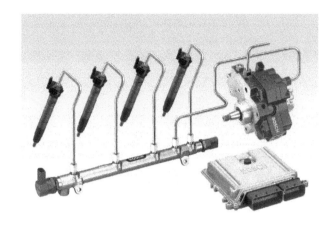

Figure 2.316 Four-cylinder system. (Source: Bosch Media)

combustion phase during the main injection period is softer and more efficient.

> **Key fact**
>
> The combustion process, with common rail injection, is improved by a pilot injection of a very small quantity of fuel, at between 40° and 90° BTDC.

Fuel injection pressures are varied throughout the engine speed and load range to suit the instantaneous conditions of driver demand and engine speed and load conditions. Data input from other vehicle system ECUs is used to further adapt the engine output, to suit changing conditions elsewhere on the vehicle. Examples are traction control, cruise control and automatic transmission gearshifts. The EDC module carries out calculations

Figure 2.315 Common rail system. (Source: Bosch Media)

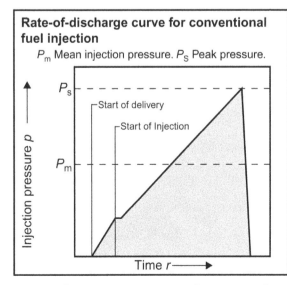

Rate-of-discharge curve for conventional fuel injection

P_m Mean injection pressure. P_S Peak pressure.

P_S
Start of delivery
Start of Injection
P_m

Injection pressure p

Time r

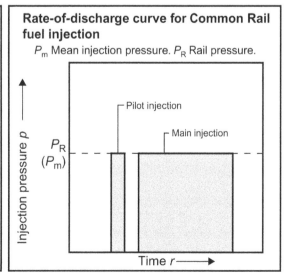

Rate-of-discharge curve for Common Rail fuel injection

P_m Mean injection pressure. P_R Rail pressure.

Pilot injection
Main injection
P_R (P_m)

Injection pressure p

Time r

Figure 2.317 Conventional system and common rail system

171

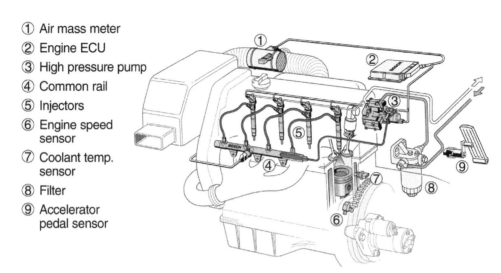

① Air mass meter
② Engine ECU
③ High pressure pump
④ Common rail
⑤ Injectors
⑥ Engine speed sensor
⑦ Coolant temp. sensor
⑧ Filter
⑨ Accelerator pedal sensor

Figure 2.318 Main components of the common rail system

to determine the quantity of fuel delivered. It also determines the injection timing based on engine speed and load conditions. The actuation of the injectors, at a specific crankshaft angle (injection advance) and for a specific duration (fuel quantity), is made by signal currents from the EDC module. A further function of the EDC module is to control the accumulator (rail) pressure.

Key fact

CR fuel injection pressures can be varied, throughout the engine speed and load range.

The Bosch CR common rail diesel fuel injection system, for light vehicles, consists of four main component areas: the low-pressure delivery components, high-pressure delivery with a high-pressure pump and accumulator (rail), the electronically controlled injectors, and the ECU and associated sensors and switches.

Definition

Accumulator: Apparatus by means of which energy can be stored (the rail on a common rail system).

The low-pressure delivery components are the fuel tank, a prefilter, presupply (low-pressure) pump, a fuel filter and the low-pressure delivery pipes to the high-pressure pump and for excess fuel return. The low-pressure pump, depending on application, can be of the roller cell type and be fitted either in the fuel tank or in-line where it is mounted to the vehicle body close to the fuel tank. Where the pump is fitted in the fuel tank, it includes a prefilter and has the fuel gauge sender unit attached to the same attachment flange on the side or top of the fuel tank.

The electrical supply to the fuel pump is made either directly or through a relay from the EDC module. An inertia switch is generally used to cut the electrical current to the pump motor in an accident. On some vehicles, a gear-type pump may be incorporated into the high-pressure pump and be driven from a common drive shaft. It can be a separate pump attached to the engine with a geared drive from the camshaft or crankshaft. The low-pressure delivery pipes connect to a fuel filter and water trap. A continuous flow of fuel runs through the filter and primes the high-pressure pump or returns to the fuel tank.

Key fact

An inertia switch is generally used to cut the electrical current to the pump motor in an accident.

Figure 2.319 High-pressure pump. (Source: Bosch Media)

The high-pressure pump is driven from the engine crankshaft through a geared drive at half engine speed and can be fitted where a conventional distributor pump would be. It can also be fitted on the end of the camshaft housing and be driven by the camshaft. It is lubricated by the diesel fuel that flows through it.

The pump has to produce all of the high pressure for fuel injection. It is a triple-piston radial pump, with a central cam for operation of the pressure direction of the pistons and return springs to maintain the piston rubbing shoes in contact with the cam. The pump has a positive displacement with inlet and outlet valves controlling the direction of flow through the pump.

The pump delivery rate is proportional to the speed of rotation of the engine so that it meets most engine speed requirements. To meet the engine load requirements the pump has a high volume. To meet the high-pressure requirements for fine atomization of the fuel on injection, the pump can produce pressures in the region of 1400 bar (scary!). A pressure control valve returns excess fuel to the fuel tank.

Safety first

A CR pump can produce pressures in the region of 1400 bar.

The pressure control valve is a mechanical and electrical unit. It is fitted on the pump or the high-pressure accumulator (rail). The mechanical part of the valve consists of a compression spring that acts on a plunger and ball valve. The electrical component is a solenoid that puts additional and variable force on the ball valve. The solenoid is actuated on signal currents from the EDC module. When the solenoid is not actuated, the ball valve opens at 100 bar against the resistance of the compression spring. This spring valve damps some of the high-frequency pressure fluctuations produced by the pump.

The solenoid in the pressure control valve is used for setting a variable mean pressure in the high-pressure accumulator (rail). The pressure in the rail is measured by a sensor and compared with a stored map in the EDC module for the current engine operating conditions. To increase the fuel rail pressure, an electrical alternating current is applied to the solenoid. The energizing current is varied by the EDC module, so that the additional force on the ball valve produces the required fuel rail pressure.

Key fact

The solenoid in the pressure control valve is used for setting pressure in the rail.

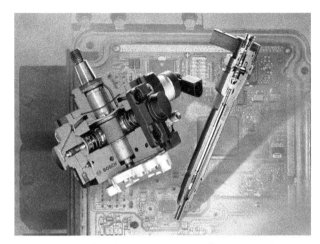

Figure 2.320 Sectioned common rail pump and injector. (Source: Bosch Media)

The high-pressure accumulator (rail) is common to all cylinders and derives its name, 'common rail', from this. This term is used in preference to fuel rail, which is used for petrol engines. The rail is an accumulator because it holds a large volume of fuel under pressure. The volume of fuel is sufficient to dampen the pressure pulses from the high-pressure pump.

The injectors on the common rail system have nozzles that are similar to all other diesel injectors for direct injection engines. The nozzle needle seats in the nozzle to obstruct the holes in the tip where the fuel is injected into the combustion chamber. The nozzle needle is held closed by a compression spring and opened by hydraulic pressure.

Opening and closing of the injector is controlled not by high-pressure fuel pulse from an injector pump, as in a conventional rotary distributor pump, but by actuation of an electrical solenoid in the injector body. This is controlled by the electronic diesel control module. A permanent high pressure is maintained in the injector at the same pressure as the rail. Operation of the injector is controllable for very small intervals of time.

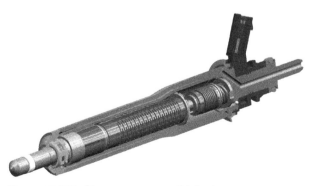

Figure 2.321 Piezo common rail injector

Figure 2.322 Common rail injection combustion

Key fact

The piezo injector reacts very quickly, thereby improving fuel control.

The electronic control of the common rail diesel injection system allows for precise control of fuelling. This results in excellent economy and very low emissions.

2.5.3.3 Diesel injection pressure

High pressure saves fuel and increases performance and torque. It is expected that most diesel engines will soon work with injection pressures of around 2,000 bar. Although 3,000 bar is not unrealistic, it will be limited to racing cars and high-performance diesel engines.

The common rail system featured Bosch's first piezo injector for passenger vehicles that works with an injection pressure of 2,500 bar. With their higher injection pressure, the new piezo models from Bosch are at the technological vanguard. The optimized fuel injection system atomizes the fuel more finely, improving combustion. Lower consumption is just one advantage of this technology.

The pressure generated by a common-rail system is roughly equivalent to the pressure that a 2,000 kg rhinoceros would exert standing on a fingernail. The compressed fuel is then finely dispersed at the speed of a supersonic jet.

A higher injection pressure generates greater specific power and increases torque. Therefore, increasing an engine's injection pressure makes it more powerful: the time available for combustion is extremely limited as soon as an engine is running at full load and high engine speed. This means the fuel must be injected into the engine very quickly at high pressure to achieve optimum power yield.

The more air there is in the combustion chamber, the higher the injection pressure must be. A large amount

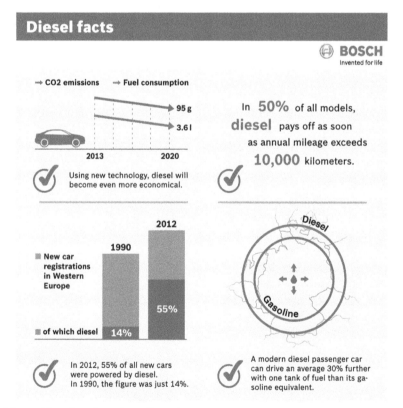

Figure 2.323 Diesel facts. The diesel engine offers the ideal combination of fuel economy and driving pleasure, particularly appreciated by business travellers and commuters.

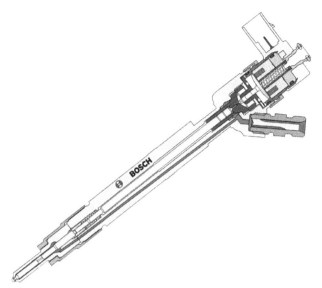

Figure 2.324 Bosch CRS3-25 with 2,500 bar pressure.

Figure 2.325 The new range of Volvo power units includes Denso i-ART technology (Source: Volvo)

of fuel must be introduced within a short space of time to achieve a combustible air–fuel mixture. Multiple turbocharged engines, particularly bi-turbo and tri-turbo models, benefit from injection pressures more than 2,000 bar. A higher injection pressure is a key factor in reducing an engine's untreated emissions. Indeed, in compact-class vehicles it can even help to avoid the need for exhaust gas treatment. The greater the injection pressure, the more finely both the injector and injection nozzle can be constructed. This improves atomization and results in a better air–fuel mixture, meaning that optimum combustion is achieved, and no soot can form.

A higher injection pressure requires more than just a re-engineered injector. With its comprehensive diesel systems competence, Bosch can assemble a finely tuned system comprising not only the control unit, but also the fuel pump, the common-rail system and the injector.

There have been some interesting developments in fuel system technology for common rail diesels. Denso has recently addressed pressure wave phenomena. This has taken the intelligence in diesel engine fuel systems to the next level with the introduction of the Intelligent Accuracy Refinement Technology (i-ART). This technology features a fuel-pressure sensor with an integrated microcomputer which monitors injection pressure based on various input data.

The whole assembly is integrated into the top of each fuel injector. The closed-loop system precisely manages injections of fuel to match specific drive cycle conditions. It replaces the single pressure sensor typically positioned in the fuel rail. Denso engineers have stated that i-ART can improve fuel efficiency by 2%, compared with open-loop systems. It was

developed to enable diesel engines to meet EURO6 regulations with a reduced after-treatment burden.

A conventional injection system could only detect an injection quantity based on indirect methods such as combustion or an engine rotation fluctuation. The i-ART system enables a direct detection of the injection quantity; each injector is equipped with a built-in fuel pressure sensor to measure injection pressure inside the injector itself. Based on the information from the built-in pressure sensor, the engine ECU reads fuel pressure values for each injection rapidly and calculates an actual injection quantity and timing for each cycle based on this information, using a rapid waveform processing technique. The learning value for the injection quantity and timing calculated with the i-ART system is applied to subsequent injections and adapted throughout its lifetime.

The actual pressure wave form generated by the i-ART pressure sensor is shown here. The system performs

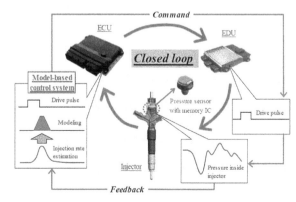

Figure 2.326 System overview – i-ART intelligent injectors and feedback data flow (Source: Denso)

175

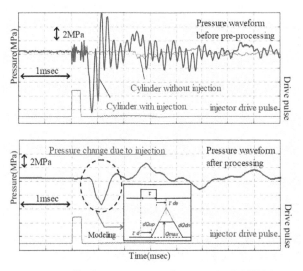

Figure 2.327 Fuel pressure waveforms at the i-ART injector (Source: Denso)

a pre-processing by comparing it to the non-injection pressure waveform to estimate the injection quantity and timing correctly. It then calculates the injection rate based on the processed pressure waveform which is optimized by filtering. The injection rate can be expressed by five parameters of a trapezoid shape. Calculating the area of the trapezoid, the injection quantity is obtained.

The i-ART system is constantly learning the injection quantity and timing while the engine is in operation and there are two advantages to using this characteristic. The first is the opportunity to use a triple pilot injection strategy, which allows a lower compression ratio to be used as less heat is needed to be able to ignite the fuel under all operating conditions. This is due to the improved mixture formation which promotes efficiency in the early stages of fuel injection/initial burning. In addition, this allows a sufficient preheating effect for the fuel with a reduced overall cylinder temperature, such that NO_x

and PM can be reduced. As a second advantage, in conjunction with cetane number detection, a stable combustion with minimized combustion noise can be achieved irrespective of the variation of cetane number with fuels in certain markets.

This technology is a big leap for common rail diesels, but also a significant step forward for measurement technology that can now be employed in production. There are significant advantages to being able to establish the fuel pressure directly at each injector, at the point of injection, as this helps considerably in being able to model the injection rate and fuel mass per stroke. The goal is to develop an injector where the rate and quantity of injection can be varied without a step and within a cycle. This would then facilitate the ability to truly control the combustion and energy release in a diesel engine with high precision on a cycle-by-cycle basis.

2.5.4 Coding components

Fitting a new component to a vehicle used to be a simple matter of obtaining the correct item and then following the appropriate procedures to 'bolt' and 'wire' it in place. In many cases it still is but, on some vehicles, certain parts must be coded. In other words, the vehicle must be 'told' what has changed. This can apply to all sorts of parts such as headlight modules, but fuel injectors are most important. Another

Figure 2.329 Injector code – a somewhat unusual one! (Source: Simon McCormac)

Figure 2.328 Bosch diesel injector

example would be fitting a tow bar – you must inform the car that it has one!

Diesel injectors were generally 'plug and play'. However, since the more widespread use of common rail diesel systems, it can be confusing as to whether an injector needs coding or not. Common rail diesel systems have several advantages compared to earlier diesel systems. For example, improved performance, lowered fuel consumption, quieter engines and reduced emissions. Coding therefore is essential if the system requires it.

An injector code is usually known as an IMA code (Bosch and Siemens) or a calibration code (Delphi). Its purpose is to ensure accurate injection control. The IMA coding is now an industry standard and stands for Injector Menge Abgleichung (injector quantity offset). When an injector is tested during manufacture, it generates an IMA code. This code identifies where the needle and nozzle assembly are in a tolerance range. When the injector is coded to the ECU it will be able to control the fuel accordingly and ensure optimal engine performance.

If injectors are not coded correctly, this can result in several issues:

▷ reduced performance, increased fuel consumption or even black smoke
▷ on some systems if you do not code in the injectors the vehicle will not start
▷ engine management light is likely to come on.

Refer to manufacturer's information but the following is a general guide:

▷ Delphi common rail injectors always require coding. Usually, the code is 16 alphanumeric characters (C2i injectors) or 20 characters for newer applications (C3i injectors). The code is usually found on the injector head.
▷ VDO/Siemens common rail injectors when used on VW applications require coding. Other vehicles will too so always check. Usual length is 6 digits and found on the injector head.
▷ DENSO common rail injectors all require coding, except for some very early ones. The usual code length is 16–24 alphanumeric characters dependent upon vehicle manufacturer. The code, as with other types, is found on the injector head. A QR code on the injector may also be used.
▷ BOSCH common rail injectors all require coding except for some very early ones. The code is usually 6–10 digits and found on the injector head.

As a rule, if an IMA code is present then the injector will need coding. A remanufactured injector should also have a new calibration code, which is usually on a sticker on the injector. Fuel pumps might need to be coded in a similar way to injectors and for the same reasons.

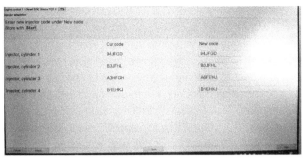

Figure 2.330 Changing injector codes (Source: Charlie Thompson)

Figure 2.331 New and old injectors (Source: Charlie Thompson)

2.6 Ignition systems

2.6.1 Ignition overview

2.6.1.1 Introduction

The purpose of the ignition system is to supply a spark inside the cylinder, near the end of the compression stroke, to ignite the compressed charge of air/fuel vapour.

Figure 2.332 Combustion taking place. (Source: Ford Media)

177

For a spark to jump across an air gap of 1.0 mm under normal atmospheric conditions (1 bar) a voltage of 4–5 kV is required. For a spark to jump across a similar gap in an engine cylinder, having a compression ratio of 8:1, approximately 10 kV is required. For higher compression ratios and weaker mixtures, a voltage of up to 20 kV may be necessary. The ignition system has to transform the normal battery voltage of 12 V to approximately 8–20 kV and, in addition, has to deliver this high voltage to the right cylinder, at the right time. Some ignition systems will supply up to 40 kV to the spark plugs.

Conventional ignition is the forerunner of the more advanced systems controlled by electronics. It is worth mentioning at this stage, however, that the fundamental operation of most ignition systems is very similar. One winding of a coil is switched on and off, causing a high voltage to be induced in a second winding. The basic types of ignition system can be classified as shown in Table 2.2.

Modern ignition systems now are part of the engine management, which controls fuel delivery, ignition and other vehicle functions. These systems are under continuous development and reference to the manufacturer's workshop manual is essential when working on any vehicle. The main ignition components are the engine speed and load sensors, knock sensor, temperature sensor and ignition coil. The ECU reads from the sensors, interprets and compares the data, and sends output signals to the actuators. The output component for ignition is the coil.

> **Key fact**
>
> Modern ignition systems now are part of the engine management.

> **Key fact**
>
> No matter how complex the electronics may seem, a high-voltage spark is produced by switching a coil on and off.

Ignition systems continue to develop and will continue to improve. However, keep in mind that the simple purpose of an ignition system is to ignite the fuel/air mixture every time at the right time. And, no matter

Figure 2.333 Electronic ignition distributor and ignition module

how complex the electronics may seem, the high voltage is produced by switching a coil on and off.

2.6.1.2 Generation of high voltage

If two coils (known as the primary and secondary) are wound on to the same iron core then any change in magnetism of one coil will induce a voltage in the other (see Chapter 3 for more details). This happens when a current is switched on and off to the primary coil. If the number of turns of wire on the secondary coil is more than on the primary a higher voltage can be produced. This is called transformer action and is the principle of the ignition coil.

The value of this 'mutually induced' voltage depends on:

▶ the primary current
▶ the turns ratio between primary and secondary coils
▶ the speed at which the magnetism changes.

The two windings are wound on a laminated iron core to concentrate the magnetism. This is how all types of ignition coil are constructed.

2.6.1.3 Ignition timing (advance angle)

For optimum efficiency the ignition advance angle should be such as to cause the maximum combustion pressure to occur about 10° after TDC. The ideal ignition timing is dependent on two main factors, engine speed and engine load. An increase

Table 2.2 Ignition systems

Type	Conventional	Electronic	Programmed	Distributorless
Trigger	Mechanical	Electronic	Electronic	Electronic
Advance	Mechanical	Mechanical	Electronic	Electronic
Voltage source	Inductive	Inductive	Inductive	Inductive
Distribution	Mechanical	Mechanical	Mechanical	Electronic

in engine speed requires the ignition timing to be advanced.

The cylinder charge of fuel/air mixture requires a certain time to burn (normally about 2 ms). At higher engine speeds the time taken for the piston to travel the same distance reduces. Advancing the time of the spark ensures that full burning is achieved.

> **Key fact**
>
> For optimum efficiency the ignition advance angle should be such as to cause the maximum combustion pressure to occur about 10° after TDC.

A change in timing due to engine load is also required as the weaker mixture used on low load conditions burns at a slower rate. In this situation further ignition advance is necessary. Greater load on the engine requires a richer mixture, which burns more rapidly. In this case some retardation of timing is necessary. Overall, under any condition of engine speed and load an ideal advance angle is required to ensure that maximum pressure is achieved in the cylinder just after TDC. The ideal advance angle may also be determined by engine temperature and any risk of detonation.

Spark advance is achieved in a number of ways. The simplest of these is the mechanical system comprising a centrifugal advance mechanism and a vacuum (load-sensitive) control unit. Manifold depression is almost inversely proportional to the engine load. I prefer to consider manifold pressure; although it is less than atmospheric pressure, the MAP is therefore proportional to engine load. Digital

Figure 2.334 First Bosch high-voltage magneto ignition system with spark plug in 1902. (Source: Bosch Media)

ignition systems adjust the timing in relation to the temperature as well as speed and load. The values of all ignition timing functions are combined either mechanically or electronically to determine the ideal ignition point.

Energy storage takes place in the ignition coil. The energy is stored in the form of a magnetic field. To ensure the coil is charged before the ignition point, a dwell period is required. Ignition timing is at the end of the dwell period as the coil is switched off.

2.6.1.4 Traditional ignition system

Very early cars used something called a magneto. For many years ignition systems were mechanically switched and distributed. Table 2.3 gives an overview of the components of this earlier system.

Table 2.3 Traditional ignition components

Component	Function
Spark plug	Seals electrodes for the spark to jump across in the cylinder. Must withstand very high voltages, pressures and temperatures
Ignition coil	Stores energy in the form of magnetism and delivers it to the distributor via the high-tension (HT) lead. Consists of primary and secondary windings
Ignition switch	Provides driver control of the ignition system and is usually also used to cause the starter to crank
Contact breakers (breaker points)	Switches the primary ignition circuit on and off to charge and discharge the coil. The contacts are operated by a rotating cam in the distributor
Capacitor (condenser)	Suppresses most of the arcing as the contact breakers open. This allows for a more rapid break of primary current and hence a more rapid collapse of coil magnetism, which produces a higher voltage output
Distributor	Directs the spark from the coil to each cylinder in a preset sequence
Plug leads	Thickly insulated wires to connect the spark from the distributor to the plugs
Centrifugal advance	Changes the ignition timing with engine speed. As speed increases the timing is advanced
Vacuum advance	Changes timing depending on engine load. On conventional systems the vacuum advance is most important during cruise conditions

Figure 2.335 Contact breaker system

Figure 2.336 Traditional ignition coil

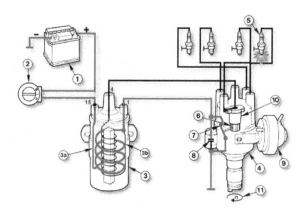

Figure 2.337 Ignition circuit of an early system: 1, battery; 2, ignition key switch; 3, coil; a, primary winding; b, secondary winding; 4, distributor body containing centrifugal (speed) advance/retard mechanism; 5, spark plug; 6, cam (with a lobe for each cylinder); 7, contact breakers (points); 8, condenser (capacitor); 9, vacuum (load) advance/retard mechanism; 10, rotor arm; 11, direction of rotation (at n rpm)

Figure 2.338 Traditional system using a distributor

All current vehicle ignition systems are electronically switched and most are now digitally controlled as part of the engine management system. However, there are many vehicles out there still using conventional electronic ignition, so the next section will give an overview of these systems.

2.6.2 Electronic ignition

2.6.2.1 Constant dwell

Mechanical ignition systems have some major disadvantages:

▶ Mechanical problems with the contact breakers occur, not least of which is the limited lifetime.
▶ Current flow in the primary circuit is limited to about 4 A or damage will occur to the contacts; or at least the lifetime will be seriously reduced.
▶ Legislation requires stringent emission limits, which means that the ignition timing must stay in tune for a long time.
▶ Weaker mixtures require more energy from the spark to ensure successful ignition, even at very high engine speed.

These problems were overcome by using a power transistor to carry out the switching function and a pulse generator to provide the timing signal.

The term 'dwell' when applied to ignition is a measure of the time during which the ignition coil is charging,

Figure 2.339 Power transistor

Figure 2.340 Electronic ignition module. (Source: Bosch Media)

in other words when primary coil current is flowing. The dwell in traditional systems was simply the time during which the contact breakers were closed, and in these early electronic systems it was the time that the transistor was switched on. Although this was a very good system in its time, constant dwell still meant that at very high engine speeds the actual time available to charge the coil would only produce a lower power spark. Note that as engine speed increases, dwell angle or dwell percentage remains the same but the actual time is reduced. All systems nowadays are known as constant energy, ensuring high-performance ignition even at high engine speed.

Definition

Dwell: A measure of the time during which the ignition coil is charging.

2.6.2.2 Constant energy

In order for a constant energy electronic ignition system to operate, the dwell must increase with engine speed. This will only be of benefit if the ignition coil can be charged up to its full capacity in a very short time (the time available for maximum dwell at the highest expected engine speed). To this end, constant energy coils are very low resistance, so a high current will flow quickly. Constant energy means that within limits, the energy available to the spark plug remains constant under all operating conditions.

This was achieved by using a pulse generator in the distributor to inform an ignition module of the engine position and speed so that the module could determine the switch-on (start of dwell) and switch-off points (end of dwell and ignition timing spark).

Owing to the high-energy nature of constant energy ignition coils, the coil cannot be allowed to remain switched on for more than a certain time. This is

Figure 2.341 Hall effect distributor

not a problem when the engine is running, as the variable dwell or current-limiting circuit prevents the coil overheating. Some form of protection must be provided, however, when the ignition is switched on but the engine is not running. This is known as stationary engine primary current cut-off.

Two types of pulse generator (sensors) were most common:

▷ Hall effect
▷ inductive.

As the central shaft of the Hall effect distributor (Fig. 2.341) rotates, the chopper plate attached under the rotor arm alternately covers and uncovers the Hall chip.

The number of vanes corresponds with the number of cylinders. In constant dwell systems the dwell is determined by the width of the vanes. The vanes cause the Hall chip to be alternately in and out of a magnetic field. The result of this is that the device will produce almost a square-wave output, which can then

181

Figure 2.342 Inductive distributor

Figure 2.343 Distributorless ignition coil in position

easily be used to switch further electronic circuits. The three terminals on the distributor are marked +, 0 and –: terminals + and – are for a voltage supply and terminal 0 is the output signal.

Typically, the output from a Hall effect sensor will switch between 0 V and about 7 V. The supply voltage is taken from the ignition ECU and on some systems is stabilized at about 10 V to prevent changes to the output of the sensor when the engine is being cranked. Hall effect distributors are very common owing to the accurate signal produced and long-term reliability. They produce a kind of square-wave output signal.

Key fact

Typically, the output from a Hall effect sensor will switch between 0 V and about 7 V (but always check specifications).

Inductive pulse generators use the basic principle of induction to produce a signal. Many forms exist, but all are based around a coil of wire and a permanent magnet. The distributor shown in Figure 2.342 has the coil of wire wound on the pick-up and as the reluctor rotates the magnetic flux varies due to the peaks on the reluctor. The number of peaks or teeth on the reluctor corresponds to the number of engine cylinders. The gap between the reluctor and pick-up can be important and manufacturers have recommended settings. These systems produce a kind of sine-wave output.

2.6.3 Distributorless ignition system (DIS)

Distributorless ignition uses a special type of ignition coil, which outputs to the spark plugs without the need for a high-tension distributor.

Key fact

Distributorless ignition outputs to the spark plugs without using a distributor.

The basic principle is that of the 'lost spark'. The distribution of the spark is achieved by using two double-ended coils, which are fired alternately by the ECU. The timing is determined from a crankshaft speed and position sensor as well as a load (MAP) sensor and other corrections such as engine temperature. When one of the coils is fired a spark is delivered to two engine cylinders, either 1 and 4 or 2 and 3. The spark delivered to the cylinder on the compression stroke will ignite the mixture as normal. The spark produced in the other cylinder will have no effect, as this cylinder will be just completing its exhaust stroke.

Because of the low compression, and the exhaust gas in the lost spark cylinder, the voltage used for the

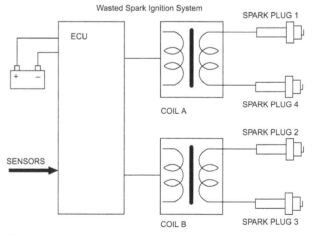

Figure 2.344 Distributorless ignition system (DIS) simplified circuit (wasted spark ignition system)

spark to jump the gap is only about 3 kV. The spark produced in the compression cylinder is therefore not affected. An interesting point here is that the spark on one of the cylinders will jump from the earth electrode to the spark plug centre. Many years ago this would not have been acceptable as the spark quality when jumping this way would not have been as good as when jumping from the hotter-centre electrode. However, the energy available from modern constant energy systems will result in a spark of high quality regardless of its polarity.

The DIS consists of three main components: the ECU, a crankshaft position sensor and the DIS coil. An MAP sensor is integrated in the module or mounted separately. The module uses an electronic spark advance system. Data on ideal dwell and timing is held in memory maps for a wide range of speed, load and voltage conditions. This can be described as an electronic spark advance (ESA) system.

The crankshaft position sensor is similar in operation to the one described in the fuel section. It is an inductive sensor and is positioned against the front of the flywheel or against a reluctor wheel just behind the front crankshaft pulley. The tooth pattern usually consists of 35 teeth. These are spaced at 10° intervals with a gap where the 36th tooth would be. The missing tooth is positioned at 90° BTDC for numbers 1 and 4 cylinders. This reference position is placed a fixed number of degrees BTDC, to allow the timing or ignition point to be calculated as a fixed angle after the reference mark.

The primary winding is supplied with battery voltage to a centre terminal. The appropriate half of the winding is then switched to earth in the module. The high-tension windings are separate and are specific to cylinders 1 and 4 or 2 and 3 (or as appropriate if a six-cylinder engine).

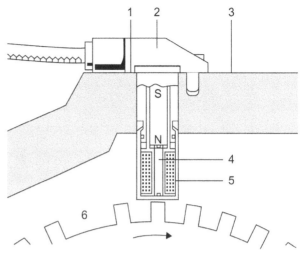

Figure 2.346 Inductive sensor: 1, magnet; 2, cover; 3, engine; 4, core; 5, winding; 6, missing tooth

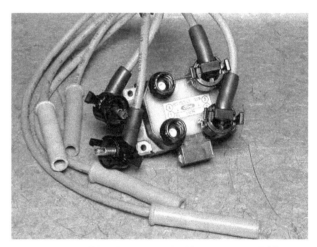

Figure 2.347 Distributorless ignition system (DIS) coil and plug leads

Key fact

The crankshaft position sensor (CPS) is similar in operation to the one described in the fuel section – and in most cases the same one is used.

On most cars now the ignition system is combined with the fuel system so that even more accurate control of outputs is possible and input data from sensors can be shared.

2.6.4 Coil on plug (COP) direct ignition system

Direct ignition is, in a way, a further improvement on distributorless ignition. This system utilizes an inductive coil for each engine cylinder. These coils are mounted directly on the spark plugs. The use of an

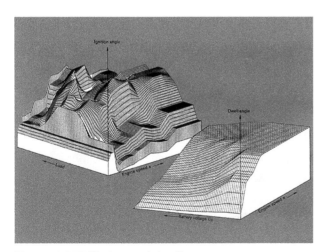

Figure 2.345 Timing and dwell maps

Figure 2.348 Six direct ignition coils in position

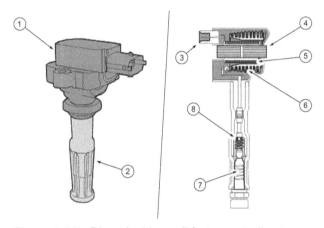

Figure 2.349 Direct ignition coil features: 1, direct ignition coil; 2, spark plug connector; 3, low-voltage connection, outer; 4, laminated iron core; 5, primary winding; 6, secondary winding; 7, spark plug; 8, high-voltage connection, inner, via spring contact

individual coil for each plug ensures that the charge time is very fast (full coil charge in a very small dwell angle). This ensures that a very high-voltage, high-energy spark is produced. This voltage, which can be in excess of 40 kV, provides efficient initiation of the combustion process under cold starting conditions and with weak mixtures.

Key fact

Direct ignition has a coil for each spark plug.

Definition

ESA: Electronic spark advance.

Ignition timing and dwell are controlled in a manner similar to the previously described ESA system. The one important addition to this on most systems is a camshaft sensor to provide information as to which cylinder is on the compression stroke. A system that does not require a sensor to determine which cylinder is on compression (engine position is known from a crank sensor) determines the information by initially firing all of the coils. The voltage across the plugs allows measurement of the current for each spark and will indicate which cylinder is on its combustion stroke. This works because a burning mixture has a lower resistance. The cylinder with the highest current at this point will be the cylinder on the combustion stroke.

A further feature of some systems is the case when the engine is cranked over for an excessive time making flooding likely. The plugs can all be fired with multisparks for a period of time after the ignition is left on to burn away any excess fuel. During difficult starting conditions, multisparking is also used by some systems during 70° of crank rotation BTDC. This assists with starting and then once the engine is running, the timing will return to its normal calculated position.

2.6.5 Spark plugs

The simple requirement of a spark plug is that it must allow a spark to form within the combustion chamber, to initiate combustion. In order to do this the plug has to withstand a number of severe conditions. It must withstand severe vibration and a harsh chemical environment. Finally, but perhaps most importantly, the insulation properties must withstand voltage pressures up to 40 kV.

Figure 2.351 shows a standard spark plug. The centre electrode is connected to the top terminal by a stud. The electrode is constructed of a nickel-based alloy. Silver and platinum are also used for some applications. If a copper core is used in the electrode this improves the thermal conduction properties. The insulating material is ceramic based and of a very high

Figure 2.350 Modern high-performance spark plug

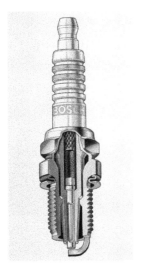

Figure 2.351 Cutaway section of a standard spark plug

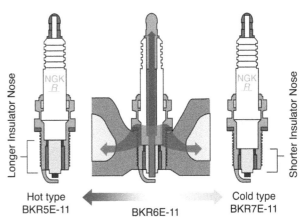

Figure 2.352 Heat-loss paths: 1, cold plug; 2, hot plug; 3, temperature (the cold plug is able to transfer heat more easily so is suitable for a hot engine)

Figure 2.353 A range of spark plugs. (Source: Bosch Media)

grade. Flash-over or tracking down the outside of the plug insulation is prevented by ribs which effectively increase the surface distance from the terminal to the metal fixing bolt, which is earthed to the engine.

Because of the many and varied constructional features involved in the design of an engine, the range of temperatures a spark plug is exposed to can vary significantly. The operating temperature of the centre electrode of a spark plug is critical. If the temperature becomes too high then pre-ignition may occur, where the fuel/air mixture may be ignited owing to the incandescence of the plug electrode. If the electrode temperature is too low, then carbon and oil fouling can occur as deposits are not burnt off. The ideal operating temperature of the plug electrode is between 400 and 900°C.

Key fact

The ideal operating temperature of a spark plug electrode is between 400 and 900°C.

The heat range of a spark plug is a measure of its ability to transfer heat away from the centre electrode. A hot running engine will require plugs with a higher thermal ability than a colder running engine. Note that hot and cold running of an engine in this sense refers to the combustion temperature, not to the cooling system.

Spark plug electrode gaps in general have increased as the power of the ignition systems driving the spark has increased. The simple relationship between plug gap and voltage required is that as the gap increases so must the voltage (leaving aside engine operating conditions). Further, the energy available to form a spark at a fixed engine speed is constant, which means that a larger gap using higher voltage will result in

a shorter duration spark. A smaller gap will allow a longer duration spark. For cold starting an engine and for igniting weak mixtures the duration of the spark is critical. Likewise, the plug gap must be as large as possible to allow easy access for the mixture to prevent quenching of the flame. The final choice is therefore a compromise reached through testing and development of a particular application. Plug gaps in the region of 0.6–1.2 mm seem to be the norm at present.

Key fact

Plug gaps of 0.6–1.2 mm are in common use.

2.7 Electric and hybrid cars

2.7.1 Types of electric vehicle

Electric vehicle (EV) refers to any vehicle that is powered, in part or in full, by a high-voltage battery. This textbook concentrates on internal

185

combustion engine (ICE) technology, but larger vehicles are similar. We will use EV as the 'catch all' phrase as this in turn includes the following technologies.

Definition

EV is used as the general description for electric vehicles

Pure-electric Vehicles (pure-EVs) are electric vehicles powered only by a battery. Range expectation of between 100 and 300 miles is currently considered acceptable for standard cars, although this technology is improving.

Plug-In Hybrid Electric Vehicles (PHEVs) have an internal combustion engine (ICE) but also a battery range in excess of 10 miles. After the battery range is utilized, the vehicle reverts to the benefits of full hybrid capability (utilizing both battery and ICE power) without compromising range.

Definition

Range anxiety: the fear that an electric vehicle has insufficient range on a fully charged battery for motorists to complete their journey.

Extended-Range Electric Vehicles (E-REVs) are like pure-EVs but with a shorter battery range of 50 miles. However, range is extended by an ICE-driven generator providing many additional miles of mobility. With an E-REV, the propulsion is always electric, unlike a PHEV where the propulsion can be electric or full hybrid.

Standard Hybrid Electric Vehicles (HEVs) are those where it is not possible to charge the battery externally – there are several variations as outlined in Table 2.4.

2.7.2 The EV experience

Cars running on electricity are easy to drive. They are also smooth, quiet and acceleration is good. Pure-EVs do not have a gearbox so are like driving an automatic. Plug-in hybrids (PHEVs) have a gearbox but this will be automatic, even though it can be controlled manually on many cars.

Electricity, when produced from sustainable sources, is easy to supply, and produces no emissions from the vehicle (often described as tailpipe emissions). EVs, therefore, have significant environmental benefits, particularly when used in urban environments. Some

Figure 2.354 Outlander PHEV (Source: Mitsubishi)

Figure 2.355 Prius HEV (Source: Toyota)

of the benefits of EVs when operating solely on battery power are:

▶ zero tailpipe emissions
▶ quiet driving
▶ easy to drive, particularly in stop-start traffic
▶ home charging avoids fuel station queues.

Electric vehicles can achieve similar speeds to ICE vehicles during normal driving. Some pure-electric cars can reach speeds over 125 mph where permitted. Power is delivered by the electric motor as soon as the vehicle begins to move which gives smooth and swift acceleration.

Key fact

Some pure-electric cars can reach speeds of over 125 mph.

Key fact

The range of an EV depends on the type and how it is driven.

Table 2.4 Summary of EVs and HEVs and their alternative names

Pure-EV, Pure-Electric Car, Battery Electric Vehicle (BEV), Fully Electric	A vehicle powered only by a battery charged from mains electricity.
Plug-In Hybrid Electric Vehicle (PHEV), Plug-In Hybrid Vehicle (PHV)	A vehicle with a plug-in battery and an internal combustion engine (ICE). Typical PHEVs will have a pure-electric range of around 30 miles. After the pure-electric range is used up, the vehicle reverts to the benefits of full hybrid capability
Extended-Range Electric Vehicle (E-REV), Range Extended Electric Vehicle (RE-EV)	A vehicle powered by a battery with an ICE-powered generator on board. E-REVs are like pure-EVs but with a shorter battery range of around 50 miles. Range is extended by an on-board generator providing additional miles of mobility. With an E-REV the vehicle is still always electrically driven and is known as a series hybrid (more on this later)
Hybrid Electric Vehicles (HEV), Full/ Normal/Parallel/Standard hybrid	A hybrid vehicle is powered by a battery and/or an ICE. The power source is selected automatically by the vehicle, depending on speed, engine load and battery charge. This battery cannot be plugged in so charge is maintained by regenerative braking supplemented by ICE generated power
Mild Hybrid	A mild hybrid vehicle cannot be plugged in, or driven solely on battery power. However, it does harvest power during regenerative braking and uses this during acceleration (current F1 cars are a type of mild hybrid)
Micro Hybrid	A micro hybrid normally employs a stop-start system and regenerative braking which charges the 12V battery
Stop-start Hybrid	A stop-start system shuts off the engine when the vehicle is stationary. An enhanced starter motor is used to support the increased number of engine starts
Alternative Fuel Vehicle (AFV)	Any vehicle that is not solely powered by traditional fuels (i.e. petrol or diesel) is referred to as alternative fuel
Internal Combustion Engine (ICE)	Petrol or diesel engine, as well as those adapted to operate on alternative fuels
Electric quadricycles	This is a four-wheeled vehicle that is categorized and tested in a similar way to a moped or three-wheeled motorcycle.
Electric motorcycles	Battery only, so full electric drive motorcycles can have a range of up to 60 miles. However, a new range of Irish-built electric bikes have a claimed range of up to 136 miles. For example, the Volt 220, which takes its name from its range of 220 km, will do up to 60 mph according to the manufacturer.

A phrase often used about EVs is 'range anxiety'. This refers to the fear about the distance an EV can drive and the worry that it may not be enough to reach our destination!

However, an interesting point to note is that the average individual journey in the UK is less than 10 miles. The average total daily distance travelled is about 25 miles. In Europe, more than 80% of drivers cover less than 63 miles in a typical day. These distances can therefore be achieved using pure-electric cars and many journeys can be made with plug-in hybrid or extended-range electric cars without using the internal combustion engine.

Key fact

The average individual journey in the UK is less than 10 miles.

EVs must comply with the same safety standards as conventional cars by obtaining 'whole vehicle type approval'. Particular attention is paid during crash testing to ensure the EV-specific safety features operate correctly. Individual components such as the battery pack are also subjected to additional impact and other mechanical tests.

EVs typically use an inertia switch or a signal from the airbag system to disconnect the traction battery if the vehicle is involved in a collision. This is very similar to conventional vehicles where an inertia switch is provided to stop the fuel supply in a crash. Battery packs are also designed with internal contactors so that if the 12 V supply is cut for any reason, the traction supply is also shut off.

Although EVs still generate tyre noise, the level of noise is much lower than an ICE car, particularly at low speed. Sight and hearing impaired people can be particularly vulnerable so drivers need to be aware of this and take extra care.

Safety first

The hierarchy in the new Highway Code places road users most at risk in the event of a collision at its top.

As with any vehicle, EV range depends on several factors, such as driving style, environmental conditions and the use of auxiliary systems in the vehicle. Manufacturers' performance claims should be seen as an indication of the capabilities of the vehicle – not what will happen in the real world! However, it is important to note how much driving style affects these performance figures and maximum range is unlikely to be achieved in a usage style based on rapid acceleration, high speeds and heavy use of auxiliary systems, such as heating and air conditioning.

For more information, please refer to another title in the series: *Electric and Hybrid Vehicles*, and there is more information from the support website: www.automotive-technology.org.

2.8 Engine systems: new developments and updates

2.8.1 Spark plug electrode designs

Every time the plug sparks, minute particles of material are worn away from the electrodes. This phenomenon is called spark erosion. Over time, this process increases the spark plug gap between the centre and the earth/ground electrode. If the gap becomes too large, misfiring will occur.

To extend the service interval of vehicles, the service life of the spark plug must be increased. Some manufacturers are fitting multi-electrode spark plugs as original equipment to achieve this. Multi-electrode spark plugs can have two, three or four ground electrodes depending on the service life requirement of the manufacturer. However, no matter how many ground electrodes the plug has, every time the spark plug fires, only one spark occurs between the centre electrode and the ground electrode, which has the lowest required voltage or the least distance to travel between the centre and the ground electrode.

> **Key fact**
> No matter how many ground electrodes the plug has, every time the spark plug fires, only one spark occurs between the centre electrode and the ground electrode

The spark plug plays a vital role in the quest to improve ignition quality, engine performance, reduce emissions, and reduce fuel consumption. Spark plugs that employ small-diameter centre and sometimes ground electrodes can offer benefits in several areas. These fine wire plugs require less voltage to create the spark, have a more consistent spark position, better gas flow around the firing position and experience less quench effect than other designs. As the electrodes erode during use it is necessary to compensate for the use of smaller electrodes by some means, otherwise the plugs would have an unacceptable (short) service life.

> **Key fact**
> Fine wire plugs require less voltage to create a spark.

By using small chips of special precious metals such as platinum or iridium which are welded to the tips of the electrodes it is possible to increase the service life significantly whilst maintaining the highest ignition performance. These metals are extremely hard and have very high melting points thus making them ideal for use in this hostile environment.

> **Safety first**
> Remember, all ignition systems are high voltage.

Some modern vehicles use a direct fuel injection system and these vehicles demand high ignition quality and extreme anti-fouling performance. NGK has developed a plug that has several special features designed to offer the required performance. Essentially, a very projected fine wire spark plug with platinum electrodes is combined with a semi-surface discharge design. The resulting plug has three ground electrodes, two of which are mostly redundant unless in extreme circumstances the plug becomes very carbon fouled. At this point the spark will discharge across the insulator nose to one of the side electrodes preventing a misfire and unburned fuel reaching the catalyst. This type of plug must only be use in the specific applications as listed in the NGK catalogues.

2.8.2 V6 diesel with electric turbocharging

Audi's (2017/18) V6 diesel includes a new electrically enhanced turbocharger system, as well as a new integrated NO_x and PM after treatment suite. It is expected that in the year 2030, more than 80% of newly registered cars and light trucks will still have an internal combustion engine on board.

> **Definition**
> **PM:** Particulate matter.
> **NO_x:** Nitrogen oxides.

Figure 2.356 Spark plug electrodes, left to right: multi-electrode, iridium wire, hybrid. (Source: NGK Plugs)

Figure 2.357 Audi V6 diesel with electric turbo. (Source: Audi Media)

It is claimed that using the e-turbo will easily gain you a lead of at least two car lengths in the first two seconds at the traffic lights! This illustrates the clear benefit of the system in that turbo lag is all but eliminated as the turbo compressor can be brought up to speed almost instantly. Other new features are to be incorporated:

▷ Piston rings have been optimized for minimal friction.
▷ The crankcase and new cylinder heads have separate coolant loops.
▷ A new thermal management system improves efficiency.
▷ The turbocharger and the fully variable-load oil pump have been updated.

The engine will include a NO_x storage catalytic converter that has been combined with a diesel particulate filter and SCR injection in a single unit.

Definition
SCR: Selective catalytic reduction.

The engine package satisfies the most stringent of emissions legislation, including Euro 6 and reduces CO_2 emissions by an average of 15 g/km. Emission and economy issues are what manufacturers are working towards with all their new developments.

2.8.3 Water injection

Even advanced petrol/gasoline engines waste roughly a fifth of their fuel. This is mostly at high engine speeds, where some of the fuel is used for cooling instead of for propulsion. Water injection means that it does not have to be that way. Particularly when accelerating quickly or driving on the motorway, the injection of additional water makes it possible to reduce fuel consumption by up to 13%.

Key fact
Even advanced petrol/gasoline engines waste roughly a fifth of their fuel.

The fuel economy offered by this Bosch technology comes especially to the fore in three- and four-cylinder downsized engines: in other words, in precisely the kind of engines to be found in any average midsize car.

This technology can make cars more powerful as well. Water injection can deliver an extra kick in any turbocharged engine. This is because earlier ignition angles mean that the engine is operated even more efficiently. On this basis, engineers can coax

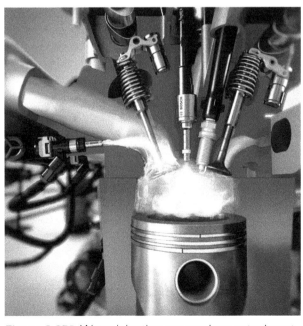

Figure 2.358 Water injection can produce extra boost for the turbocharged engine. (Source: Bosch Media)

189

additional power out of the engine, even in powerful sports cars.

The basis of this innovative engine technology is a simple fact: an engine must not be allowed to overheat. To stop this happening, additional fuel is injected into nearly every SI engine on today's roads. This fuel evaporates, cooling parts of the engine block. With water injection, Bosch engineers have exploited this physical principle. Before the fuel ignites, a fine mist of water is injected into the intake duct. Water's high heat of vaporization means that it provides effective cooling.

> **Key fact**
>
> Water's high heat of vaporization means that it provides effective cooling.

This is also the reason only a small additional volume of water is needed: for every one hundred kilometres driven, only a few hundred millilitres are necessary. As a result, the compact water tank that supplies the injection system with distilled water only has to be refilled every few thousand kilometres. If the tank should run empty, the engine will still run smoothly, albeit without the higher torque and lower consumption provided by water injection.

2.8.4 Cylinder deactivation

The purpose of cylinder deactivation technology is to reduce fuel consumption on petrol/gasoline engines. Large-capacity engines were renowned for poor fuel consumption under low speed and torque conditions. This inefficiency is known as pumping loss and is used to describe a situation where an engine that is unable to draw in a sufficient quantity of air and fuel mixture on the inlet stroke, to produce high cylinder pressure during compression.

Low pressure in an engine cylinder results in low efficiency because more fuel is required to compensate for the lack of pressure and the ability to draw fuel into the cylinder. To deactivate a cylinder of a running engine whilst retaining balance, driveability and emission targets is a challenge. The best method of deactivation, therefore, is to disconnect the valve operating mechanism. If the inlet and exhaust valves of the deactivated cylinder are closed, no gases can enter or exit. This also prevents an increase in exhaust emissions. Fuel injection to the deactivated cylinder is also switched off.

> **Key fact**
>
> Low pressure in an engine cylinder results in low efficiency because more fuel is required to compensate for the lack of pressure

Figure 2.359 Valves operating normally (left), valve closed (right) and cylinder deactivated. (Source: Volkswagen Media)

Volkswagen use a cylinder deactivation technology on smaller engines (1.4 L. for example). Their method employs a system of cam lobes that are splined to the camshafts. In the normal working position the cam lobes are aligned with the valves. To deactivate the cylinder a pin is engaged with a scroll that moves the cams out of alignment with the rockers and prevents the valve from opening. The pin is solenoid operated. In this state the valve rockers run on a concentric part of the shaft. After the sliding component has moved, the pin is retracted. To re-activate the valves a second pin is engaged with the scroll and the cams move back to their normal position. The switchover process occurs during half of a camshaft rotation.

The VW Active Cylinder Technology (ACT) system shuts down the second and third cylinders during low and medium loads and therefore reduces fuel consumption. It is active over an engine speed range of 1400 to 4000 rpm a torque output range of about 25 to 100 Nm. When the driver demands acceleration, this is detected by the pedal sensor and both cylinders are re-activated without any noticeable change. The system has no detrimental effects on smooth running of the engine. Changes are made to ignition timing and the throttle valve position to ensure a smooth transition. If the system detects irregular driving, cylinder shut-off is deactivated.

Honda use a different deactivation system. This employs rockers that are connected and disconnected by hydraulically operated locking pins. One rocker always remains in contact with the profile of the cam and, when connected by the pin, operates a second rocker that in turn operates the valve. When deactivation is required, oil pressure, controlled by a solenoid valve, forces the locking pin out of engagement with the second rocker. This prevents the valve from opening. This technology has been used in hybrid models since the year 2000.

The Bosch cylinder deactivation system can facilitate the deactivation of cylinders that are not required

Figure 2.360 Honda cylinder deactivation mechanism

within part-load operations in almost every gasoline engine. This saves fuel and thus reduces CO_2 emissions. The Bosch system deactivates the hydraulic valve lifters to prevent the valves opening.

This section has outlined three methods of cylinder deactivation:

- locking pins and a second rocker
- splined moveable cams
- deactivation of hydraulic lifters.

It is quite likely that more methods will be introduced because the system has great potential for reducing consumption and emissions.

2.8.5 Dynamic skip fire

Tula Technology's Dynamic Skip Fire (DSF) is an interesting extension of cylinder deactivation technology in which any of the cylinders for automobile engines are fired or skipped (deactivated) on a continuously variable basis. The engine control system commands the appropriate number and sequence of fired cylinders to deliver the instantaneous torque demanded from the

Figure 2.361 DSF Operation tracking engine torque demand. (Source: Tula, www.tulatech.com)

engine. This enables fully optimized combustion and reduced engine pumping losses, thereby increasing fuel efficiency.

Operating the engine in a dynamic skip fire manner alters the torque excitations on the vehicle powertrain, which could lead to unacceptable NVH characteristics. Tula's novel firing decision and control algorithms manage noise and vibration algorithmically to maintain a high-quality driving experience.

Definition

NVH: Noise vibration and harshness.

This interesting technology from Tula integrates advanced signal processing with sophisticated powertrain controls to create the ultimate variable displacement engine. The result is optimal fuel efficiency at the lowest cost.

2.8.6 Oil filter modules

The company Mahle has developed an oil filter module that is already installed in over a million vehicles. The unit includes an integrated oil/water heat exchanger. However, this does not require plumbing into the coolant as it is designed to connect via its baseplate,

Figure 2.362 Oil filter module. (Source: Mahle)

191

which is sealed with O rings, and held in place by four simple self-tapping screws.

The housing is a glass fibre polyamide and has a built-in pressure relief valve. There is also a drain screw on the bottom to prevent spills as the unit is changed.

Another interesting development in oil changing is a complete module that contains the oil and filter, and can be changed as one complete unit in just a few minutes. An electric oil pump is used to fill the unit with the old oil before it is removed, and then to return the new oil to the sump after replacement.

2.8.7 Miller cycle

A traditional reciprocating internal combustion engine uses four strokes (or the Otto cycle), of which two can be considered high-power: the compression stroke (high power flow from crankshaft to the charge) and power stroke (high power flow from the combustion gases to crankshaft).

In the Miller cycle, the intake valve is left open longer than it would be in a normal four-stroke engine. This effectively makes the compression stroke into two discrete cycles:

1 First part, when the intake valve is open
2 Second part, when the intake valve is closed.

> **Key fact**
>
> In the Miller cycle, the intake valve is left open longer than it would be in an Otto-cycle engine.

This two-stage intake stroke creates the so-called 'fifth' stroke. As the piston initially moves upwards (20–30%) in what is traditionally the compression stroke, the charge is partially expelled back out through the still-open intake valve. This loss of charge air would normally result in a loss of power. However, in the Miller cycle, this is compensated for by the use of a positive displacement supercharger (Roots or screw) as these work at low engine speed. The final 70–80% of the compression stroke (after the inlet valve has closed) is where the mixture is further compressed.

Efficiency is increased by having the same effective compression ratio but a larger expansion ratio. This allows more work to be extracted from the expanding gases as they are expanded to almost atmospheric pressure.

Audi's A4 was the first model to receive a new 2.0 litre engine that starts to bring spark-ignition technology much closer to the fuel consumption and torque of compression-ignition engines. It does so via technology that is mostly on the inlet side of the cylinder. It is described as being 'comparable to the Miller cycle'.

Focusing on the inlet side of the engine, intake timing has been greatly reduced by adopting a 140° crank angle compared to typical 190–200° timing. As mentioned above, Miller-cycle engines traditionally employ a Roots or screw-type positive-displacement supercharger, but Audi is using a turbo. The engine gets a higher boost pressure on the inlet side to provide optimal cylinder charges, despite shorter intake timing.

> **Key fact**
>
> Miller-cycle engines traditionally employ a Roots or screw-type positive-displacement supercharger.

To help achieve further performance targets, in part-load range there is an additional injection upstream from the intake valve. This results in an efficient mixture formation that is already complemented by the direct injection in the combustion chamber as well as in the intake manifold. The Audi Valvelift System ensures a short intake time at part-load and a longer at higher loads.

Other features of the Audi engine include controlled coolant flow control to reduce engine warm-up

Figure 2.363 Audi engine using features of the Miller cycle

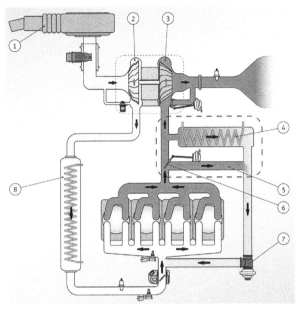

Figure 2.364 Cooled gas EGR block diagram. 1, Air filter; 2, Turbocharger compressor; 3, Turbocharger turbine; 4, EGR cooler; 5, Bypass duct; 6, Bypass flap (vacuum controlled in this example); 7, EGR valve; 8, Charge air cooler. (Source: Pierburg)

Figure 2.365 EGR cooler module with integrated EGR valve and bypass flap. (Source: Pierburg)

time. Another interesting feature is that the exhaust manifold is integrated in the cylinder head. Reduction of friction, as well as the use of low-friction engine oil (0W-20), also contributes to enhanced engine efficiency. Maximum torque of the engine is claimed to be 320 Nm (236 lb ft), available from 1450 rpm to 4400 rpm.

2.8.8 Spark Controlled Compression Ignition (SPCCI)

The SPCCI engine developed by Mazda (known as the Skyactiv X) combines the processes of compression and spark ignition to achieve a highly efficient gasoline engine. Using the spark plugs only in certain circumstances, such as a cold start, the technology makes more power, but produces fewer emissions and uses less fuel than a conventional petrol engine.

The engine switches between spark ignition and a compression ignition mode to achieve a more efficient combustion of the air–fuel mixture. By compressing a very dense mixture, it raises the temperature within the cylinder, causing the remaining fuel to be unignited after the conventional spark to ignite under pressure just like a diesel.

Key fact

Normal petrol engines use an air–fuel ratio of 14.7:1, the SPCCI engine uses a ratio of 40:1

To pressurize the air during SPCCI mode, there is a small, Roots-style supercharger, but the function is not to add power in the conventional sense. To further enhance the economy of the power plant, it is also available with Mazda is hybrid ranges. Mazda is the first manufacturer to bring this type of technology to mass market although other manufacturers have attempted to achieve improved efficiency using a petrol compression concept before them, such as Lotus in 2001. One of the reasons Mazda has enjoyed success with bringing the concept to mass production is that engine control processors are fast enough to monitor and control the pressure in each cylinder. As ambient conditions change the control system is able to actively adjust the spark ignition to compensate. In addition, electric variable valve timing uses stepper motors for faster valve timing and a water cooled EGR to help prevent premature combustion.

A potential problem for the SPCCI engine is that higher compression ratios increase the potential for knock, which is why they normally require a higher-octane fuel. It is important to check fuel compatibility for all vehicles using manufacturers' information, or the government fuel checker resource.

SPCCI is a type of Homogenous Charge Compression Ignition (HCCI), but strictly speaking an HCCI engine uses compression alone, so bringing in a spark plug to a compression engine might seem odd. However, the spark needed for the highly compressed fuel mix is very small, the extra pressure ensures all the mixture is ignited very quickly. Previous attempts at HCCI have used exhaust gas recirculation, igniting the fuel with the hot emissions gases, but this concept proved difficult for cold start conditions and under a full load.

Electrical systems

3.1 Electrical and electronic principles

3.1.1 Electrical fundamentals

To understand electricity properly we must start by finding out what it really is. This means we must think very small. The molecule is the smallest part of matter that can be recognized as that particular matter. Subdivision of the molecule results in atoms. The atom is a basic unit of matter and consists of a central nucleus made up of protons and neutrons. Around this nucleus electrons are in orbit, like planets around the sun. The neutron is a very small part of the nucleus. It has an equal positive and negative charge. It is therefore neutral and has no polarity. The proton is another small part of the nucleus. It is positively charged. As the neutron is neutral and the proton is positively charged, this means the nucleus of the atom is positively charged.

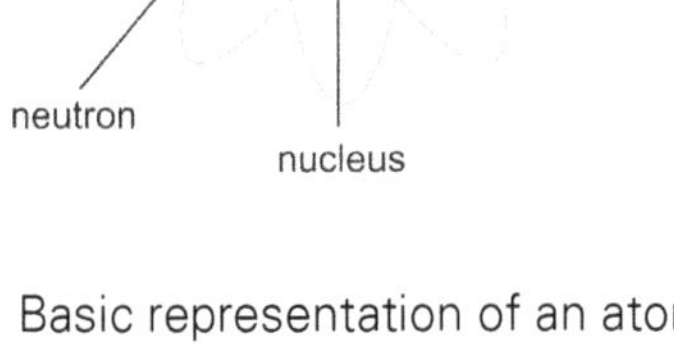

The electron is an even smaller part of the atom, and is negatively charged. It is held in orbit around the nucleus by the attraction of a positively charged proton. When atoms are in a balanced state the number of electrons orbiting the nucleus equals the number of protons. The atoms of some materials have electrons that are easily detached from the parent atom and join an adjacent atom. In so doing they move an electron (like polarities repel) from this atom to a third atom and so on through the material. These are called free electrons.

Materials are called conductors if the electrons can move easily. However, in some materials it is difficult to move the electrons. These materials are called insulators.

If an electrical pressure (voltage) is applied to a conductor, a directional movement of electrons will take place. There are two conditions for electrons to flow: a pressure source, e.g. from a battery or generator, and a complete conducting path for the electrons to move, e.g. wires.

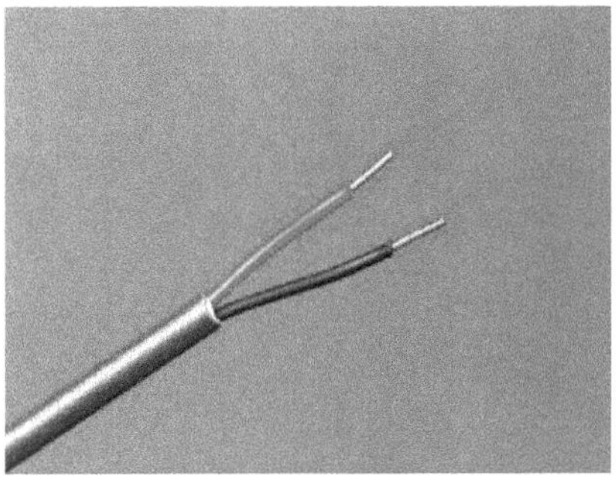

Figure 3.1 Basic representation of an atom

Figure 3.2 Insulated conductors

Automobile Mechanical and Electrical Systems. 978-0-415-72578-1 © 2018 Tom Denton.
Published by Taylor & Francis. All rights reserved.

DOI: 10.1201/9781003299066-3

An electron flow is termed an electric current. The battery positive terminal is connected, through a switch and lamp, to the battery negative terminal. With the switch open, the chemical energy of the battery will remove electrons from the positive terminal to the negative terminal via the battery. This leaves the positive terminal with fewer electrons and the negative terminal with a surplus of electrons. An electrical pressure exists between the battery terminals. With the switch closed, the surplus electrons on the negative terminal will flow through the lamp back to the electron-deficient positive terminal. The lamp will therefore light until the battery runs down.

Definition

Electron: A subatomic particle that carries a negative electric charge. It has a mass that is approximately 1/1836 that of the proton (not very big then!).

Figure 3.3 Heating effect

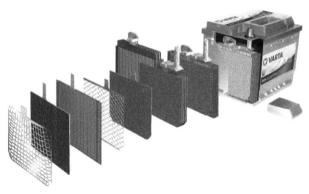

Figure 3.4 Chemical effect (Source: Varta Batteries)

The movement from negative to positive is called the electron flow. However, it was once thought that current flowed from positive to negative. This convention is still followed for practical purposes. Therefore, even though it is not correct, the most important point is that we all follow the same convention. We say that current flows from positive to negative.

Definition

Current flow: Conventional current flow is from positive to negative.

When a current flows in a circuit, it can produce only three effects: heat, chemical and magnetic. The heating effect is the basis of electrical components such as lights and heater plugs. The chemical effect is the basis for electroplating and battery charging. The magnetic effect is the basis of relays, motors and generators. The three effects are reversible. For example, electricity can make magnetism, and magnetism can be used to make electricity.

The number of electrons through a circuit in every second is the rate of flow. The cause of electron flow is the electrical pressure. A lamp, for example, produces an opposition to the rate of flow set up by

Figure 3.5 Magnetic effect

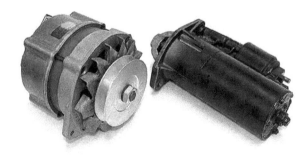

Figure 3.6 Reversibility: motor and generator

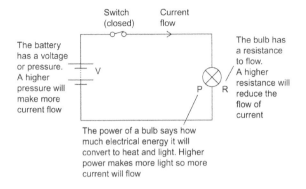

The battery has a voltage or pressure. A higher pressure will make more current flow

Switch (closed)

Current flow

The bulb has a resistance to flow. A higher resistance will reduce the flow of current

The power of a bulb says how much electrical energy it will convert to heat and light. Higher power makes more light so more current will flow

Figure 3.7 Basic circuit

the electrical pressure. Power is the rate of doing work or changing energy from one form to another. If the voltage applied to the circuit was increased but the lamp resistance stayed the same, then current would increase. If the voltage was kept constant but the lamp was changed for one with a higher resistance, then current would decrease.

Definition

Rate of flow: The number of electrons passing through a circuit in 1 second.

Any one value can be calculated if the other two are known. This relationship is stated in a law called Ohm's law.

Ohm's law states that in a closed circuit the current is proportional to the voltage and inversely proportional to the resistance. This means:

$$\text{Voltage} = \text{Current} \times \text{Resistance}$$
$$(V = IR) \text{ or } (R = V/I) \text{ or } (I = V/R)$$

When voltage causes current to flow, energy is converted. This is described as power. The unit of power is the watt. As with Ohm's law, any one value can be calculated if the other two are known.

$$\text{Power} = \text{Voltage} \times \text{Current}$$
$$(P = VI) \text{ or } (I = P/V) \text{ or } (V = P/I)$$

WWW

Check the book website for the Ohm's law and power 'magic' triangles.
www.automotive-technology.org

Electrical units are summarized in Table 3.1.

All metals are conductors. Silver, copper and aluminium are among the best and are frequently used. Liquids that will conduct an electric current are called electrolytes. Insulators are generally non-metallic and include rubber, porcelain, glass, plastic, cotton, silk, wax, paper and some liquids. Some materials can act as either insulators or conductors depending on conditions. These are called semiconductors. They are used to make transistors and diodes.

The amount of resistance offered by a conductor is determined by four factors:

▷ **Length** – The greater the length the greater the resistance.

Figure 3.8 Semiconductor transistors

Table 3.1 Electrical units summary

Name	Definition	Name	Symbol	Abbreviation
Electrical flow or current	The number of electrons past a fixed point in one second	Ampere	I	A
Electrical pressure	A pressure of 1 volt applied to a circuit will produce a current flow of 1 amp if the circuit resistance is 1 ohm	Volt	V	V
Electrical resistance	The opposition to current flow in a material or circuit when a voltage is applied across it	Ohm	R	Ω
Electrical power	When a voltage of 1 volt causes a current of 1 amp to flow the power developed is 1 watt	Watt	P	W

197

▷ **Cross-sectional area** – The larger the area the smaller the resistance.

▷ **The material** – The resistance offered by a conductor will vary according to the material from which it is made.

▷ **Temperature** – Most metals increase in resistance as temperature increases.

When resistors are connected so that there is only one path for the same current to flow through each resistor they are connected in series. In a series circuit:

▷ Current is the same in all parts of the circuit.

▷ Applied voltage equals the sum of the volt drops around the circuit.

▷ The total resistance of the circuit equals the sum of the individual resistance values.

When resistors are connected such that they provide more than one path for the current to flow in, and have the same voltage across each component, they are connected in parallel. In a parallel circuit:

▷ Voltage across all components of a parallel circuit is the same.

▷ Total current from the source is the sum of the current flowing in each branch. The current splits up depending on each component's resistance.

▷ The total resistance of the circuit is the sum of the reciprocal (one divided by the resistance) values.

Magnetism can be created by a permanent magnet or by an electromagnet. The space around a magnet in which the magnetic effect can be detected is called the magnetic field. Flux lines or lines of force represent the shape of magnetic fields in diagrams. Electromagnets are used in motors, relays and fuel injectors, to name just a few. Force on a current carrying conductor in a magnetic field results from two magnetic fields interacting. This is the basic principle of how a motor works.

> **Key fact**
> Magnetism can be created by a permanent magnet or by an electromagnet.

When a conductor cuts or is cut by magnetism a voltage is induced in the conductor. The direction of this voltage depends on the direction of the magnetic field and the direction in which the field moves relative to the conductor. The size is proportional to the rate at which the conductor cuts or is cut by the magnetism. This effect of induction,

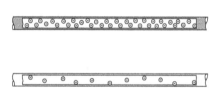

Figure 3.9 Conductor resistance

Figure 3.10 A simple series circuit

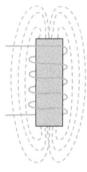

Figure 3.12 Electromagnet showing the field or lines of flux

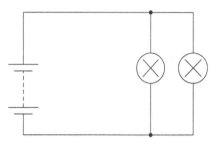

Figure 3.11 A simple parallel circuit

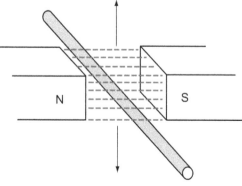

Figure 3.13 Induction

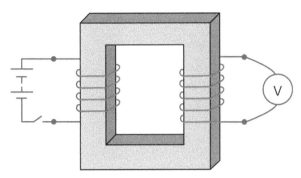

Figure 3.14 Transformer: as the switch is turned on and off the voltmeter needle will move

meaning that voltage is made in the wire, is the basic principle of how generators such as the alternator in a car work. A generator is a machine that converts mechanical energy into electrical energy.

If two coils, primary and secondary, are wound on to the same iron core, any change in magnetism of one coil will induce a voltage in the other. This happens when the primary current is switched on and off (Fig. 3.14). If the number of turns of wire on the secondary coil is more than on the primary, a higher voltage can be produced. This is called transformer action and is the principle of the ignition coil.

3.1.2 Electrical components and circuits

A switch is a simple device used to break a circuit; that is, it prevents the flow of current. A wide range of switches is used. Some switches are simple on/off devices such as an interior light switch on the door pillar. Other types of switch are more complex. They can contain several sets of contacts to control, for example, a vehicle's indicators, headlights and horn. These are described as multifunction switches.

A relay is a very simple device. It can be thought of as a remote-controlled switch. A very small electric current is used to magnetize a small winding. The magnetism then causes some contacts to close,

which in turn can control a much heavier current. This allows small, delicate switches to be used, to control large current users, such as the headlights or the heated rear window.

Some form of circuit protection is required to protect the electrical wiring of a vehicle and to protect the electrical and electronic components. It is now common practice to protect almost all electric circuits with a fuse. A fuse is the weak link in a circuit. If an overload of current occurs then the fuse will melt and disconnect the circuit before any serious damage is caused. Automobile fuses are available in three types: glass cartridge, ceramic and blade.

The blade type is now the most popular choice owing to its simple construction and reliability. Fuses are available in a number of rated values (Table 3.2). Only the fuse recommended by the manufacturer should be used.

A fuse is used to protect the device as well as the wiring. A good example of this is a fuse in a wiper motor circuit. If a value were used that was much too high then it would still protect against a severe short-circuit. However, if the wiper blades froze to the screen, a large value fuse might not protect the motor from overheating.

Figure 3.16 Simple 'cube' relay

Figure 3.15 Multifunction switch

Figure 3.17 Blade fuses

199

Table 3.2 Blade type fuse ratings

Continuous current (A)	Colour
3	Violet
4	Pink
5	Clear/beige
7.5	Brown
10	Red
15	Blue
20	Yellow
25	Neutral/white
30	Green

> **Key fact**
>
> A fuse is used to protect the device as well as the wiring.

Fusible links in the main output feeds from the battery protect against major short-circuits in the event of an accident or error in wiring connections. These links are simply heavy-duty fuses and are rated in values such as 50, 100 or 150 A.

Many types of terminal are available. These have developed from early bullet-type connectors into the high-quality waterproof systems now in use. A popular choice for many years was the spade terminal. This is still a standard choice for connection to relays, for example, but is now losing ground to the smaller blade terminals. Circular multipin connectors are used in many cases; the pins vary in size from 1 to 5 mm. With any type of multipin connector an offset slot or similar is used to prevent incorrect connection.

Protection against corrosion of the connector is provided in a number of ways. Earlier methods included

Figure 3.19 Wiring and connectors on a vehicle

applying suitable grease to the pins to repel water. It is now more usual to use rubber seals to protect the terminals, although a small amount of contact lubricant can still be used. Many multiconnectors use some kind of latch not only to prevent individual pins working loose but also to ensure that the complete plug and socket is held securely.

Cables or wires used for motor vehicle applications are usually copper strands insulated with polyvinyl chloride (PVC). Copper, besides its very low resistance, has ideal properties such as ductility and malleability. This makes it the natural choice for most electrical conductors. For the insulation, PVC is ideal. It not only has very high resistance, but also is very resistant to fuel, oil, water and other contaminants.

> **Key fact**
>
> High voltage cables are brightly coloured.

The choice of cable size depends on the current it will have to carry. The larger the cable used then the

Figure 3.18 Fusible links

better it will be able to carry the current and supply all of the available voltage. However, it must not be too large or the wiring becomes cumbersome and heavy. In general, the voltage supply to a component must not be less than 90% of the system supply. Cable is available in stock sizes, but a good rule-of-thumb guide is that one strand of 0.3 mm diameter wire will carry 0.5 A safely.

The selection of symbols shown in Figure 3.22 is intended as a guide to some of those in use on circuit diagrams. Many manufacturers use their own variation.

The idea of a symbol is to represent a component in a very simple but easily recognizable form.

The conventional type of diagram shows the electrical connections of a circuit but does not attempt to show

Figure 3.20 High voltage cables are easy to spot

Figure 3.21 Failed connecting plug

the various parts in any particular order or position (Fig. 3.23).

A layout circuit diagram attempts to show the main electrical components in a position similar to those on the actual vehicle (Fig. 3.24). Owing to the complex circuits and the number of individual wires, some manufacturers now use two diagrams, one to show electrical connections, and the other to show the actual layout of the wiring harness and components.

A terminal diagram shows only the connections of the devices and not any of the wiring (Fig. 3.25). The terminal of each device, which can be represented pictorially, is marked with a code. This code indicates the device terminal designation, the destination device code and its terminal designation, and in some cases the wire colour code.

The diagram in Figure 3.26 is laid out so as to show current flow from the top of the page to the bottom. These diagrams often have two supply lines at the top marked 30 (main battery positive supply) and 15 (ignition controlled supply). At the bottom of the diagram are the earth or chassis connections (marked 31).

Three descriptive terms are useful when discussing electric circuits:

▷ **Short-circuit** – A fault has caused a wire to touch another conductor and the current uses this as an easier way to complete the circuit (Fig. 3.27).
▷ **Open circuit** – The circuit is broken and no current can flow (Fig. 3.28).
▷ **High resistance** – A part of the circuit has developed a high resistance (such as a dirty connection), which will reduce the amount of current that can flow.

The complexity of modern wiring systems (Fig. 3.21) has increased dramatically in recent years. The number of separate wires required on a top of the range vehicle can be thousands. The wiring loom required to control all functions in or from the driver's door alone can require up to 50 wires. This is clearly becoming a problem as, apart from the obvious issue of size and weight, the number of connections and number of wires increase the possibility of faults developing.

A data bus or multiplex system solves many of these problems. The data bus and the power supply cables must 'visit' all areas of the vehicle electrical system. To illustrate the operation of this system, consider the events involved in switching the sidelights on and off. First, in response to the driver pressing the light switch, a unique signal is placed on the data bus. This signal is only recognized by special receivers built as

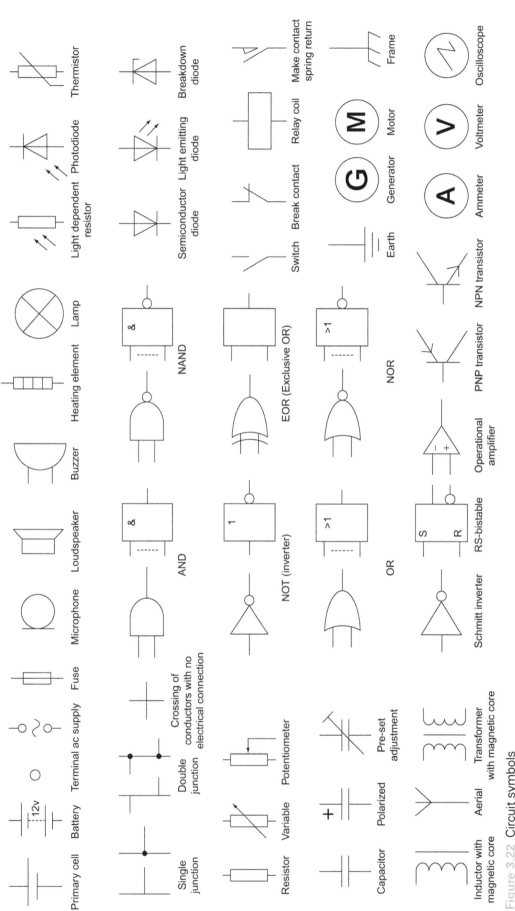

Figure 3.22 Circuit symbols

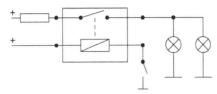

Figure 3.23 Conventional diagram with a resistor, a relay, two bulbs and a switch

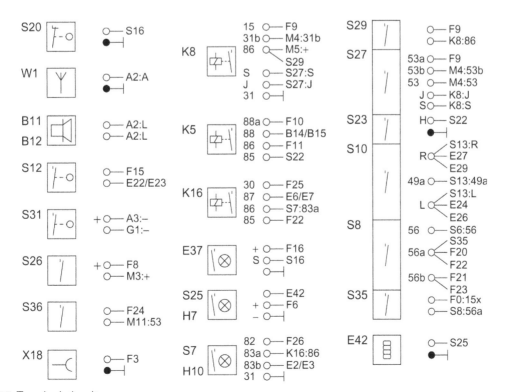

Figure 3.24 Layout circuit

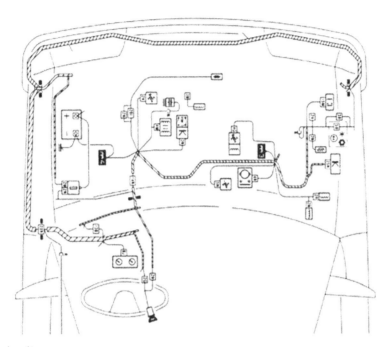

Figure 3.25 Terminal circuit

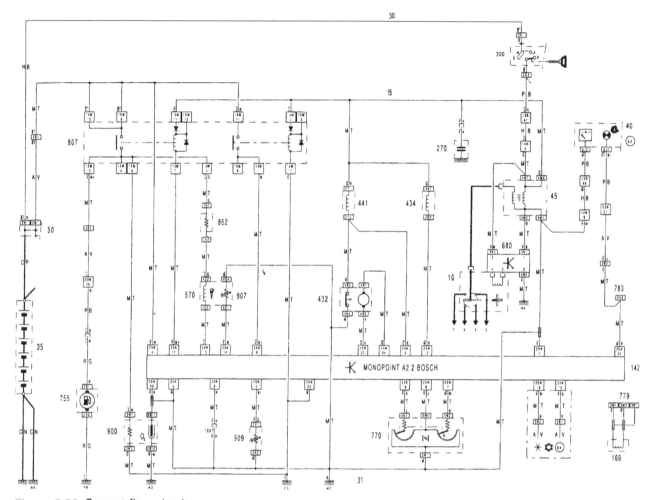

Figure 3.26 Current flow circuit

Figure 3.27 Short-circuit

Figure 3.28 Open circuit

Figure 3.29 Coming in after someone else's repairs can be problematic!

part of each light unit assembly, which in turn will make a connection between the power supply cable and the lights. The events are similar in turning off the lights, except that the code placed on the data bus will be different and will be recognized only by the appropriate receivers as an off code. A data bus can also carry information from sensors and allows electronic control units (ECUs) to communicate with each other.

Key fact

A data bus can carry information from sensors and allows ECUs to communicate with each other.

Bosch has developed a data bus protocol known as the controller area network (CAN) (Fig. 3.30). CAN is suitable for transmitting data in the area of driveline components, chassis components and mobile communications. It is a compact system, which will make it practical for use in many areas. Two variations on the physical layer are available which suit different transmission rates. One variation is for data transmission of between 100 kbits and 1 Mbits per second. It is used for rapid control devices. The other variation transmits data between 10 kbits and 100 kbits per second. It is used for switching and control operations.

The CAN message signal consists of a sequence of binary digits or bits. A high voltage present indicates the value 0, a low or no voltage indicates 1. The actual message can vary between 44 and 108 bits in length. This is made up of a start bit, name, control bits, the data itself, bits used for error detection, a confirmation signal and finally several stop bits.

Key fact

The CAN message signal consists of a sequence of binary digits or bits.

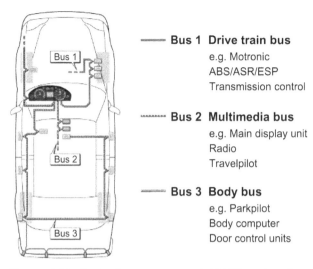

Bus 1 **Drive train bus**
e.g. Motronic
ABS/ASR/ESP
Transmission control

Bus 2 **Multimedia bus**
e.g. Main display unit
Radio
Travelpilot

Bus 3 **Body bus**
e.g. Parkpilot
Body computer
Door control units

Figure 3.30 Controller area network (CAN) system for instrumentation. (Source: Bosch Media)

A binary format message can be something like:

1000101010001010111110000111010111101010101010001
1111010111100110011000001111110101010100001111111111
000000001

Errors in a message are recognized by an error detection scheme in which all the bits in a block of data are divided by a predetermined binary number. A check character, known to the transmitter and receiver, is determined by the remainder. If an error is recognized the message on the bus is destroyed. This in turn is recognized by the transmitter, which then sends the message again. This technique, when combined with additional tests, makes it possible to discover all faulty messages.

Because each node in effect monitors its own output, interrupts disturbed transmissions and acknowledges correct transmissions faulty stations can be recognized and uncoupled (electronically) from the bus. This prevents other transmissions from being disturbed.

Key fact

A CAN message may vary between 44 and 108 bits in length.

A controller area network (CAN) signal is made up of voltage pulses that represent ones and zeros, in other words, binary signals. The data is applied to two wires known as CAN-high and CAN-low.

It is possible to verify that data is being continuously exchanged along the CAN bus. It is also possible to check that the peak-to-peak voltage levels are correct and that a signal is present on both CAN lines. CAN uses a differential signal, and the signal on one line should be a mirror image (the signals should line up) of the data on the other line. This means that interference is cancelled out.

Definition

DLC: Diagnostic/Data link connector.

CAN: Controller area network.

LIN: Local interconnect network.

The usual reasons for examining the CAN signals is where a CAN fault has been indicated by OBD, or to check the CAN connection to a suspected faulty CAN node. Manufacturers' data should be referred to for precise waveform parameters.

Key fact

CAN uses a differential signal so the signal on one line should be a mirror image of the data on the other line.

A local interconnect network (LIN) is a serial bus system especially suited for networking 'intelligent' devices, sensors and actuators within a subsystem. LIN enables the implementation of a hierarchical vehicle network.

The LIN standard includes the specification of the transmission protocol, the transmission medium, the interface between development tools, and the interfaces for software programming. LIN guarantees the interoperability of network nodes from the viewpoint of hardware and software, and predictable electromagnetic compatibility (EMC) behaviour.

Definition

EMC: Electromagnetic compatibility requirements stipulate that a device shall not cause interference within itself or in other devices, or be susceptible to interference from other devices.

LIN provides a cost-efficient bus communication where the bandwidth and versatility of CAN are not required. It is used for non-critical systems.

The FlexRay Communications System has a dual channel data rate of 10 Mbps for advanced in-vehicle control applications. The communications protocol is designed to provide high-speed deterministic distributed control for advanced automotive applications. FlexRay's dual-channel architecture offers system-wide redundancy that meets the reliability requirements of emerging safety systems, such as brake-by-wire. The FlexRay system can also

be employed as a vehicle-wide network backbone, working in conjunction with already well-established systems, such as CAN and LIN. It can drive down costs by reducing the number of parallel CAN networks that have been used to solve bandwidth bottlenecks.

MOST is a high-speed multimedia network technology. It can be used for applications inside or outside the vehicle. The serial MOST bus uses a daisy-chain topology or ring topology and synchronous data communication. Depending on the specific version, the signals are sent via plastic optical fibre or electrical conductors. These are known as the physical layers. MOST technology is used in almost every car brand worldwide. It is mostly used for audio, video, voice and data.

Network applications will soon need bit rates of 5 to 10 Mbps. CAN is limited to 1Mbps so new protocols are under development by Bosch and others. One answer instead of using event-triggered transmission is to trigger the communication at a precise time. Because of this, speeds close to 'real time' can be achieved. TT-CAN is an extension of the existing CAN protocol so the system will still be able to use event triggered communications. An additional advantage of TT-CAN is that missing messages are detected immediately, and there is much better protection against unauthorized bus access.

Byteflight is a protocol designed for use with safety-critical systems such as air bags and seat belt tensioners. It has a high fault tolerance and allows high-speed data. It is known as a flexible time division multiple access (FTDMA) protocol and because of the flexibility can also be used for body and convenience functions such as locking, windows and seat control.

Star topography is used but the protocol also works on linear systems. Communication is by plastic optic fibre (POF) and the data rate, depending on loads is 5 to 10 Mbps. One example of its use is the BMW 1 series where it links the airbag crash sensor satellite ECUs. The protocol was not designed for X-by-wire but it is high performance and has many of the required features.

Ethernet is a family of computer networking technologies commonly used in local area networks,

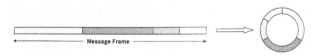

Figure 3.31 TT-CAN signal with the beginning and end joined to make a continuous cycle

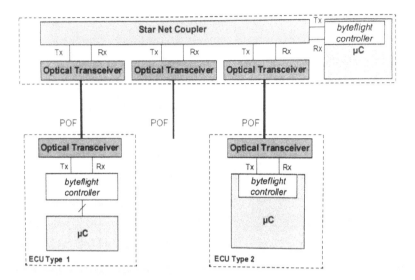

Figure 3.32 Byteflight uses a star topography as this ensures that a node failure cannot bring down the network

Figure 3.33 MOST fibre optics

Figure 3.34 Ethernet cables (Source: Delphi)

metropolitan area networks and wide area networks. It was commercially introduced in 1980 and first standardized in 1983 as IEEE 802.3. Automotive Ethernet does more than support mobile connectivity and complex, high-bandwidth automotive applications; it is a key enabler for the fully connected autonomous vehicle (CAV). NXP, for example, has designed a broad portfolio of robust, flexible and cost-effective automotive Ethernet products to connect vehicle systems faster and more efficiently.

In an automotive Ethernet set-up multiple vehicle systems simultaneously access high bandwidth over a single cable (a twisted pair). The Ethernet is the backbone of the vehicle network and supports higher levels of data processing and more communication types. Each port receives dedicated bandwidth, and the entire backbone is capable of IP connectivity. Instead of supporting individual high-bandwidth functions, the architecture supports all the high-bandwidth functions that reside on the same physical network but use logically separated virtual networks.

Automotive Ethernet will connect the vehicle's internal digital devices, connect the vehicle with other vehicles and even make the car a productive part of the IoT. It supports the high bandwidth and real-time processing required for today's infotainment and ADAS systems

and provides a platform for the development of truly autonomous driving.

Vehicle networks are usually a combination of several different protocols and systems. Figure 3.35 shows a typical example used on and Audi A8 (c2015). The networks on this vehicle are:

▷ CAN
▷ LIN
▷ MOST

The CAN, on this vehicle, is further divided into these separate networks:

▷ drivetrain CAN
▷ convenience CAN (comfort)
▷ diagnosis CAN
▷ dash panel insert CAN (info)
▷ adaptive cruise control CAN

A useful feature has been added to a number of scanners, which is the ability to show the CAN topology. Other scanners are available, but I used a TopDon Phoenix Pro to get the results shown in Figure 3.36. The vehicle was a Volkswagen Golf GTE (2018). Figure 3.37 shows the same thing but with the fault codes cleared. The colour coded topology display shows all vehicle

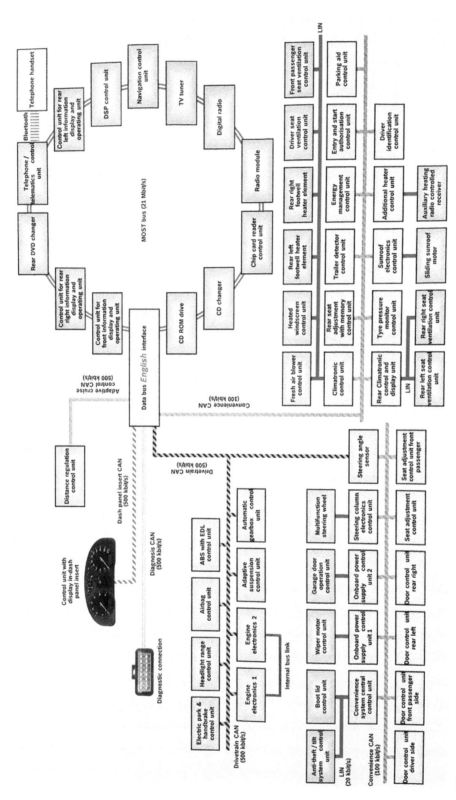

Figure 3.35 Vehicle network structure example (Source: Bosch media)

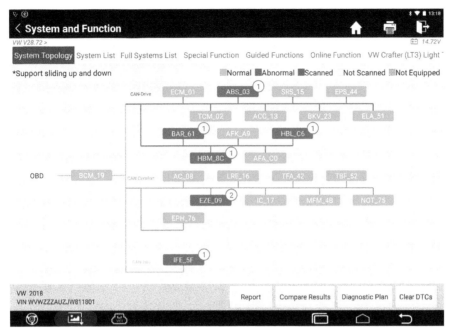

Figure 3.36 System topology showing possible DTCs on some of the systems (Source: www.diagnosticconnections.com)

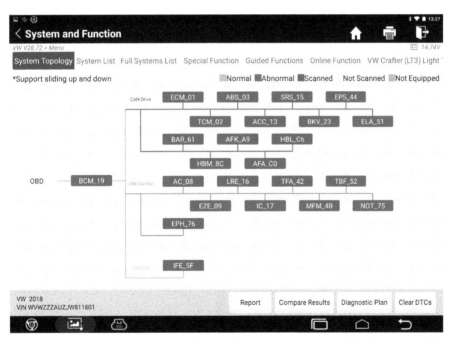

Figure 3.37 System topology showing no DTCs after the faults have been rectified or cleared (Source: www.diagnosticconnections.com)

systems and their status on one page. Note how it also shows the relationship between the different nodes and the different networks:

▷ CAN-Drive
▷ CAN-Comfort
▷ CAN-Info.

The display shows the structure of communication between modules, and you can then press on a system module to display any DTCs and access the system.

The benefit of this topology display is that as well as showing the relationship between different networks, you can also see which modules are on which network so that, for example, several errors on one network may indicate a common cause. It is also very convenient to tap on a node where an error is shown and open full details of that part.

As vehicle systems become ever more complex and interlinked, diagnostic tools are developing new ways to display network data. The 'topology view' is a very welcome and useful improvement.

3.1.3 **Electronic components**

Resistors are probably the most widely used component in electronic circuits (Fig. 3.38). Two factors must be considered when choosing a suitable resistor, namely the ohmic value and the power rating. Resistors are used to limit current flow and provide fixed voltage drops. Most resistors used in electronic circuits are made from small carbon rods; the size of the rod determines the resistance. A thermistor is a resistor that changes resistance with temperature.

A capacitor is a device for storing an electrical charge (Fig. 3.39). In its simple form, it consists of two plates separated by an insulating material. One plate can have excess electrons compared with the other. In vehicles, its main uses are for reducing arcing across contacts and for radio interference suppression circuits, as well as in ECUs.

Figure 3.40 DLC socket – pin 6 is CAN-high and pin 14 is CAN-low

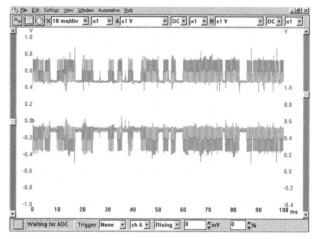

Figure 3.41 CAN high and low signals on a dual trace scope

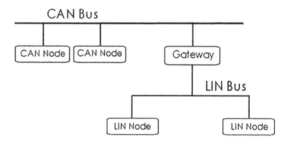

Figure 3.42 Structure using CAN and LIN

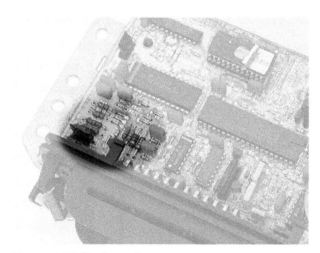

Figure 3.38 Resistors in use

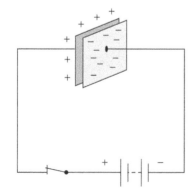

Figure 3.39 Capacitor operation

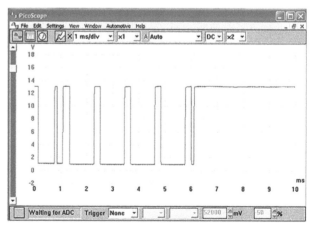

Figure 3.43 LIN waveform

Diodes can be described as one-way valves (Fig. 3.44). For most uses, this is a good description. The diode is made from two types of silicon (N type and P type). Electrons can flow from negative (N-type) to positive (P-type) material, but not the other way. Zener diodes are very similar in operation except that they are designed to conduct in the reverse direction at a preset voltage. They can be thought of as a type of pressure-relief valve.

Definition

Zener diode: A type of electrical pressure-relief valve.

Transistors are the devices that have allowed the development of today's complex and small electronic systems (Fig. 3.45). The transistor is used either as a switch or as an amplifier. Transistors are constructed from the same materials as diodes but with three terminals. A small voltage (about 0.7 V) supplied to the base terminal of a transistor known as an NPN will cause it to fully switch on, joining the collector and emitter. It is sometimes useful to think of a transistor as a type of relay. However, with a transistor, a smaller voltage will partially switch the collector–emitter circuit on and hence the component works as an amplifier.

Inductors are most often used as part of an oscillator or amplifier circuit. In these applications, it is essential for the inductor to be stable and of reasonable size. The basic construction of an inductor is a coil of wire wound on a former.

It is the magnetic effect of the changes in current flow that gives this device the properties of inductance. The inductor is also used as a filter because it tends to prevent changes in signals.

Figure 3.44 Diodes are one-way valves for electricity

Figure 3.45 Transistor in an electronic control unit

Figure 3.46 Electronic control unit. (Source: Bosch Media)

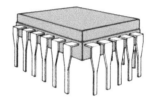

Figure 3.47 Integrated circuit (IC) package

Integrated circuits (ICs) are constructed on a single slice of silicon (Fig. 3.47). Combinations of thousands or even millions of the components mentioned previously can be combined to carry out various tasks. These tasks can range from a simple switching action to the operation of a microprocessor of a computer. The components required for these circuits can be made directly on to one slice of silicon. The advantage of this is not just the size of the ICs (which can be very small) but the speed at which they can be made to work.

3.2 Engine electrical

3.2.1 Batteries

A reliable supply of electric power is necessary for vehicles. Internal combustion engines require a large current to operate the starter motor and many other systems are electrically powered. Alternative fuel vehicles still require a 12 V supply in most cases to bring high voltage systems online. Modern cars use a 12 V system. Many vehicle batteries are of conventional design, using lead plates in a dilute sulphuric acid electrolyte. This feature leads to the common description of 'lead-acid' batteries. The output from a lead-acid battery is direct current (DC).

Electricity is not 'stored' in a battery, a rechargeable battery is an electrochemical unit that converts an

211

Figure 3.48 Automotive batteries (Source: Varta)

electric current into a modified chemical compound. This chemical reaction can be reversed to release an electric current. The modified chemical compound in the battery allows the change of state to create energy on demand and then use it for the items that require power that are connected to the battery's circuit.

The first function of a car battery is to use a large current to start the car. Many other electrical systems depend on the battery whilst the car is running, and for a modern vehicle – there can be numerous systems! Most vehicles depend on a 12 V battery, this includes high-voltage cars which (in most cases) use 12 V batteries to operate the relay that brings the high voltage system online. Without a stable and reliable battery, vehicles can suffer from many problems, so it is important to understand the different types and how they work.

Definition

Direct current (d.c.): Current that moves in a single direction in a steady flow.

Definition

Alternating current (a.c.): This is often used to indicate an alternating potential rather than a current, as in 240 V or 110 V a.c.

A vehicle 12 V battery is made up from six cells. Each lead-acid cell has a nominal voltage of 2.1 V, which gives a value of 12.6 V for a fully charged battery under no-load conditions. The six cells are connected in series, internally in the battery, with lead bars. The cells are formed in the battery case and are completely separate from each other. Each cell has a set of interleaved positive and negative plates kept apart by porous separators. The separators prevent contact of the plates, which would give an internal short-circuit and affect the chemical reaction in the battery cell. The cell plates are supported above the bottom of the case. This leaves a sediment trap below the plates so that any loose material that falls to the bottom does not cause a short-circuit between the plates.

Key fact

A lead-acid cell has a nominal voltage of 2.1 V, which gives a value of 12.6 V for a normal car battery.

The cell plates are formed in a lattice grid of lead–antimony or lead–calcium alloy (Fig. 3.49). The grid carries the active material and acts as the electrical conductor. The active materials are lead peroxide for the positive plate and spongy lead for the negative plate.

When a battery is in a charged state the positive plates of lead peroxide (PbO_2) are reddish brown in colour, and the negative plates of spongy lead (Pb) are grey in colour (Fig. 3.50). When the battery is discharging, a chemical reaction with the electrolyte changes both plates to lead sulphate ($PbSO_4$) (Fig. 3.51).

Applying an electric current to the battery reverses the process (Fig. 3.52). The charged battery stores chemical energy. This can be released as electrical energy when the battery is connected into a circuit.

The electrolyte is dilute sulphuric acid, which reacts with the cell plate material during charging and discharging of the battery. Sulphuric acid (H_2SO_4) consists of hydrogen, sulphur and oxygen. These

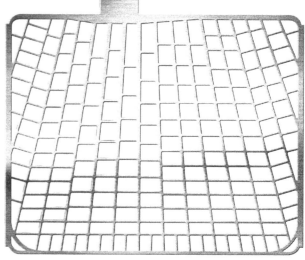

Figure 3.49 Power frame (Source: Varta)

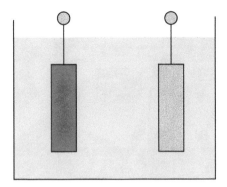

Figure 3.50 Charged state (see text)

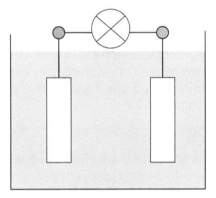

Figure 3.51 Discharged state (see text)

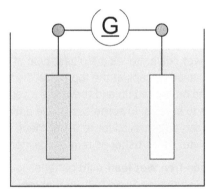

Figure 3.52 Chemical changes reverse when current is applied (see text)

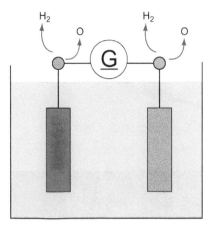

Figure 3.53 Discharge of hydrogen and oxygen (see text)

Near the fully charged state some hydrogen (H_2) and oxygen (O_2) may be lost as gas from the battery vent (Fig. 3.53). Some water (H_2O) can also be lost by vaporization in hot weather. With older batteries, this meant that the battery electrolyte needed regular inspection and topping up.

Only water is lost from the battery and therefore only water should be used for topping up. Any contaminants will affect the chemical reactions in the battery and, therefore, the performance. Only distilled or specially produced topping-up water should be used. Tap water is not suitable for topping up a battery. Acid should never be used, as this would strengthen the acid solution and alter the chemical reactions. Most modern batteries, however, include vapour traps and other features to minimize water loss and therefore need little or no attention.

The electrolyte chemical composition changes with the state of charge. It is possible to measure this change using a hydrometer. Sulphuric acid is denser and provides greater buoyancy than water. This property is called specific gravity or relative density. Water, which is used as the base for measurement of all liquids, is given a value of 1 for readings at 15°C (60°F).

The dilute sulphuric acid of the electrolyte of a fully charged battery cell has a reading of 1.280. The reading for a half-charged battery cell is 1.200 and for a fully discharged battery is 1.150. A reading below 1.140 may indicate a cell that can no longer be recharged. It is common to write these values with three decimal places but to just say the significant digits (e.g. twelve eighty).

A hydrometer consists of a calibrated float in a glass cylinder (Fig. 3.54). A bulb on the top of the cylinder is depressed so that it acts as a vacuum pump when it is released. A small rubber tube is attached to the bottom

chemicals separate during the charge and discharge process and attach to the cell plate active material or return to the electrolyte.

During discharge, the sulphate (SO_4) combines with the lead to form lead sulphate ($PbSO_4$). The oxygen in the positive plate is released to the electrolyte and combines with the hydrogen that is left, to form water (H_2O). During charging, the reverse process occurs with the sulphate (SO_4), leaving the cell plates to reform with the hydrogen in the electrolyte to produce sulphuric acid (H_2SO_4). Oxygen in the electrolyte is released to reform with the positive cell plate material as lead peroxide (PbO_2).

213

Figure 3.54 Construction of a hydrometer

Figure 3.55 Battery testing

of the cylinder and is inserted into the electrolyte in the battery cell. A sample of the electrolyte can, therefore, be drawn into the cylinder.

The sample of electrolyte in the hydrometer lifts the float in proportion to the buoyancy of the liquid. The higher it floats, the greater the relative density. Calibrated marks on the float align with the top of the liquid to give the actual reading. This is compared with standard data and all cells are compared with each other to check the general condition of the battery. There should be very little difference between the cells.

Key fact

The relative density reading for a fully charged battery cell is 1.280, for a half-charged cell 1.200 and for a fully discharged cell 1.150.

Figure 3.56 Computerized battery testing (Source: Varta)

However, as most batteries are sealed the voltage of the battery can be used to indicate battery condition. A fully charged battery will give a reading of 12.6 V and a discharged battery 12.0 V.

Key fact

A fully charged battery will give a reading of 12.6 V and a discharged battery 12.0 V (when not under load).

3.2.2 The main types of 12 V car batteries

Lead-acid wet cell batteries are oldest type of car battery, identified by caps at the top that allow the electrolyte to be topped up and they have a vent to allow gases to escape. In some cases, the entire top of the battery can be removed for refurbishment. They can spill electrolyte. These batteries require maintenance.

Maintenance-free wet lead-acid batteries look remarkably like the lead-acid battery that requires maintenance but are more modern in their design. They do not have caps as they do not require topping up. They have a liquid electrolyte and need to vent, in the 1970s these were given the misnomer of 'sealed', but this is not accurate due to the control venting devices needed to prevent the battery from exploding during stressful charge or rapid discharge.

Calcium batteries are lead-acid batteries with a calcium component added to both the positive and negative plates. This helps to reduce water loss through gassing and allows the battery to be fully sealed and maintenance free. Rather than being different to the lead-acid battery, calcium batteries are the next stage in lead-acid battery evolution. Calcium reduces self-discharge, but the positive lead calcium plate has the side effect of growing when over-charged.

Enhanced Flooded Battery (EFB) is also known as Advanced Flooded Battery (AFB) and Enhanced Cycling Mat (ECM). EFB batteries are another development of lead-acid batteries designed to provide a cost-effective solution for more demanding use, such as for stop-start technology or cars that perform frequent starts such as taxis.

Sealed lead-acid batteries use a gel instead of a liquid electrolyte. The battery is completely sealed and does not have a vent. This battery does not 'slosh' when given a shake like other lead-acid batteries

Absorbed Glass Mat (AGM) batteries can look very similar to a gel-filled sealed lead-acid battery, but inside they are very different. AGM batteries use a fine network of glass fibres that create a mesh inside the battery.

AGM are higher performance than the common wet or flooded type. Instead of submerging the plates into liquid electrolyte, the electrolyte is absorbed in a mat of fine glass fibres. This makes the battery spill-proof and therefore can be transported without hazardous material restrictions. The plates can be made flat like the standard flooded lead-acid and placed in a rectangular case or wound into cylinders.

AGM batteries have a very low internal resistance so they can deliver high currents and offer long service even if occasionally deep-cycled. These batteries are lower weight and more reliable than the conventional lead-acid type. They also perform at high and low temperatures and have a low self-discharge. Other advantages over regular lead-acid are a better specific power rating (high load current) and faster charge times (up to five times faster). The disadvantages are that they have a slightly lower specific energy (capacity) and higher manufacturing costs.

AGM batteries are found in high-end vehicles to run power-hungry accessories such as heated seats, steering wheels, mirrors and wipers. NASCAR and other auto racing leagues choose AGM products because they are vibration resistant. Start-stop batteries are almost exclusively AGM because the conventional types do not respond well to repeated micro cycling.

All sealed units such as gel and AGM batteries are sensitive to overcharging. They can be charged to 2.40 V/cell (and higher) without problem. However, the float charge, used to keep the battery at 100%, should be reduced to between 2.25 and 2.30 V/cell. Vehicle charging systems for conventional lead-acid batteries often have a fixed voltage setting of 14.40 V (2.40 V/cell). A direct replacement with a sealed unit could therefore cause the battery to be overcharged. As always, the correct battery is the one specified by the vehicle or battery manufacturer.

AGM and gel electrolyte batteries do not like heat. Manufacturers recommend stopping charge if the battery core reaches 49°C. Many batteries on modern vehicles now include a temperature and current sensor of some sort that is linked to a smart charging system.

Almost all vehicles are now fitted with maintenance-free batteries, and a hydrometer cannot be used to find the state of charge. This can, however, be determined from the voltage of the battery, as given in the table. An accurate voltmeter is required for this test.

Battery Volts at 20°C	State of Charge
12.0 V	Discharged (20% or less)
12.3 V	Half charged (50%)
12.7 V	Charged (100%)

Traditionally batteries were tested with a voltmeter and a high-rate discharge tester. This monitored the battery voltage while causing a current discharge of 100s of amps. Some batteries exploded during the process! The device (Fig. 3.54) will test 12 V batteries rated from 100 to 2000 CCA. It uses advanced conductance measurement technology to show battery health status including voltage, cranking power, state of charge and more. It displays one of five easy-to-understand results that can be printed out for the customer:

▷ Good battery
▷ Good battery, need recharge
▷ Replace, life cycle
▷ Replace, bad cell
▷ Charge, retest.

Built in reverse-connection protection prevents damage to the battery or tester. It can be used for batteries on or off the vehicle.

Measurement of battery condition by heavy-duty discharge should only be carried out on a fully charged battery. Heavy-duty discharge is carried out with a specialist item of equipment. The test consists of connecting a low-value resistor in circuit with the battery for a few seconds and measuring the voltage drop. This test applies a load similar to that of the starter circuit. In general, the voltage should not fall below 10 V. Figure 3.56 shows test equipment for this purpose; as always with this type of thing, refer to the manufacturer's instructions before use.

Battery performance can be determined by its ampere-hour (Ah) rating, but this is also affected by the ambient temperature. A cold battery will have a poor performance in comparison to a warm battery. A more useful measure of battery performance is its 'reserve capacity' rating. This refers to the time taken in minutes for a battery to discharge to a cell voltage of 1.75 V, when supplying a constant current of 25 A. This test reflects a typical current draw for a vehicle that would be needed if the charging system failed during night driving. The reserve capacity for a 40 Ah battery will be approximately 60 minutes.

Safety first

A discharge test applies a load similar to that of the starter circuit, which results in a high current and heat.

Another measure is the 'cold cranking amps' rating, which is a measure of the maximum current that can be supplied for a period of 30 seconds before the battery voltage falls below 7.2 V. This test is carried out at minus 18°C (0°F) so that it represents the most severe conditions of cold engine starting. The cold-cranking amps (CCA) rating of a battery is an important measure in regions that suffer very cold winter temperatures.

Some maintenance-free batteries incorporate a built-in hydrometer to indicate the state of charge and condition of the battery. The hydrometer is colour coded (Fig. 3.57). A green colour indicates that the battery is charged and serviceable. A green–black or black colour indicates that the battery requires recharging. A yellow colour indicates that the battery is faulty. Where a yellow hydrometer is showing, the battery should not be recharged or tested, and the use of jump leads for starting should not be carried out. A new battery should be fitted and the alternator checked for correct operation.

Battery charging can be described as slow or fast. Slow charging is best but in an emergency a fast charge is acceptable. When recharging, a battery should ideally be disconnected from the vehicle electrical systems.

As batteries have become more sophisticated to meet the demands of modern vehicles, so has the equipment to look after them. Computerized testing,

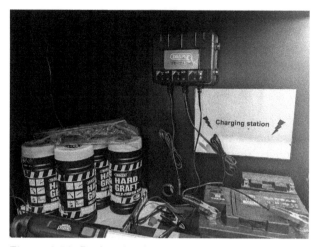

Figure 3.58 Recharging batteries on a bench

Figure 3.59 Compact start pack

clever start packs and battery support devices are now part of the modern workshop in addition to traditional bench-top chargers.

The bench charger should be situated in a well-ventilated area of the workshop. Smoking should also be prohibited, if it is not already. Bench chargers with voltage control have high initial current outputs, which fall as the battery charges. Chargers with current control can be adjusted to suit individual batteries.

The usual rule for battery charging is that the charge current should be set to about one-tenth of the ampere-hour (Ah) rating of the battery. Alternatively, about one-sixteenth of the reserve capacity, or one-fortieth of the CCA figure gives a good guide. A fully discharged battery will take about 12 hours to fully recharge. When recharging partially charged batteries, it is recommended that they should be checked at regular intervals. Always switch off the charger before disconnecting the leads and carrying out any tests.

Figure 3.57 Maintenance-free battery hydrometer colours

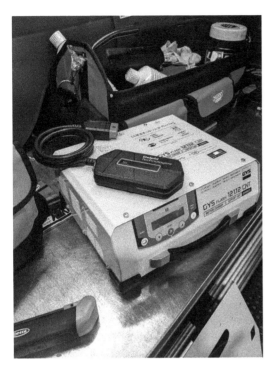

Figure 3.60 Battery support unit

Figure 3.61 AGM Battery (Source: Varta)

> **Key fact**
>
> Ideal charge rates: one-tenth of the ampere-hour rating of the battery, one-sixteenth of the reserve capacity, or one-fortieth of the cold cranking amp rating.

> **Key fact**
>
> Battery cables must have sufficient cross-sectional area to carry the starter motor and electrical systems current.

Battery cables must have sufficient cross-sectional area to carry the starter motor and electrical systems current. The feed to the starter motor is a heavy-duty

Figure 3.62 Correct polarity is important: red cables are used for the positive connection

insulated cable, and the earth or ground cable is of similar construction.

There are two ways of charging batteries in the workshop; one is a slow or trickle charge and the other a fast charge. These require two different types of charger. Most batteries can be fast charged, but this should only be carried out infrequently. If a high charge is used this can cause some deterioration of the battery's active materials.

A slow charger or bench charger uses mains electricity. Inside is fitted a transformer to reduce the voltage to 6, 12 or 24 V, to suit the battery or batteries on charge. Also fitted is a rectifier to change the a.c. volts of the mains supply to the d.c. volts needed for charging batteries. The charger is connected to the battery terminals with the correct polarity (Fig. 3.62). After setting the control switches, the charger is then turned on at the main switch.

There are several different types of charger and these should be used in accordance with the manufacturers' instructions. The recommended charge rate for a battery is one-tenth of the ampere-hour capacity. A 40 Ah battery should be charged at 4 A. If the ampere-hour capacity is not known, set the rate to one-sixteenth of the reserve capacity. Where the charge current can be adjusted, this facility should be used to set the rate.

> **Key fact**
>
> **Smart charger:** A device which will revive, charge, condition and maintain lead-acid batteries. It determines a battery's charge and then sets the appropriate rate. Intended for unsupervised use they are spark proof and reverse-polarity protected. The clean voltage and current supplied means they are safe to use on a battery which is still connected to the vehicle. The charger can be left attached to a battery for an indefinite time without risk of damage.

Figure 3.63 Older style fast charger with time clock

It is important that the charger is switched off before it is disconnected from the battery. For further safety, leave the batteries for about 5 minutes before the charger leads are disconnected. This will allow any flammable gas to dissipate to the atmosphere.

An older-style fast charger can be connected to a battery on a vehicle, to give a quick boost when a battery has a low charge. Some of these chargers have an engine start facility. Always follow the equipment manufacturer's instructions when using this type of charger. Some batteries are not suitable for fast charging; therefore, always refer to the vehicle or battery manufacturer's data for recommendations.

Safety first

Always familiarize yourself with battery equipment. If in doubt, ask for help!

3.2.3 Starting system

The engine starting system consists of a heavy-duty motor, with a drive pinion that engages with a gear on

Figure 3.64 Pre-engaged starter

Figure 3.65 Inertia motors with a solenoid switch were used on older cars but have now been replaced by the pre-engaged drive

the engine flywheel, and an electrical control circuit to operate the motor (Figs 3.64 and 3.65).

The starter motor power output has to be able to crank a cold engine at sufficient speed to start it. A 2 litre petrol engine will have a starter motor of about 1 kW, which will spin the engine at about 150 rpm. A similarly sized diesel engine will require double the power and possibly twice the cranking speed to start.

Key fact

The main components of the starter motor are the magnetic fields, armature, drive pinion and solenoid.

The main components of the starter motor are the magnetic fields, armature, drive pinion and solenoid (Fig. 3.66). The circuit consists of a battery supply, earth cables and the starter switch.

The starter motor is a direct current (d.c.) electromagnetic unit that usually has two pairs of magnetic pole shoes arranged at opposite positions inside the motor

Figure 3.66 Starter internal components

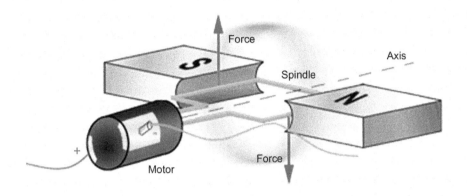

Figure 3.67 Direct current motor operating principle

casing. The casing acts as the yoke for the magnetic poles. The magnetic pole shoes can be strong permanent magnets or electromagnets using a winding.

The armature, which consists of a series of wire conductor loops wound around a laminated iron core, is mounted on the motor spindle. The conductor loops are terminated into segments of a commutator. Carbon or composite brushes conduct the motor electrical supply through the commutator segments to the individual conductor loops. The construction of a simple d.c. motor is shown in Figure 3.67. The magnetic force between the poles is from north to south. A loop conductor inside the magnetic field is provided with a d.c. electrical supply through a split slip ring, which forms a simple commutator.

When an electrical current is passed through a conductor a magnetic field is formed around that conductor. The magnetic field direction depends on the direction of the current flow. When the conductor is placed inside a fixed magnet, the magnetic field distorts to produce a repelling magnetic force, which pushes on the conductor.

In practice, it requires a large series of loop conductors to provide a motor with continuous rotation and good

torque characteristics. In order to supply each loop of the winding, when it is in alignment with the field magnets, and to maintain the current in the proper direction, a commutator is fitted. Current is passed to the commutator segments through spring-loaded brushes held in position by brush holders on the motor end plate (Fig. 3.68).

A starter motor requires strong magnetic forces to produce the speed and torque to crank an engine at sufficient speed for starting. For this, the armature is made with soft iron cores to make strong electromagnets, which are able to change polarity with the direction of current flow in the loop conductors. Laminations of soft iron are used for the cores to reduce magnetization losses. They are insulated from each other and assembled as a single unit on the armature.

The magnetic strength of the field magnetic poles is usually determined by using an electrical winding around the pole shoe. The wire coil is wound around one pole shoe and then the other in the opposite direction, so that the opposing field poles are produced opposite each other in the casing.

Figure 3.68 A series of windings on the armature gives a continuous rotary motion in conjunction with the windings on the fields, which produce strong magnetism

Figure 3.69 Starter in position to mesh with a ring gear on the engine flywheel

219

The drive from the motor is taken from a pinion gear on the spindle to the large diameter starter ring gear on the engine. The starter ring gear is fitted to the outside of the flywheel on manual transmission vehicles, or the torque converter drive plate on automatic transmission vehicles. The pinion meshes with the ring gear only during starting and is made to slide axially on or with the spindle to engage the drive when operated (Fig. 3.69).

> **Key fact**
>
> A pre-engaged starter motor moves the drive pinion into mesh by the action of a solenoid.

In a pre-engaged starter motor, the drive pinion is brought into mesh by the action of an electromagnetic solenoid mounted on the starter motor casing (Fig. 3.70). The solenoid has a soft iron plunger, which is drawn into the magnetic field that is produced inside the solenoid when an electrical current is passed through the solenoid windings. Connected to the plunger is a lever, which is pivoted so that, as one end is pulled into the solenoid, the opposite end pushes the pinion into mesh with the starter ring gear. The pinion is mounted on a unidirectional clutch, which is fitted to a sleeve with an internal spline to take the drive from the starter spindle. On the outside of the sleeve is a radial groove to take the fork of the engagement lever.

At the other end of the solenoid are the electrical contacts that form the switch to pass the electrical current to the motor. The solenoid on many pre-engaged starter motors has two windings. These are the 'closing' and 'holding' windings.

The closing winding or pull-in coil operates as soon as the solenoid is energized. This winding has an earth, or ground, return through the motor windings. This passes a current into the motor so that it rotates slowly during the engagement phase. Once the switch contacts are fully engaged, the holding winding holds the switch in place. The closing winding does not conduct once the motor current has been switched on.

Figure 3.71 One-way clutch behind the drive pinion

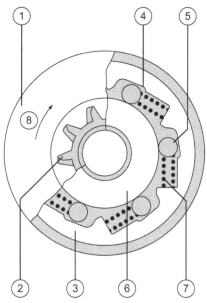

Figure 3.72 Clutch operation: 1, casing; 2, pinion; 3, clutch shell; 4, roller race; 5, roller; 6, pinion shaft; 7, spring; 8, direction of rotation

A holding coil is wound around the solenoid. This creates the magnetic field required to hold the solenoid in the engaged position during starting. When the starter switch is released, a spring returns the solenoid plunger to its 'off' position. If the engine were to start under these conditions, it would drive the motor spindle at an excessive speed. To prevent this occurring in pre-engaged drive starter motors, a unidirectional overrun clutch is fitted on the pinion (Fig. 3.71). This allows the motor to drive the engine but stops the engine driving the motor. A roller-type overrun clutch is a popular method, although a few other types are used. These clutch units are sealed for life and require replacement if they fail in service.

Clutch operation is summarized in Figure 3.72. On early inertia-type starter motors, a spiral or helical sleeve carried the pinion, which slid into mesh because of the forward drive from the motor spindle, and out of mesh by the engine spinning. A spring

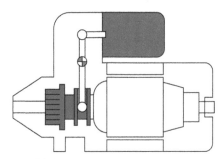

Figure 3.70 Pinion movement

Figure 3.73 Heavy-duty starter motor. (Source: Bosch Media)

Figure 3.74 Intermediate transmission gears

inside the pinion barrel held the gears out of mesh when the starter was not in operation.

Most small, modern starter motors use permanent magnets for the field poles. These motors have high-speed and low-torque characteristics and are suitable, without additional gearing, for engines up to about 2 litres. A heavy-duty starter motor is shown in Figure 3.73.

On some permanent magnet starter motors, for light diesel engines and petrol/gasoline engines up to 5 litres, an intermediate planetary gear set between the motor and drive pinion may be used (Fig. 3.74). This intermediate gearing modifies the speed and torque characteristics of the motor and makes it possible to construct starter motors that can be 40% lower in weight. The starter electrical current is passed through the armature only. On a planetary gear motor, the spindle is fitted with a sun wheel, and the motor casing with the annulus. The output to the drive pinion is made from the planetary gear carrier.

Key fact

Small, modern starter motors tend to use permanent magnet field poles.

Key fact

On a planetary gear starter motor, the armature drives the sun wheel, the motor casing holds the annulus and the output to the drive pinion is from the planetary gear carrier.

Starter motor control circuits use a heavy-duty electrical relay, called a solenoid, to switch the large starter current to the motor. The solenoid is an electromagnetic switch and, on modern pre-engaged starter motors, is attached to the top of the motor, where it performs the switching function, and is also used to slide the motor drive pinion into mesh with the starter ring gear on the engine flywheel.

A basic starter circuit is shown in Figure 3.75. The main components are the battery, starter switch, which is usually part of the ignition switch, the solenoid and motor, connecting cables and the earth, or ground, return circuit. The battery and starter cables are of heavy-duty construction to carry a large current to the motor. The control cables are standard low-current cable sizes. If any of these cables have to be replaced, cables of the same size, or as specified by the vehicle manufacturer, should always be used.

Starter motor circuits may have additional automatic switching to prevent the engine being started in particular situations. Automatic transmission systems incorporate an inhibitor switch on the gear selector, which allows engine starting in the park and neutral positions only. This prevents the engine being started with the transmission in gear, which could result in the vehicle pulling away unexpectedly. The inhibitor switch must be carefully checked and adjusted so that there is no risk of incorrect operation.

In the Ford diagram shown in Figure 3.76, the powertrain control module (PCM) allows the engine to start only when the passive anti-theft system (PATS) reads a key that transmits a valid code. In a key-free

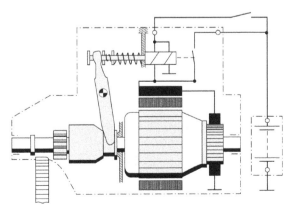

Figure 3.75 Starter circuit

221

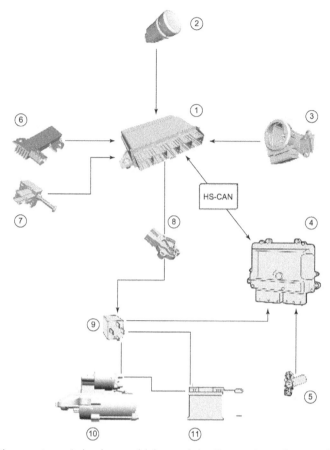

Figure 3.76 Keyless starting system: 1, keyless vehicle module; 2, start/stop button; 3, electronic steering lock; 4, powertrain control module; 5, crank sensor; 6, keyless vehicle antenna; 7, vehicles with manual transmission: clutch pedal position switch/vehicles with automatic transmission: stoplamp switch; 8, transmission range sensor; 9, starter relay; 10, starter motor, 11, battery. (Source: Ford Motor Company)

vehicle, the passive key is recognized by the key-free module and if the key is valid the permission to start is issued directly. In vehicles with a manual transmission it is necessary to depress the clutch pedal; on those with automatic transmission the brake pedal must be pressed. In a key-free system the key-free module switches on the control voltage for the starter relay.

The PCM switches the ground in the control circuit of the starter relay, which then connects power through to the starter solenoid. As soon as the speed of the engine has reached 750 rpm or the maximum permitted start time of 30 seconds has been exceeded, the PCM switches off the starter relay and, therefore, the starter motor. This protects the starter. If the engine does not turn or turns only slowly, the starting process is aborted by the PCM.

> **Key fact**
>
> In a key-free system the key-free module switches on the control voltage for the starter relay.

An electronically controlled starting system differs from a normal system in that the driver is not in direct control. The driver sends a 'start' request to an ECU which then in turn can carry out several different checks. For example:

▷ Is the driver authorized to start the engine?
▷ Is the engine stationary?
▷ Is battery state of charge adequate at the current temperature?
▷ Is the clutch disengaged or is the automatic transmission in neutral?

The checks only take a fraction of a second after which the ECU initiates the start. Once a set engine speed has been reached the starter is switched off. This always ensures the shortest possible start time and reduces starter wear. A similar process can be used for start-stop operation.

The benefit of this system is 5% or more reduction in CO_2 and fuel consumption in urban traffic. Almost all

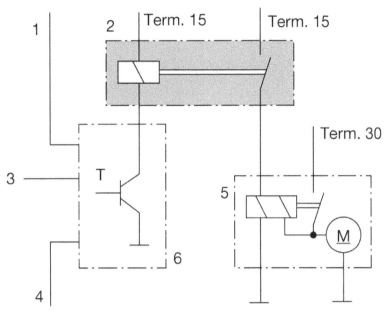

Figure 3.77 Automatic starting circuit: 1, start signal from driver; 2, relay; 3, other input signals; 4, park/neutral or clutch signal; 5, starter; 6, ECU (terminal 15 is ignition live and 30 is battery live)

European automakers are now integrating stop-start technology into compact cars, premium sedans, light trucks and even in powerful sports cars. To ensure reliable operation a high degree of control of other systems is necessary. For example, electrical energy management that incorporates battery charge detection. Engine systems must also be optimized to ensure a fast start, and finally, the starter and ring gear components must be enhanced to not wear out prematurely.

Stop-start systems, which largely use existing components such as the starter motor and engine control unit, can be quickly and cost-effectively adapted to different engines and vehicles. Therefore, this starter motor-based method has now become very popular. Efficient generators that charge the battery as fast as possible are an ideal addition to these systems since they allow the function to be used more often.

In vehicles fitted with automatic transmission, the stop-start function is very easy to use. The driver presses the brake pedal, and as soon as the car has stopped, the engine stops automatically. It starts again when the brake pedal is released. This is very convenient for the driver, who uses the accelerator and brake as normal. The system works completely in the background and evaluates a wide array of sensor information before being activated. For instance, a battery sensor determines the battery's level of charge, and the engine will only be switched off if a quick restart is guaranteed. Another example is the vehicle's cabin temperature. If the cabin temperature is too cold or too hot, the engine will continue to run until the occupants' desired temperature is reached.

To further reduce consumption, and CO_2 emissions, engineers are continually working to extend engine shutdown times. This will apply initially to the time when the vehicle is coasting to a stop and then it will also apply to periods while driving, when the driver is not accelerating. The effectiveness of all safety and comfort functions will be guaranteed, even during these longer periods with the engine switched off. The enhanced stop-start systems could save up to an additional 10% of fuel.

3.2.4 Charging system

The electrical generator on modern vehicles is an alternator (Fig. 3.79). Older vehicles used a dynamo, which gives a direct current without the need for a rectifier.

There are two main parts of an alternator: the rotor and the stator (Fig. 3.80). Together, they produce an a.c. voltage output. An electric current is induced

Figure 3.78 Starter motor

223

Figure 3.79 Alternator

Figure 3.80 Stator and rotor

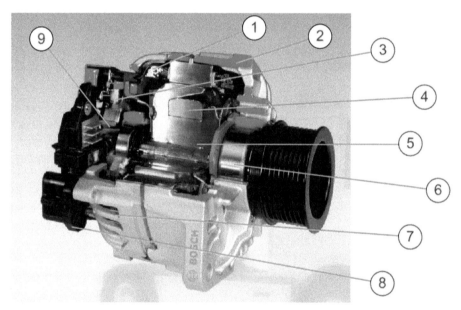

Figure 3.81 Cutaway alternator: 1, stator winding; 2, case; 3, rectifi er diodes; 4, rotor (field) winding; 5, claw pole rotor; 6, drive end bearing; 7, ventilation slots; 8, regulator, brush box and slip rings; 9, slip ring end bearing. (Source: Bosch Media)

or generated in the stator by the magnetic fields produced in the rotor.

Key fact

The main parts of an alternator are the rotor and stator.

A rectifier changes the a.c. voltage to a d.c. voltage, because that is what is needed for battery charging. Diodes in a bridge formation are used to route the electric current in such a way as to convert the a.c. voltage to a d.c. voltage.

A voltage regulator senses the alternator output voltage. It then controls the rotor magnetic field strength to maintain the voltage at the correct level.

All the main components of modern alternators are enclosed in a lightweight aluminium casing (Fig. 3.81). The vehicle engine provides power to the alternator, through a drive belt and pulleys to the rotor, which is mounted on bearings in the end covers of the alternator casing.

Key fact

A voltage regulator controls the rotor magnetic field strength to maintain the voltage at the correct level.

In a light vehicle, alternator magnetic fields are produced around magnetic poles on the rotor by an electrical current passing through coil windings. The poles are made from iron and shaped like claws with six fingers. There are two of these, one for each pole, facing each other, and set at each end of the rotor. Wound inside the poles is the rotor winding, which is connected to slip rings at one end of the rotor. Carbon brushes are used to conduct an electrical current to the rotor windings through the slip rings. The rotor rotates inside the stator and induces a current into the stator windings.

When an electrical current is passed through the rotor windings, they become 'excited' and a magnetic field is produced. The strength of the magnetic field is proportional to the voltage in the windings. The voltage in the windings is provided by the alternator during the charging phase and then controlled to regulate the alternator output voltage.

The initial electrical current to excite the windings can be provided through the ignition or generator warning light circuit. The light acts as an indicator that the generator field is being provided with an initial current to excite the rotor windings, and as a warning when the voltage from the stator is less than battery voltage. Under normal conditions, the light should go out as soon as the engine is running.

The stator, which is fitted inside the alternator casing, is made from soft iron laminations wound with three sets of windings (Fig. 3.82). The three sets of windings give three separate outputs, or phases, of alternating current. The electrical current induced in the alternator flows in the stator because of the changing magnetic fields produced by rotation of the rotor. The speed of rotation and the magnetic strength of the rotor determine the value of the voltage that is produced.

The three-phase stator windings are enamel-coated copper wire of a heavy gauge and, for light vehicle

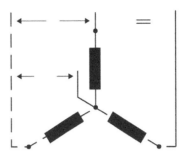

Figure 3.83 Star stator windings showing main output and a tapping from the centre

applications, are usually connected in a 'star' formation (Fig. 3.83). The windings can also be connected in a 'delta' (Greek letter Δ) formation (Fig. 3.84). The voltage and current outputs from the two formations are different for the same magnetic field strength and alternator speed. The voltage is higher and current lower for the star formation, in comparison with the delta formation.

Key fact

Alternator rectifiers use semiconductor diodes in a bridge formation.

Alternator rectifiers use semiconductor diodes, in a bridge formation, to provide rectification of the alternating current to the direct current required to charge the vehicle battery (Fig. 3.85).

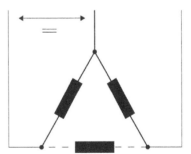

Figure 3.84 Delta stator windings showing main output

Figure 3.82 Stator construction

Figure 3.85 Rectifier

225

To achieve full wave rectification, a series of diodes is connected in a 'bridge' arrangement, so that the current flow is routed through open paths created by the bias of the diodes (Fig. 3.86). There are two open paths in the bridge rectifier, one for each direction of current flow. The output current flow is always in the same direction, and this gives a direct current flow.

The three phases of the alternator output require six diodes arranged in the circuit as shown in Figure 3.87. The arrows of the diode symbols show the current flow for each phase and direction of flow of the alternating current. The output from the rectifier is connected to the vehicle battery and the circuit

completed by an earth, or ground return connection through the alternator casing.

Three additional diodes are often fitted in the circuit so that part of the output from the stator can be passed to the rotor. This is to increase the magnetic field strength when the stator voltage increases as the engine speed rises (Fig. 3.88). Without a control system, the voltage would rise to high levels and cause extensive damage to the alternator and electrical systems on the vehicle. The regulator is connected into the rotor field circuit to control the rotor winding voltage and therefore the rotor magnetic field strength.

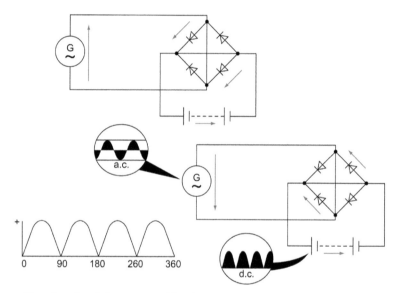

Figure 3.86 Bridge rectifier showing full wave rectification

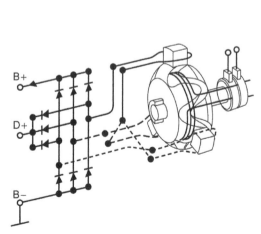

Figure 3.87 Alternator rectifier and current flow paths

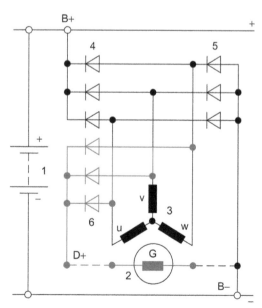

Figure 3.88 Field diodes and current flow to rotor (via a regulator)

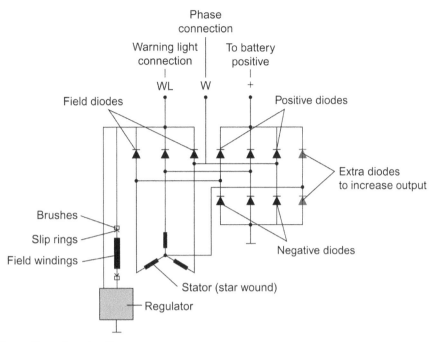

Figure 3.89 Modern alternator circuit

Key fact

Voltage from the stator increases as the engine speed increases.

As the alternator output is dependent on speed and rotor magnetic field strength, it is necessary to reduce the magnetic field strength as the speed increases. The regulator does this and therefore maintains a constant alternator output voltage. This is usually at about 14.2 V, which is sufficient to charge the battery without causing excessive gassing and, for maintenance-free batteries, is the optimum voltage level for correct charging.

The regulator consists of a small electronic circuit built around a zener diode. A zener diode conducts an electrical current only when its rated voltage is exceeded. This is used to control transistors that in turn switch the rotor field current on and off. As the voltage in the rotor windings is switched off, the voltage induced in the stator reduces until it falls below the rated voltage of the zener diode. The zener becomes non-conductive and allows the current to return once again to the rotor windings. This cycle occurs very quickly in the regulator to maintain a constant voltage output.

A typical circuit for a modern alternator is shown in Figure 3.89. The extra diodes from the centre of the stator help to improve the overall efficiency.

The ratio between the power supplied to a machine and the power it actually produces is known as its efficiency. In an alternator there are five key areas

where power is lost as it is converted from mechanical (or kinetic) energy to electrical energy:

▷ Copper losses: These are caused by the electrical resistance of the stator and rotor windings. It is proportional to the square of the current ($P = I^2R$)
▷ Iron losses – stator: These occur in the iron of the stator core. They are caused by alternating magnetic fields and hysteresis, which results in eddy currents that produce heat. They can be reduced by laminating the stator core.
▷ Iron losses – rotor: Eddy currents are produced on the surface of the rotor because of the fluctuations of magnetic flux caused by the slits in the stator core.
▷ Rectifier losses: Rectifier diodes cause a voltage drop. This results in heat and hence the need for the rectifier heat sink. High-efficiency diodes (HEDs) can reduce but not eliminate this loss
▷ Mechanical losses: These are caused by friction in the bearings and sliding contacts, but mostly by air resistance, sometimes called windage. This is caused by the rotor and fan and is particularly noticed at higher speeds.

Ongoing developments aim to optimize all these losses.

Alternator noise is caused by the cutting of magnetic flux. Interestingly, on some alternators, it is possible to determine if they are working by listening for the 'whine' they produce. Of course proper measurements are needed too! One area of research that may reduce the noise is the use of a 5-phase,

227

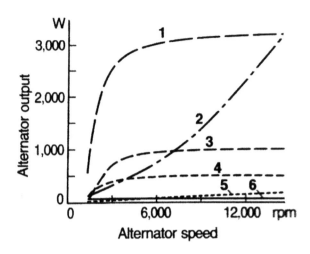

Figure 3.90 Comparison of losses in a 220 A alternator. 1, power output; 2, iron losses; 3, copper losses (stator); 4, rectifier losses; 5, friction losses; 6, copper losses (rotor)

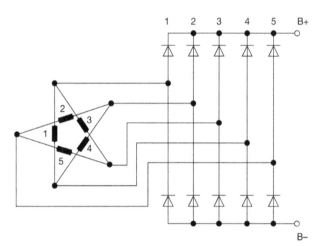

Figure 3.91 Pentagram connected 5-phase alternator with a bridge rectifier.

pentagram-connected stator winding as shown here. Another method is to use two 3-phase windings offset by 30 degrees.

Bosch 'Efficiency Line' (EL) series alternators are designed for vehicles that feature several safety and comfort functions and therefore require a greater amount of electrical energy. By further improving the electrical design of this alternator and adding new diodes, its efficiency is as much as 77%. This means that the alternator needs less mechanical energy to generate the same electrical power. In real driving conditions, fuel consumption and CO_2 emissions can be reduced by as much as 2%. These alternators can also withstand temperatures of up to 125°C. Series production of these alternators started in 2010.

The EL alternators are available in three sizes, covering a range between 130 and 210 A. They ensure that the battery is well charged even when the engine is idling or running at low rpm. A well-charged battery is a condition for start-stop systems. This means that if this fuel-saving system is used, consumption is reduced even further. The combination of a Bosch EL alternator and a start-stop system can reduce fuel consumption (and therefore CO_2 emissions) in urban driving by as much as 10%.

More fuel can be saved if the electrical energy is generated as far as possible during overrun (when rolling or braking). This calls for intelligent alternator control, which is provided by an interface such as the local interconnected network (LIN). The variable control of charging voltage lowers fuel consumption and extends the battery's life cycle.

The standard (simplified) method of testing a charging system is to measure battery voltage with the engine off and then again when running at about 3000 rpm with the lights on. If the voltage increases to about 14.2 V then the system is almost certainly OK. See the section on testing charging for more details.

Smart charging alternators, however, are different. The alternator is controlled by the PCM (power control

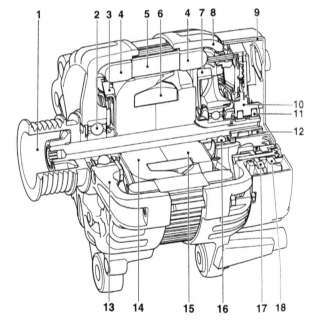

Figure 3.92 Efficiency Line alternator components:1, belt pulley; 2, A-side (drive-end) ball bearing; 3, A-side fan; 4, stator winding heads; 5, laminated stator core; 6, rotor winding (excitation winding); 7, B-side fan; 8, B-side end shield; 9, protective cap; 10, brush holder; 11, carbon brush; 12, collector ring; 13, A-side end shield; 14, A-side claw pole; 15, B-side claw pole; 16, B-side ball bearing; 17, negative heat sink of rectifier; 18, positive heat sink of rectifier. (Source: Bosch)

Figure 3.93 Alternator 28 V regulator with a LIN interface

module) which monitors certain parameters such as the engine temperature, battery temperature and electrical demand. If the alternator does not receive a signal from the PCM, the battery light is illuminated on the vehicle. This can sometimes be misdiagnosed as an alternator failure.

3.2.5 48 V technology

The benefits of 48 V systems have been under discussion for many years but at last they seem to be gaining ground.

It is thought that 48 V architectures will transform and enhance just about every major subsystem of conventional vehicles. However, they will also be a bridge technology that will allow current combustion engines to evolve into the all-electric future. One area that will be important for combustion engine power and efficiency is the use of electrically accelerated turbochargers. The higher power that 48 V can deliver also enables use of electromechanically actuated valves.

Electrification of mechanical components also allows additional digital control and all the benefits that brings. For example, Audi are working on an electromechanical rotary-damper system to replace conventional hydraulic dampers. This would have been much more difficult with 12 V. New possibilities for adjusting the suspension will now exist as the response characteristics of a component (or system) will be controlled by software.

The 48 V micro-hybrids system employs a starter/generator, hence it can be used to charge the batteries as well as start the vehicle. It is a 3-phase device and includes two sensors, position and current, which provide information back to the ECU. The CAN and LIN connections make data available across the vehicles as well as data from other sensors available to the micro-hybrid system. Infineon is one of the leading companies that offers a micro-hybrid 48 V control system.

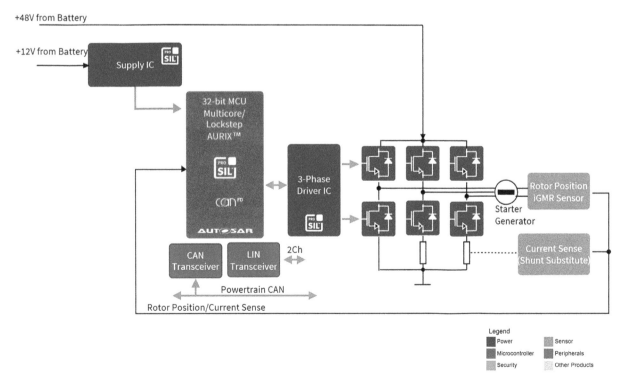

Figure 3.94 Micro-hybrid 48 V control system block diagram (Source: Infineon, www.infineon.com)

229

3.3 Lighting and indicators

3.3.1 Lighting systems

Vehicle lighting systems are very important, particularly where road safety is concerned. If headlights were suddenly to fail at night and at high speed, the result could be serious. Remember that lights are to see with and to be seen by. Lights are arranged on a vehicle to meet legal requirements and to look good. Headlights, sidelights and indicators are often combined on the front (Fig. 3.95). Taillights, stoplights, reverse lights and indicators are often combined at the rear (Fig. 3.96).

> **Safety first**
>
> Lights are to see with AND to be seen by.

The number, shape and size of bulbs used on vehicles are increasing all the time. A common selection is shown in Figure 3.97. Most bulbs used for vehicle lighting are generally either conventional tungsten filament bulbs or tungsten halogen.

In the conventional bulb, the tungsten filament is heated to incandescence by an electric current. The temperature reaches about 2300°C. Tungsten, or an alloy of tungsten, is ideal for use as filaments for electric light bulbs. The filament is normally wound into a 'spiralled spiral' to allow a suitable length of thin wire in a small space, and to provide some mechanical strength.

Almost all vehicles now use tungsten halogen bulbs for the headlights (Fig. 3.98). The bulb will not blacken and,

Figure 3.96 Rear light

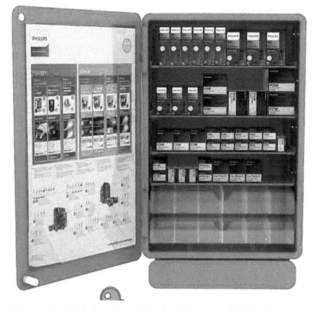

Figure 3.97 Selection of bulbs (Source: Philips)

therefore, has a long life. In normal gas bulbs, about 10% of the filament metal evaporates. This is deposited on the bulb wall. Design features of the tungsten halogen bulb prevent this deposition. The gas in halogen bulbs is mostly iodine and the glass envelope is made from fused silicon or quartz. This allows higher temperatures and the shape of the envelope causes the metal to condense back on the filament.

The object of the headlight reflector is to direct the random light rays produced by the bulb into a beam of concentrated light, by applying the laws of reflection.

Figure 3.95 Front light

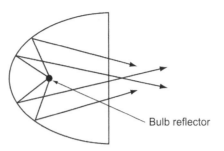

Figure 3.101 Light source in front of the focal point

Figure 3.98 Headlight bulbs

Bulb filament position relative to the reflector is important if the desired beam direction and shape are to be obtained. A reflector is a layer of silver, chrome or aluminium deposited on a smooth and polished surface such as brass or glass. Consider a mirror reflector that 'caves in': this is called a concave reflector (Fig. 3.99). The centre point on the reflector is called the pole, and a line drawn perpendicular to the surface from the pole is known as the principal axis.

Key fact

A headlight reflector directs the random light rays produced by the bulb into a beam.

If a light source is moved along the principal axis, a point will be found where the radiating light produces a reflected beam parallel to the axis. This point is known as the focal point, and its distance from the pole is known as the focal length. If the filament is between the focal point and the reflector, the reflected beam will diverge; that is, spread outwards along the principal axis (Fig. 3.100).

Key fact

A headlight lens redistributes the reflected light beam.

If the filament is positioned in front of the focal point, the reflected beam will converge towards the principal axis (Fig. 3.101).

A common type of bulb arrangement is shown in Figure 3.102, where the dip filament is shielded. This gives a nice sharp cut-off line when positioned in a reflector like that of Figure 3.101 on dip beam (the other filament is at the focal point to give a main beam).

A good headlight should have a powerful, far-reaching central beam, around which the light is distributed both horizontally and vertically to illuminate as great an area of the road surface as possible. The beam formation can be considerably improved by passing the reflected light rays through a transparent block of lenses. It is the function of the lenses to partially redistribute the reflected light beam and any stray light rays. This gives better overall road illumination.

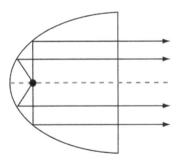

Figure 3.99 Concave reflector (light source at the focal point)

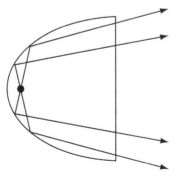

Figure 3.100 Light source behind the focal point

Many headlights are now made with clear lenses, such that all the redirection of the light is achieved by the reflector (Fig. 3.103). A clear lens does not restrict the light in any way. This makes the headlights more efficient as well as attractive.

Sidelights, taillights, brake lights and others are relatively straightforward. Headlights present the most problems.

231

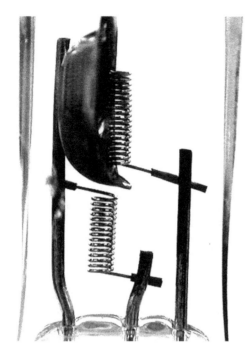

Figure 3.102 Shielded dip filament (top) and main beam filaments in a tungsten halogen bulb

Figure 3.103 Headlights with a clear lens and complex reflector

Figure 3.104 The beam setter is correctly aligned before use

This is because on dipped beam they must provide adequate light for the driver, but not dazzle other road users (Fig. 3.104). The conflict between seeing and dazzling is very difficult to overcome. The main requirement is that headlight alignment must be set correctly. Some cars have a headlight adjuster that the driver can control. The adjuster is connected to levelling actuators. The function of levelling actuators is to adjust the dipped or low beam in accordance with the load carried by the car. This will avoid dazzling oncoming traffic.

3.3.2 Stoplights and reverse lights

Stoplights, or brake lights, are used to warn drivers behind that you are slowing down or stopping. Reverse lights warn other drivers or pedestrians that you are reversing, or intend to reverse. The circuits are quite simple; one switch in each case operates two or three bulbs via a relay.

The circuits for these two systems are similar. Figure 3.106 shows a typical stoplight or reverse light circuit. Most incorporate a relay to switch on the lights, which in turn is operated by a spring-loaded switch on the brake pedal or gearbox. Links from the stoplight circuit to the cruise control system may be found. This is to cause the cruise control to switch off as the brakes are operated. A link may also be made to the antilock brake system.

The circuits are operated by the appropriate switch. The stoplight switch (Fig. 3.107) is usually fitted so that it acts on the brake pedal. The reverse light switch (Fig. 3.108) is part of the gearbox or gear change linkage.

Figure 3.105 Headlamp bulbs should be changed in pairs, especially if a different colour

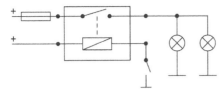

Figure 3.106 Stoplight or reverse light circuit

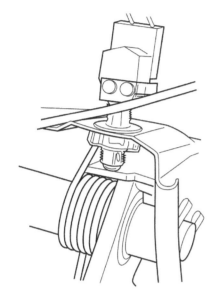

Figure 3.107 Stoplight switch)

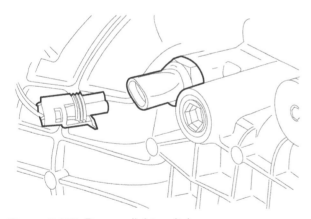

Figure 3.108 Reverse light switch

Light emitting diodes (LEDs) are more expensive than bulbs. However, the potential savings in design costs due to long life, sealed units being used and greater freedom of design could outweigh the extra expense. LEDs are ideal for stoplights because they illuminate more quickly than ordinary bulbs. This time is approximately the difference between 130 ms for the LEDs and 200 ms for bulbs. If this is related to a vehicle brake light at motorway speeds, then the increased reaction time equates to about a car length. This is potentially a major contribution to road safety.

Figure 3.109 Switch positioned in the door pillar

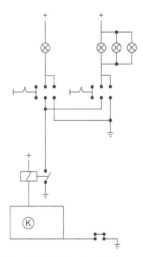

Figure 3.110 Circuit for interior lights

Safety first

LEDs are ideal for stoplights because they illuminate more quickly than ordinary bulbs.

3.3.3 Interior lighting

Interior lighting consists of several systems, the main ones being courtesy lights, map lights and panel illumination lights. Features such as delay and fade-out are now common. This requires some electronic control. Map lights are an extra feature to assist with reading a map in the dark. Many types are available: some are small spotlights, which form part of the interior light assembly, while others are positioned on the centre console of the vehicle.

Lights are designed to illuminate the vehicle interior when the doors are opened. Most cars have one central interior light above the rear-view mirror, or two lights, on the sides above the driver's and passenger's shoulders. Door switches are simple spring-loaded contacts that are made as the door opens (Fig. 3.109). The contacts

233

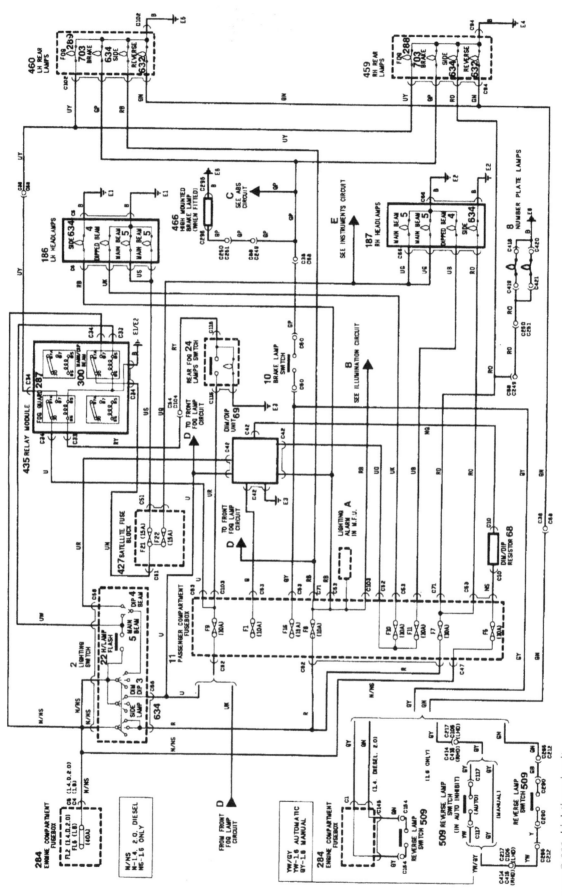

Figure 3.111 Lighting circuit

are broken again as the door closes. Rubber seals are sometimes used to keep water out. The same switches may also be used for the alarm system.

The circuit shown in Figure 3.110 is typical of many in common use. The sliding switches have three positions: 'off', 'on' and 'door operated'. The control module is to allow delay operation. In this case, it is also used for the central locking system.

Panel and instrument lights are illuminated when the vehicle sidelights are switched on. Most cars also incorporate a dimmer switch so the level of illumination can be set.

Interior lights are important for passenger comfort. Most now operate via some type of electronic control. One enhancement is a switching-off delay, after the doors are closed. Some manufacturers are linking functions such as interior lights with other systems, by a central control module.

3.3.4 Lighting circuits

Lighting circuits can appear complex at first view. However, if you concentrate on just one part of the circuit at a time you will find it easier to understand. Relays are often used because they take load off the control switches. They are still simple switches, so don't panic! Take your time and you will find electrical circuits an interesting challenge.

The diagram shown in Figure 3.111 is the complete lighting circuit of a vehicle. Operation of the switch allows the supply on the N or N/S wire to pass to fuses 7 and 8 on the R wire. The two fuses then supply left lights on an RB wire, and right lights on an RO wire. The number plate lights are also supplied from here. When the dip beam is selected, a supply is passed, on a U and UR wire, to the dim dip unit, which is de-energized. This then allows a supply to fuses 10 and 11 on the OU wire. This supply is then passed to the left light on a UK wire and the right light on a UB wire. Selecting main beam allows a supply on the UW wire to the main/dip relay, thus energizing it. A supply is therefore placed on fuses 21 and 22 and hence to each of the headlight main beam bulbs.

Key fact

Relays take the load off control switches.

When sidelights are on, there is a supply to the dim dip unit on the RB wire. If the ignition supplies a second feed on the G wire from fuse 1, the unit will allow a supply from fuse 5 to the dim dip resistor on the NS wire. This continues on to the dim dip unit on an NG wire. The dim dip unit links this supply to fuses 10 and 11. These are the dip beam fuses. The supply is therefore passed to the left light on a UK wire and the right light on a UB wire. When the headlights are switched on, a supply is made from the light switch to fuse 9 on a U wire. From this fuse, a supply is sent to the fog light relay contacts on a U wire, and the rear fog lamp switch on a UR wire. When the fog switch is operated, it sends a supply on the RY wire to close the relay. The main supply is now fed from the relay on a UY wire to both rear fog lamps.

Following a circuit diagram is easy after a bit of practice. Think of it as a railway map that is used to get from A to B. Electricity will only complete the 'journey' if the path is a circuit, i.e. it has a return ticket!

3.3.5 Indicators and hazard lights

Direction indicators (turn signals) have a number of statutory requirements. The light produced must be amber (or red on the rear of some American cars), but they may be grouped with other lamps. The flashing

Figure 3.112 Indicator bulb

Figure 3.113 Light cluster that uses LEDs

Figure 3.114 Stoplight and rear light twin filament bulb

Figure 3.116 Indicator warning light

rate must be between one and two per second with a relative 'on' time of between 30% and 57%. If a fault develops this must be apparent to the driver by the operation of a warning light on the dash. The fault can be indicated by a distinct change in frequency of operation or by the warning light remaining on. If one of the main bulbs fails then the remaining lights should continue to flash perceptibly.

Legislation exists as to the mounting position of the exterior lamps. The rear indicator lights must be within a set distance of the rear lights and within a set height. The wattage (power) of indicator bulbs is normally 21 W (Fig. 3.112). These lights often come under the heading of auxiliaries or signalling. A circuit is examined later in this section. The bulbs are often combined with the rear lights (Figs 3.113 and 3.114).

The operation of the flasher unit is usually based around an integrated circuit. The electronic type shown in Figure 3.115 can operate at least four 21 W bulbs (front and rear) and two 5 W side repeaters when operating in hazard mode. This will continue for several hours if required. Flasher units are rated by the number of bulbs they are capable of operating. When towing a trailer the unit must be able to

operate at a higher wattage. Most units use a relay for the actual switching as this provides an audible signal. Thermal-type flasher units are still used on older vehicles. LEDs are now also being used in place of indicator bulbs. An indicator warning appears on the dash (Fig. 3.116).

The diagram shown in Figure 3.117 is the complete indicator circuit of a vehicle. You can follow the circuit by looking for the labels on the wires. 'G', for example, means 'Green', but this has no effect on how it works.

Note how the hazard switch, when operated, disconnects the ignition supply from the flasher unit and replaces it with a constant supply. The hazard system will therefore operate at any time, but the indicators will only work when the ignition is switched on. When the indicator switch is operated left or right, the front, rear and repeater bulbs are connected to the output terminal of the flasher unit. This is what makes it operate and causes the bulbs to flash.

When the hazard switch is operated, five pairs of contacts are made to open or close. Two sets connect left and right circuits to the output of the flasher unit. One set disconnects the ignition supply and another set connects the battery supply to the unit. The final set of contacts causes a hazard warning light to be operated. On this and most vehicles, the hazard switch is illuminated when the sidelights are switched on.

With the ignition switched on, fuse 1 provides a feed to the hazard warning switch on the G wire. Provided the hazard warning switch is in the off position, the feed crosses the switch and supplies the flasher unit on the LGK wire. When the switch control is moved for a right turn, the switch makes contact with the LGN wire from the flasher unit, which is connected to the GW wire. This allows a supply to pass to the right-hand front and rear indicator lights, and then to earth on the B wire.

When the switch control is moved for a left turn, the switch makes contact with the GR wire, which allows

Figure 3.115 Electronic flasher unit

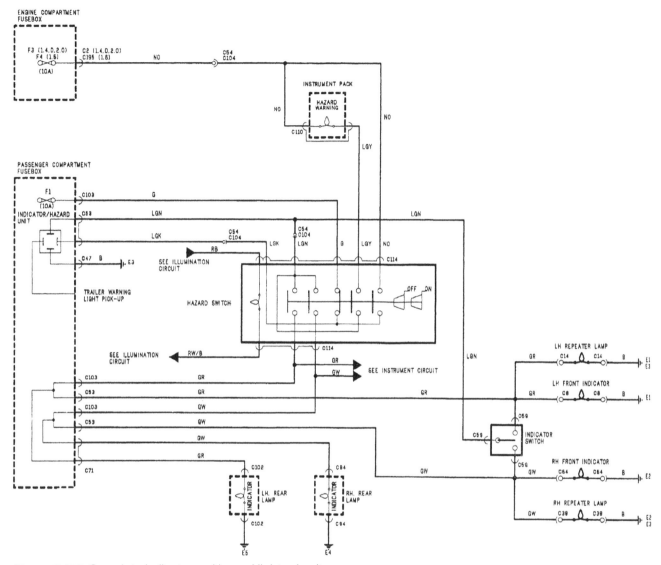

Figure 3.117 Complete indicator and hazard lights circuit

the supply to pass to the left-hand front and rear indicator lights, and then to earth on the B wire. The action of the flasher unit causes the circuit to make and break.

When the hazard warning switch is switched, a battery supply on the N0 wire from fuse 3 or 4 crosses the switch and supplies the flasher unit on the LGK wire. At the same time, contacts are closed to connect the hazard warning light and the flasher unit to both the GW and GR wires. These are the right-hand and left-hand indicators. The warning light and the main lights flash alternately.

Indicators and hazard lights are interesting circuits. Hazard lights are intended to show a hazard such as a broken-down car. They also seem to be used to prevent parking tickets when in a restricted area; my experience is that this does not work!

3.4 Body electrical and electronic systems

3.4.1 Washers and wipers

The requirements of the wiper system are simple. The windscreen must be clean enough to provide suitable visibility at all times. To do this, it must:

▷ achieve efficient removal of water and snow
▷ achieve efficient removal of dirt
▷ operate at temperatures from −30 to 80°C
▷ pass the stall and snow load test
▷ have a service life in the region of 1.5 million wipe cycles
▷ be resistant to corrosion from acid, alkali and ozone.

Wiper blades are made of a rubber compound (Fig. 3.118) and are held on to the screen by a spring in the

237

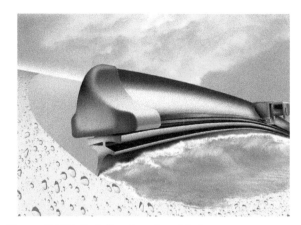

Figure 3.118 Details of a wiper blade

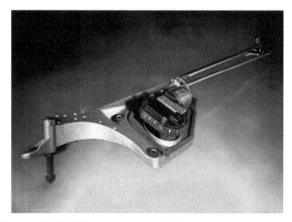

Figure 3.119 Wiper motor and linkage

wiper arm. The aerodynamic property of the wiper blades has become increasingly important. The strip on top of the rubber element is often perforated to reduce air drag. A good quality blade will have a contact width of about 0.1 mm. The lip wipes the surface of the screen at an angle of about 45°. The pressure of the blade on the screen is also important.

Key fact

The aerodynamic property of the wiper blades has become increasingly important.

Most wiper linkages consist of a series or parallel mechanism (Fig. 3.119). Some older types use a flexible rack and wheel boxes similar to the operating mechanism of many sunroofs. One of the main considerations for the design of a wiper linkage is the point at which the blades must reverse. This is because of the high forces on the motor and linkage at this time. If the reverse point is set so that the linkage is at its maximum force transmission angle, then the reverse action of the blades puts less strain on the system. This also ensures smoother operation.

All modern wiper motors are permanent magnet types. The drive is taken via a worm gear to increase torque and reduce speed. Three brushes may be used, or some form of electronic control, to allow two-speed operation. In three-brush motors, the normal speed operates through two brushes placed in the usual positions opposite to each other. For a fast speed, the third brush is placed closer to the earth brush. This reduces the number of armature windings between them, which reduces resistance, hence increasing current and therefore speed (Fig. 3.120).

Wiper motors or the associated circuit must have some kind of short-circuit protection. This is to protect the motor in the event of stalling, if the wiper blade is frozen to the screen for example. A thermal trip of some type is often used, or a current sensing circuit in the wiper ECU if fitted.

Key fact

Wiper motors or the circuit must have short-circuit protection to protect the motor in the event of stalling.

A windscreen washer system consists of a simple d.c. permanent magnet motor (Fig. 3.121), which drives a centrifugal water pump. The water, preferably with a cleaning additive, is directed onto an appropriate part of the screen by two or more jets. A non-return valve is often fitted in the line to the jets to prevent water siphoning back to the tank. This also allows 'instant' operation when the washer button is pressed. The washer circuit is normally linked in to the wiper circuit, such that when the washers are operated, the wipers start automatically and will continue, for several more sweeps, after the washers have stopped.

Figure 3.122 shows a circuit for fast, slow and intermittent wiper control. The switches are shown in the off position and the motor is stopped and in

Figure 3.120 Motor brushes and armature

Figure 3.121 Washer motors. (Source: MadeInChina.com)

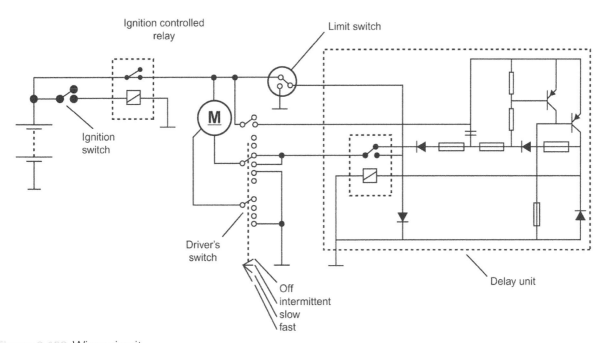

Figure 3.122 Wiper circuit

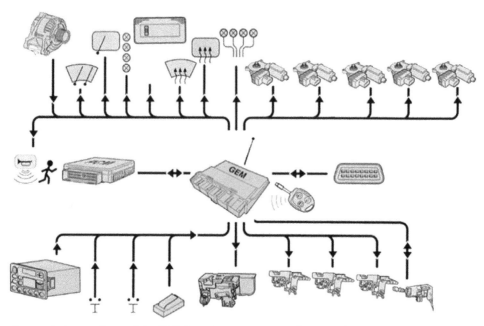

Figure 3.123 General electronic module (GEM) and components. (Source: Ford Motor Company)

its park position. Note that the two main brushes of the motor are connected together via the limit switch, delay unit contacts and the wiper switch. This causes regenerative braking because of the current, generated by the motor due to its momentum, after the power is switched off. Being connected to a very low resistance loads up the 'motor/generator' and, when the park limit switch closes, it stops instantly.

When either the delay contacts or the main switch contacts are operated, the motor will run at slow speed. When fast speed is selected, the third brush on the motor is used. On switching off, the motor will continue to run until the park limit switch changes over to the position shown in Figure 3.122. This switch is only in the position shown when the blades are in the parked position.

Many vehicles use a system with more enhanced facilities. This is regulated by what may be known as a central control unit (CCU), a multifunction unit (MFU) or a general electronic module (GEM) (Fig. 3.123). These units often control other systems as well as the wipers, thus allowing reduced wiring bulk under the dash area. Electric windows, headlights and heated rear window, to name just a few, are now often controlled by a central unit.

Using electronic control, a CCU allows the following facilities for the wipers:

- front and rear wash/wipe
- intermittent wipe
- time delay set by the driver
- reverse gear selection rear wipe operation
- rear wash/wipe with 'dribble wipe' (an extra wipe several seconds after washing)
- stall protection.

3.4.2 Horns

Regulations in most countries state that the horn (or audible warning device) (Fig. 3.124) should produce a uniform sound. This makes sirens and melody-type fanfare horns illegal. Most horns draw a large current so are switched by a relay.

The standard horn operates by simple electromagnetic switching. Current flow causes an armature, which is attached to a tone disc, to be attracted to a stop. A set of contacts is then opened. This disconnects the current, allowing the armature and disc to return under spring tension. The whole process keeps repeating when the horn switch is on. The frequency of movement, and hence the tone, is arranged to

Figure 3.124 Vehicle horns

Figure 3.125 Ultrasonic reversing aid

Figure 3.126 Cruise control switch

lie between 1.8 and 3.5 kHz. This note gives good penetration through traffic noise.

Twin horn systems, which have a high and low tone horn, are often used. This produces a more pleasing sound, but is still very audible in both town and higher speed conditions.

Definition

Hertz: One hertz (1 Hz) is one cycle or oscillation per second.

3.4.3 Obstacle avoidance

The principle of radar as a reversing aid is illustrated in Figure 3.125. This technique is in effect a range-finding system. The output can be audible and/or visual, the former being perhaps more appropriate, as the driver is likely to be looking backwards. The audible signal is a 'pip pip pip' type sound, the repetition frequency of which increases as the car comes nearer to the obstruction, becoming almost continuous as impact is imminent. Some cars have 'all-round' obstacle avoidance and also use the car speakers in a way such that the sound comes from the direction of the obstacle.

The units fitted in the rear bumper as shown in Figure. 3.125 transmit a radio signal and also receive the reply if the signal bounces off a nearby object. The time it takes to receive the signal tells the system the distance.

3.4.4 Cruise control

Cruise control is the ideal example of a closed loop control system. The purpose of cruise control is to allow the driver to set the vehicle speed and let the system maintain it automatically. The system reacts to the measured speed of the vehicle and adjusts the

throttle accordingly. The reaction time is important so that the vehicle's speed does not feel to be surging up and down. Other facilities are included such as allowing the speed to be gradually increased or decreased at the touch of a button (Fig. 3.126). Most systems also remember the last set speed. They will resume this speed at the touch of a button. The main switch switches on the cruise control and this, in turn, is ignition controlled. Most systems do not retain the speed setting in memory when the main switch has been turned off. Operating the 'set' switch programs the memory, but this normally will only work if conditions similar to the following are met:

- Vehicle speed is greater than 40 km/h.
- Vehicle speed is less than 120 km/h.
- Change of speed is less than 8 km/h/s.
- Automatics must be in 'drive'.
- Brakes or clutch are not being operated.
- Engine speed is stable.

Definition

Closed loop control system: A system in which the output of the system has a direct effect on the input because of a feedback loop.

Once the system is set, the speed is maintained to within 1–2 mph until it is deactivated by pressing the brake or clutch pedal, pressing the resume switch or turning off the main control switch. The last set speed is retained in memory except when the main switch is turned off. If the cruise control system is required again, then either the set button will hold the vehicle at its current speed or the resume button will accelerate the vehicle to the previous set speed. When cruising at a set speed, the driver can press and hold a button to accelerate the vehicle until the desired speed is reached. If the driver accelerates from the set speed to overtake, for example, then when the throttle is released, the vehicle will slow down again.

Several methods are used to control the throttle position. Vehicles fitted with drive-by-wire systems allow the cruise control to operate the same throttle actuator. A motor can be used to control the throttle cable or, in some cases, a vacuum-operated diaphragm is used.

The brake switch is very important, as braking would be dangerous if the cruise control system was still trying to maintain the vehicle speed. This switch is normally of superior quality and is fitted in place or as a supplement to the brake light switch activated by the brake pedal. Adjustment of this switch is important. The clutch switch is fitted in a similar manner to the brake switch. It deactivates the cruise system to prevent the engine speed increasing, if the clutch is pressed. The automatic gearbox switch will only allow the cruise to be engaged when it is in the 'drive' position. This is to prevent the engine overspeeding if the cruise tried to accelerate to a high road speed with the gear selector in position '1' or '2'.

Key fact

Vehicles fitted with drive-by-wire systems allow the cruise control to operate the throttle actuator.

The speed sensor will often be the same sensor that is used for the speedometer. If not, several types are available, the most common producing a pulsed signal, the frequency of which is proportional to the vehicle speed.

Conventional cruise control has now developed to a high degree of quality. It is, however, not always very practical on many roads as the speed of the general traffic is constantly varying and often very heavy. The driver has to take over from the cruise control system on many occasions to speed up or slow down.

Adaptive cruise control can automatically adjust the vehicle speed to the current traffic situation. The system has three main aims:

▷ to maintain a speed as set by the driver
▷ to adapt this speed and maintain a safe distance from the vehicles in front
▷ to provide a warning if there is a risk of collision.

The operation of an adaptive cruise system is similar to a conventional system. However, when a signal from the headway sensor (Fig. 3.127) detects an obstruction, the vehicle speed is decreased. If the optimum stopping distance cannot be achieved by just backing off the throttle, a warning is supplied to the driver. The more complex system can take control of the vehicle transmission and brakes. It is important to note that adaptive cruise control is designed to relieve the burden on the driver, not take full control of the vehicle!

Key fact

The operation of an adaptive cruise system is similar to a conventional system except that when a headway sensor detects an obstruction, the vehicle speed is decreased.

3.4.5 Seats, mirrors, sunroofs, locking and windows

3.4.5.1 Seat adjustment

Electrical movement of seats, mirrors (Fig. 3.128) and the sunroof is included in one area as the operation of each system is quite similar. The operation of electric windows and central door locking is also much the same. Fundamentally, all the systems discussed in this

Figure 3.127 Headway sensor. (Source: Bosch Media)

Figure 3.128 Mirrors

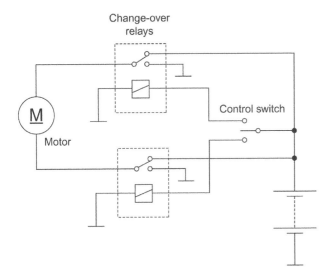

Figure 3.129 Motor reverse circuit

Figure 3.130 Seat motors

topic operate using one or several permanent magnet motors, together with a supply reversing circuit.

A typical motor reverse circuit is shown in Figure 3.129. When the switch is moved, one of the relays operates and changes the polarity to one side of the motor. If the switch is moved the other way then the polarity of the other side of the motor is changed. When at rest, both sides of the motor are at the same potential. This has the effect of regenerative braking, so that when the motor stops it will do so instantly. Further refinements are used to enhance the operation of these systems. Limit switching, position memory and force limitation are the most common.

Adjustment of a seat is achieved by using a number of motors to allow positioning of different parts of the seat. Movement is possible in the following ways:

▷ front to rear
▷ cushion height rear
▷ cushion height front

Figure 3.131 Mirror switch

Figure 3.132 Mirror motors

▷ backrest tilt
▷ headrest height
▷ lumbar support.

Three seat operation motors are shown in Figure 3.130. Each motor is operated by a simple rocker type switch that controls two relays as described previously. When the seat position is set, some vehicles have memories, to allow automatic repositioning if the seat has been moved. This is often combined with electric mirror adjustment.

3.4.5.2 Mirrors

Many vehicles have electrical adjustment of mirrors (Fig. 3.131). The system used is much the same as has been discussed previously in relation to seat movement. Two small motors are used to move the mirror vertically or horizontally (Fig. 3.132). Many mirrors also contain a small heating element on the rear of the glass. This is operated for a few minutes when the ignition is first switched on. The circuit may include feedback resistors for position memory.

3.4.5.3 Sunroof

The operation of an electric sunroof is once again based on a motor reverse circuit (Fig. 3.133).

243

Figure 3.133 Sunroof

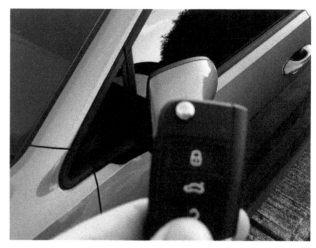

Figure 3.135 Remote door locking

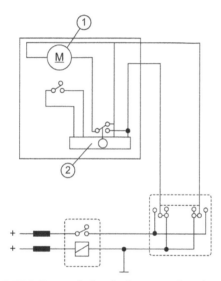

Figure 3.134 Sunroof circuit: 1, motor; 2, microswitch

However, further components and circuitry may be needed to allow the roof to slide, tilt and stop in the closed position. The extra components used in many cases are a microswitch and a latching relay. A latching relay works in much the same way as a normal relay except that it locks into position each time it is energized. The mechanism used to achieve this is much like that used in ballpoint pens that have a button on top.

A microswitch is mechanically positioned such as to operate when the roof is in its closed position. A rocker switch allows the driver to adjust the roof. The switch provides a supply to the motor to run it in the chosen direction. The roof will open or tilt. When the switch is operated, to close the roof, the motor is run in the appropriate direction until the microswitch closes (when the roof is in its closed position). This causes the latching relay to change over, which stops the motor. The control

switch now has to be released. If the switch is pressed again, the latching relay will once more change over and the motor will be allowed to run (Fig. 3.134).

3.4.5.4 Locking

When the key is turned in the driver's door lock (or the remote button is pressed), all the other doors on the vehicle should also lock (Fig. 3.135). Motors or solenoids in each door achieve this. If the system can only be operated from the driver's door key, then an actuator is not required in this door. If the system can be operated from either front door or by remote control then all the doors need an actuator. Vehicles with built-in alarm systems lock all the doors as the alarm is set.

A simplified door locking circuit is shown in Figure 3.136. The main control unit contains two change-over relays (reverse circuit), which are actuated by either the door lock switch or, if fitted, the remote key. The motors for each door lock are wired in parallel and all operate at the same time. Most door actuators are small motors or solenoids.

Remote-control central door locking is controlled by a small hand-held transmitter and a sensor receiver unit. When the key is operated, by pressing a small switch, a complex code is transmitted. Trillions of different code combinations are used in modern systems. The sensor in the car picks up this code and sends it in an electrical form to the main control unit. If the received code is correct, the relays are triggered and the door locks are either locked or unlocked. If an incorrect code is received on three consecutive occasions when attempting to unlock the doors, then some systems will switch off until the

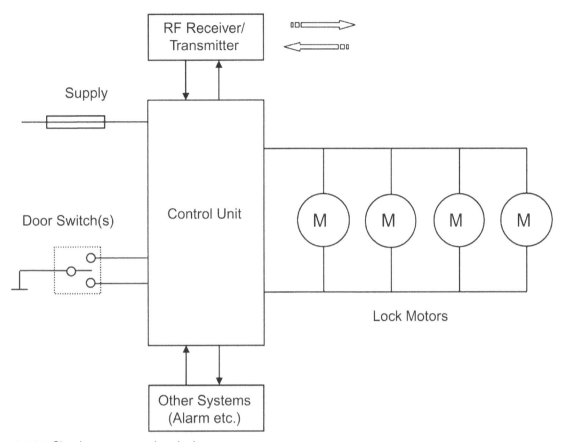

Figure 3.136 Circuit to operate door locks

Figure 3.137 After replacing window regulators it is important to reseal the compartment.

door is opened by the key. This technique prevents a scanning-type transmitter unit from being used to open the doors.

Key fact

Trillions of different code combinations are used in modern locking systems.

3.4.5.5 Windows

The basic form of electric window operation is similar to many of the systems discussed so far in this module; that is, a motor reversing system either using relays or directly by a switch. More sophisticated systems are now popular for reasons of safety as well as improved comfort. The following features are now available from many manufacturers:

▷ one touch up or down
▷ inch up or down
▷ lazy lock
▷ back off or bounce back.

The complete system consists of an electronic control unit containing the window motor relays, switch packs and a link to the door lock and sunroof circuits.

Key fact

A bounce-back feature is used to prevent people and things being trapped in windows.

When a window is operated in 'one-touch' mode, the window is driven in the chosen direction until the switch position is reversed, the motor stalls or the ECU receives

245

a signal from the door lock circuit. The problem with one-touch operation is that if a child, for example, should become trapped in the window there is a serious risk of injury. To prevent this, a bounce-back feature is used. An extra commutator is fitted to the motor armature. This produces a signal, via two brushes, proportional to the motor speed. Hall sensors are used on some systems. If the rate of change of speed of the motor is detected as being below a certain threshold, the ECU reverses the motor until the window is fully open.

By counting the number of pulses received, the ECU can also determine the window position. This is important, as the window must not reverse when it stalls in the closed position. In order for the ECU to know the window position, it must be initialized. This is often done simply by operating the motor to drive the window first fully open, and then fully closed. If this is not done then the one-touch feature and bounce-back will not operate.

A 'lazy-lock' feature allows the car to be fully secured by one operation of a remote key. This is done by linking the door lock ECU and the window and sunroof ECUs. A signal is supplied and causes all the windows to close in turn, and then the sunroof, and finally locks the doors. The alarm will also be set if required. The windows close in turn to prevent the excessive current demand that would occur if they all tried to operate at the same time.

A circuit for electric windows is shown in Figure 3.138. Note the rear window isolation switch. This is commonly fitted to allow the driver to prevent rear window operation, for child safety for example.

Most window lift motors (Fig. 3.139) are permanent magnet types and drive through a worm gear. This reduces speed and greatly increases the torque.

All of the systems examined in this section are based on motor reverse circuits. Door locks, windows, sunroofs, mirrors and seats all operate in this way. Most of the systems are designed to improve driver and passenger comfort.

3.4.6 Screen heating

Electrical heating is used for screens, windows, seats and mirrors. Some heavy vehicles also incorporate cab heaters, which use fuel from the tank. As far back as the 1920s, when vehicle heaters were not fitted, electrically heated gloves were available. Beware of short-circuits!

Heating of the rear screen involves a circuit with a relay and usually a timer. The heating elements are thin metallic strips bonded to, or built inside, the glass (Fig. 3.140). When a current is passed through

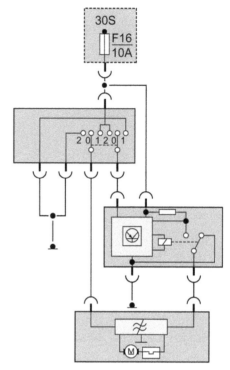

Figure 3.138 Circuit for electric windows

Figure 3.139 Electric window lift motor

the elements, heat is generated and the window will defrost or demist.

This circuit can draw high current, 10–15 A being typical. Because of this, the circuit often contains a timer relay to prevent the heater being left on too long (Fig. 3.141). The timer will switch off after 10–15 minutes. The rear screen elements are usually shaped to defrost the area of the rear wiper blade resting position, if fitted.

Front windscreen heating is used on some vehicles (Fig. 3.142). This presents more problems than with the rear screen, as vision must not be obscured. The technology used is drawn from the aircraft industry and involves very thin wires cast into the glass. As with the heated rear window, this device can consume a large current and uses a timer relay.

Figure 3.140 Rear screen heater elements

Figure 3.141 Timer relay

Figure 3.142 Front screen heater elements

3.4.7 Security systems

Huge numbers of cars are reported missing each year and many are never recovered; even when returned many are damaged. Most car thieves are opportunists, so even a basic alarm system serves as a deterrent (Fig. 3.143). Car and alarm manufacturers are constantly fighting to improve security.

Building the alarm system as an integral part of the vehicle electronics has made significant improvements. Even so, retrofit systems can still be very effective. The main types of intruder alarm used are:

▷ switch operated on all entry points
▷ trembler operated

▷ battery voltage sensing
▷ volumetric sensing.

> **Key fact**
>
> When the alarm system is an integral part of the vehicle electronics it is more effective.

There are four main ways to disable the vehicle:

▷ ignition circuit cut-off
▷ fuel system cut-off
▷ starter circuit cut-off
▷ engine ECU code lock (now the most common).

A separate switch or transmitter can be used to set an alarm system. Often, such a device is set automatically when the doors are locked. Some types have electronic sirens and give an audible signal when arming and disarming. They are all triggered when the car door opens and will automatically reset after a period of time, often 1 or 2 minutes. The alarms are triggered instantly when an entry point is breached. Most systems are in two parts, with a separate control unit and siren.

Most systems now come with remote 'keys' (Fig. 3.144) that use small button-type batteries and may have an LED that shows when the signal is being sent; they operate with one vehicle only. When operating with flashing lights most systems draw about 5 A. Without flashing lights (siren only) the current draw is less than 1 A. The sirens produce a sound level of about 95 dB, when measured 2 m in front of the vehicle.

Most factory-fitted alarms are combined with the central door locking system. This allows the facility mentioned in a previous section known as 'lazy lock'. Pressing the button on the remote unit, as well as setting the alarm, closes the windows and sunroof, and locks the doors.

A security code in the engine electronic control unit is a powerful deterrent. This can only be 'unlocked' to allow the engine to start when it receives a coded signal.

Figure 3.143 Alarm switch type sensor

247

Figure 3.144 Inside a remote key

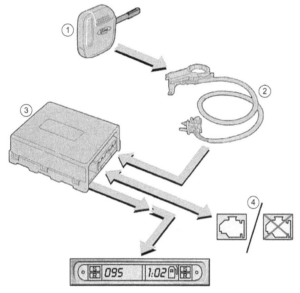

Figure 3.145 Coded key: 1, coded key; 2, harness; 3, ECU; 4, malfunction indicator light (MIL). (Source: Ford Motor Company)

Many manufacturers use a special ignition key, which is programmed with the required information (Fig. 3.145). Even the correct 'cut' key will not start the engine.

3.4.8 Safety systems

Active safety relates to any development designed to actively avoid accidents. It can be considered under four general headings:

▶ handling safety
▶ physiological safety
▶ perceptual safety
▶ operational safety.

Definition

Passive and active safety features: Passive safety features are used during an unavoidable accident. Active safety features work to avoid the accident in the first place.

Passive safety relates to developments that protect the occupants of the vehicle in the event of an accident (Fig. 3.146). Air bags are a good example of this.

A seat belt, seat belt tensioner and an air bag are at present the most effective restraint system in the event of a serious accident. At speeds in excess of 40 km/h, the seat belt alone is no longer adequate. Research after a number of accidents has determined that in 68% of cases an air bag provides a significant improvement. It is suggested that if all cars in the world were fitted with an air bag then the number of fatalities annually would be reduced by well over 50,000. Some air bag safety issues have been apparent in the USA, where air bags are larger and more powerful. This is because in many areas the wearing of seat belts is less frequent.

The method becoming most popular for an air bag system is that of building most of the required components into one unit (Fig. 3.147). This reduces the amount of wiring and connections, thus improving reliability. An important aspect is that some form of system monitoring must be built in, as the operation cannot be tested; it only ever works once.

The sequence of events in the case of a frontal impact at about 35 km/h is as follows (Fig. 3.148):

1 Driver in normal seating position prior to impact.
2 About 15 ms after the impact the vehicle is strongly decelerated and the threshold for triggering the air bag is reached. The igniter ignites the fuel tablets in the inflater.
3 After about 30 ms, the air bag unfolds and the driver will have moved forwards as the vehicle's crumple zones collapse. The seat belt will have locked or been tensioned, depending on the system.
4 At 40 ms after impact, the air bag will be fully inflated and the driver's momentum will be absorbed by the air bag.

Figure 3.146 Passive safety is used only in an emergency. (Source: Saab Media)

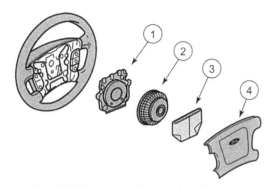

Figure 3.147 Air bag unit: 1, container; 2, gas generator; 3, airbag; 4, cover

5 About 120 ms after impact, the driver will be moved back into the seat and the air bag will have almost deflated through the side vents, allowing driver visibility.

Safety first

Taking the 'slack' out of a seat belt in the event of an impact contributes to passenger safety.

Passenger air bag deployment events are similar to the previous description. The position is different but the basic principle of operation is the same. The positions of the side and rear air bags are

shown in Figures 3.149 and 3.150, and the ECU in Figure 3.151.

A block diagram of an air bag circuit is shown in Figure 3.152. A digitally based system, using electronic sensors, has about 10 ms at a vehicle speed of 50 km/h, to decide whether the restraint systems should be activated. In this time, about 10,000 computing operations are necessary. Data for the development of these algorithms is based on computer simulations, but digital systems can also remember the events during a crash, allowing real data to be collected.

Taking the 'slack' out of a seat belt in the event of an impact is a good contribution to vehicle passenger safety. The decision to take this action is the same as for the air bag. The two main methods are:

▷ spring tension
▷ pyrotechnic.

When the explosive charge is fired, the cable pulls a lever on the seat belt reel, which in turn tightens the belt (Fig. 3.153). The unit must be replaced once deployed. This feature is sometimes described as anti-submarining.

3.4.9 Mobile multimedia

It would be almost unthinkable to not have a quality music system in our vehicles. Shown here is a

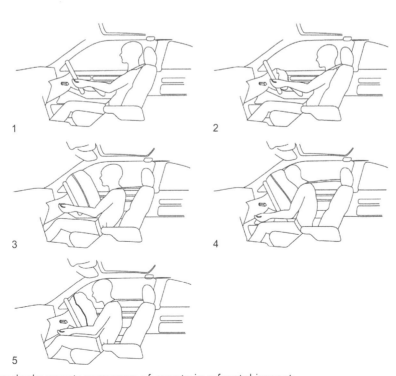

Figure 3.148 Air bag deployment: sequence of events in a frontal impact

249

Figure 3.149 Side air bag position

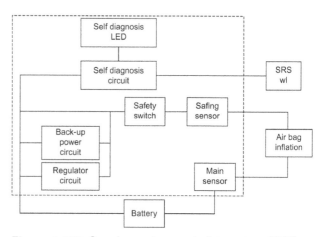

Figure 3.152 Supplementary restraint system (SRS) block diagram

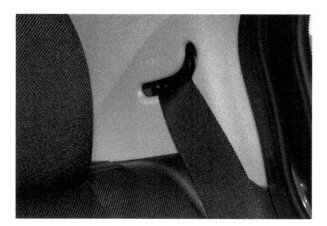

Figure 3.150 Rear air bag position

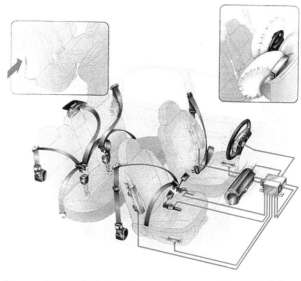

Figure 3.153 Belt tensioners. (Source: Ford Media)

Figure 3.151 Electronic control unit (ECU)

factory-fitted, in car entertainment (ICE) system. It has auxiliary inputs for either a 3.5 mm jack plug, USB or Bluetooth. It also has a feature that mirrors smartphone apps onto its touch screen, an iPhone in this example.

Linking a phone to the car allows most of the common features to synchronize. Phone address book, music and sat nav for example. This means that some cars do not have features built in because they would only duplicate those that we already have available.

ICE systems usually include at least six speakers, two larger speakers in the rear to produce good low frequency reproduction, two front door speakers for mid-range and two front door tweeters for high frequency notes.

Figure 3.154 Multimedia options

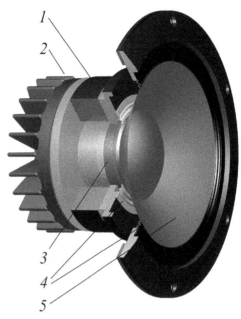

Figure 3.156 Cutaway view of a dynamic midrange speaker: 1, magnet; 2, cooler (sometimes present); 3, voicecoil; 4, suspension diaphragm; 5, cone

Figure 3.155 Satellite navigation on the multimedia screen from an iPhone

▷ Automatic tuning finds the best available signal for the chosen radio station.
▷ Traffic information and news broadcasts can be identified, and a setting made so that whatever you are listening to at the time can be interrupted.

Frequency modulation (FM) is generally a better source of high-fidelity sound. This is because the quality of AM reception is limited by the narrow bandwidth of the signal. FM does, however, present problems with reception when mobile. As most vehicles use a rod aerial, which is omni-directional, it will receive signals from all directions. Because

Speakers are a very important part of a sound system. No matter how good the receiver or CD player is, the sound quality will be reduced if inferior speakers are used. Equally, if the speakers are of a lower power output rating than the set, distortion will result at best and damage to the speakers at worst. Speakers fall into the following categories:

▷ tweeters – high frequency reproduction
▷ mid-range – middle range frequency reproduction (treble)
▷ woofers – low frequency reproduction (bass)
▷ sub-woofers – very low frequency reproduction.

Radio data system (RDS) has become standard on many radio sets. It is an extra inaudible digital signal which is sent with FM broadcasts in a similar way to how text is sent with TV signals. RDS provides information so a receiver can appear to act intelligently. The possibilities available when RDS is used are as follows:

▷ The station name can be displayed in place of the frequency.

Figure 3.157 RDS display on an older set with a CD player

Figure 3.158 Digital broadcast radio set

Fuel level is measured using a variable resistor that is moved by a float (Fig. 3.160). The position of the float is determined by how much fuel is in the tank. The resistance value is varied by a contact sliding over a resistor.

The most common temperature measurement is that of the engine coolant. However, outside air, cabin, air intake and many other temperatures are measured (Fig. 3.161). Most applications use a thermistor, which is a special material that changes its resistance with temperature. Most types are described as negative temperature coefficient (NTC). This means that, as the temperature increases, their resistance decreases.

A reed switch consists of two small strips of steel. When these become magnetized, they join and make a circuit. Bulb failure circuits often use a reed relay to monitor the circuit. In the circuit shown in Figure 3.159, the contacts of the reed switch will only close when electricity is flowing to the bulb being monitored.

Road speed is often sensed using an inductive pulse generator. This sensor produces an a.c. output with a frequency that is proportional to speed. It is like a small generator that is driven by a gear on the gearbox output shaft. This type of sensor is also used to sense engine speed from the flywheel or crankshaft.

Engine speed can be sensed in a number of ways. An inductive sensor as described in the fuel section is most common. The Hall effect sensor is also a popular choice, as it is accurate and produces a square-wave output with a frequency proportional to engine speed.

Fluid levels, such as washer fluid or radiator coolant, are often measured or sensed using a float and reed switch assembly. The float has a magnet attached that causes the contacts to join when it is in close proximity. The float moves up or down depending on the fluid level.

Oil pressure may be measured and displayed on a gauge or, as is most common, by using a simple warning light. For this purpose, a diaphragm switch

of this, reflections from buildings, hills and other vehicles can reach the set all at the same time and distort the signal.

Digital audio broadcast (DAB) is designed to provide high quality digital radio broadcasting for reception by stationary and mobile receivers. It is designed to operate at any frequency up to 3GHz. The system uses digital techniques to remove redundancy and perceptually irrelevant information from the audio source signal. All transmitted information is then spread in both the frequency and the time domains (multiplexed), so a high-quality signal is obtained in the receiver, even under poor conditions.

Mobile multimedia has changed significantly over the years. On earlier cars it was common for them not to be fitted with a radio. Radio cassette players became standard equipment followed by radio and CD or even multi-CD players. Quality in general of the systems has increased significantly too. Almost all modern cars now have a multimedia unit that does include a radio but allows connection of a smartphone as a minimum to be a music player but now more often full multimedia facilities.

3.5 Monitoring and instrumentation

3.5.1 Sensors

Sensors are used on vehicles for many purposes (Fig. 3.159). For example, the coolant temperature thermistor is used to provide information to the engine management system as well as to the driver. The information to the driver is provided by a display or gauge. Sensors convert what is being measured into an electrical signal. This signal can then be used to operate a display, such as a gauge or warning light, on the instrument panel.

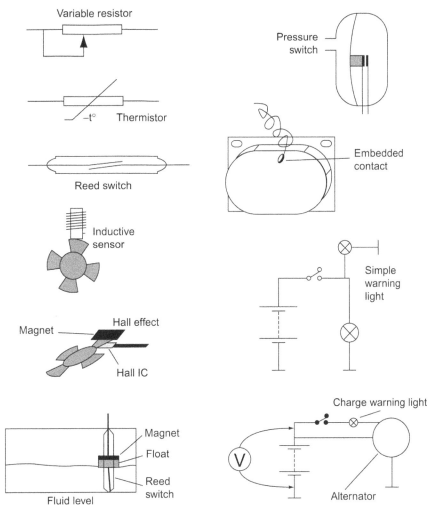

Figure 3.159 Sensors used for instrumentation

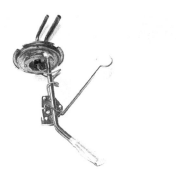

Figure 3.160 Tank sender unit

Figure 3.161 Temperature sensor

is used. As oil pressure increases, it is made to act on a diaphragm. Once it overcomes spring pressure, the contacts are operated. The contacts can be designed to open or close as the pressure reaches a set level.

Brake pad wear is sensed by using a simple embedded contact wire (Fig. 3.162). When the friction material wears down, the embedded contact makes contact with the disc to complete a circuit. Some systems use a loop of wire that is broken when the pad wears out.

A wide range of sensors is used to operate instrument displays. Sensors convert what is being measured into an electrical signal. This may be by a simple on/off operation, a changing voltage output or a change in resistance.

3.5.2 Gauges

By definition, an instrumentation system can be said to convert a variable into a readable or usable display

253

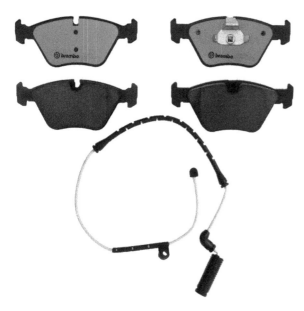

Figure 3.162 Brake pads with sensor wires

(Fig. 3.163). For example, a fuel level instrument system will display a representation of the fuel in the tank using an analogue gauge.

Instrumentation is not always associated with a gauge or a read-out type display. In many cases, a system can be used just to operate a warning light. However, it must still work to certain standards. For example, if a low outside temperature warning light did not illuminate at the correct time, a dangerous situation could develop.

Thermal gauges, which are ideal for fuel and engine temperature indication, have been in use for many years (Fig. 3.164). This will continue because of their simple design and inherent 'thermal' damping. The gauge works by utilizing the heating effect of electricity and the widely adopted benefit of the

Figure 3.163 Instrument panel

Figure 3.164 Fuel gauge display

bimetal strip. As a current flows through a simple heating coil wound on a bimetal strip, heat causes the strip to bend. The bimetal strip is connected to a pointer on a suitable scale. The amount of bending is proportional to the heat, which in turn is proportional to the current flowing. Provided the sensor can vary its resistance in proportion to the fuel level or temperature, the gauge will indicate a suitable representation (Fig. 3.165).

The inherent damping is due to the slow thermal effect on the bimetal strip. This causes the needle to move very slowly to its final position. This is a particular advantage for displaying fuel level, as the variable resistor in the tank will move as the fuel moves, owing to vehicle movement. If the gauge reacted quickly, it would be constantly moving. The movement of the fuel, however, is in effect averaged out and an accurate display can be obtained. Thermal gauges are used with a variable resistor. This is either a float in the fuel tank or a thermistor in the engine water jacket. The sender resistance is usually at a maximum when the tank is empty or the engine is cold.

A constant voltage supply is required to prevent changes in the system voltage affecting the reading.

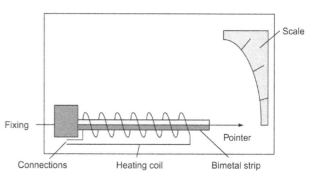

Figure 3.165 Thermal gauge operation

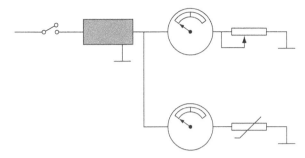

Figure 3.166 Fuel and temperature gauge circuit with a voltage stabilizer

Figure 3.167 Air-cored gauge details

This is because if system voltage increased, the current flowing would increase and the gauges would read higher. Most voltage stabilizers are simple zener diode circuits as shown in Figure 3.166.

Air-cored gauges work on the same principle as a compass needle lining up with a magnetic field (Fig. 3.167). The needle of the display is attached to a very small permanent magnet. Three or more coils of wire are used and each produces a magnetic field. The magnet, and therefore the needle, will line up with the resultant of the three fields. The current flowing through each coil is, therefore, the key to moving the needle position.

When an air-core gauge is used as a temperature gauge and the thermistor resistance changes, the current in all three coils is made to change. This moves the needle from cold to hot, or the other way, as needed.

The air-cored gauge has a number of advantages. It has almost instant response and, as the needle is held in a magnetic field, it will not move as the vehicle changes position. The gauge can be arranged to continue to register the last position even when

switched off. If a small 'pull-off' magnet is used, it will return to its zero position. A change in system voltage would affect the current flowing in all three coils. Variations are therefore cancelled out so that a voltage stabilizer is not needed.

A variation of any of the above types of gauge can be used to display other required outputs such as voltage or oil pressure. Gauges to display road or engine speed, however, need to react very quickly to changes. Many systems now use a stepper motor or another type of electrical gauge for this purpose.

A few cars still use conventional cable-driven speedometers. The head units usually work by either friction or magnetism. The frictional or magnetic drag increases as speed increases and this is used to move a needle. The flexible cable is driven from the gearbox output. It has square ends to transfer the rotation (Fig. 3.168).

> **Key fact**
> Many systems now use a stepper motor to drive speedometers and tachometers.

The system for driving most rev-counters (or tachometers) is similar to the electronic speedometer system. Pulses from the ignition primary circuit are often used to drive the gauge. The rev-counter needle response is damped to give a steady reading (Fig. 3.169).

Various gauges are used for instrumentation displays. The most common for fuel and temperature display are thermal and air-cored. Speedometers and tachometers use stepper motors (Fig. 3.170), electrical gauges or mechanical systems.

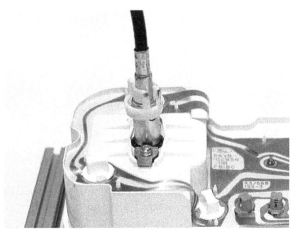

Figure 3.168 Speedometer cable and head

Figure 3.169 Tachometer and speedometer display

Figure 3.171 GPS display in a Jaguar

Figure 3.170 Stepper motor tachometer

Figure 3.172 GPS III Space Vehicle 05 launched June 2021 orbiting 12,550 miles above Earth (Source: Lockheed Martin)

3.5.3 Global Positioning System (GPS)

From 1974 to 1979, a trial using six satellites allowed navigation in North America for just four hours per day. This trial was extended worldwide by using eleven satellites until 1982, at which time it was decided that the system would be extended to 24 satellites, in six orbits, with four operating in each. There are now over 30 satellites in use. They are set at a height of about 21,000 km (13,000 miles), inclined 55° to the equator, and take approximately 12 hours to orbit the Earth. The orbits are designed so that there are always six satellites in view from most places on the Earth (Fig. 3.172).

The system was developed by the American Department of Defense. Using an encrypted code allows a ground location to be positioned to within a few centimetres. The signal employed for civilian use is artificially reduced in quality so that positioning accuracy is in the region of 50 m. Some systems, however, now improve on this and can work down to about 15 m.

GPS satellites send out synchronized information 50 times a second. Orbit position, time and identification signals are transmitted. A modern GPS receiver will typically track all of the available satellites, but only a selection of them will be used to calculate position. The times taken for the signals to reach the vehicle are calculated and from this information the computer can determine the distance from each satellite. The current vehicle position can then be worked out using three coordinates. Imagine the three satellites forming a triangle (represented in Figs 3.173–3.175 as A, B and C). The position of a vehicle within that triangle can be determined if the distance from each fixed point (satellite) is known. This is called triangulation.

Key fact

Position can be worked out using three coordinates; this is known as triangulation.

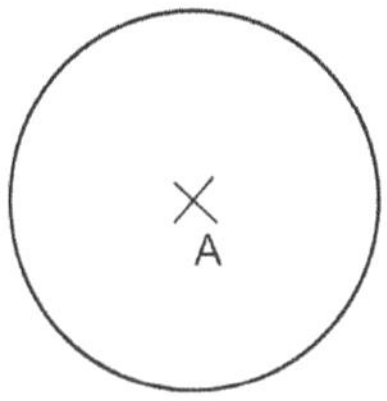

Figure 3.173 At a known distance from a fixed point A you could be anywhere on a circle

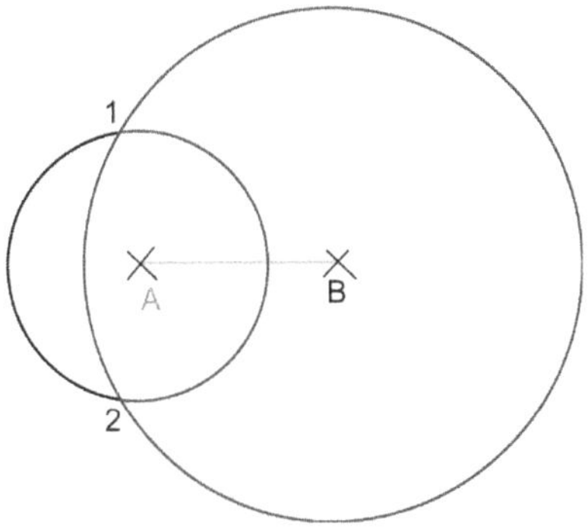

Figure 3.174 At a known distance from two fixed points A and B you must be at position 1 or 2

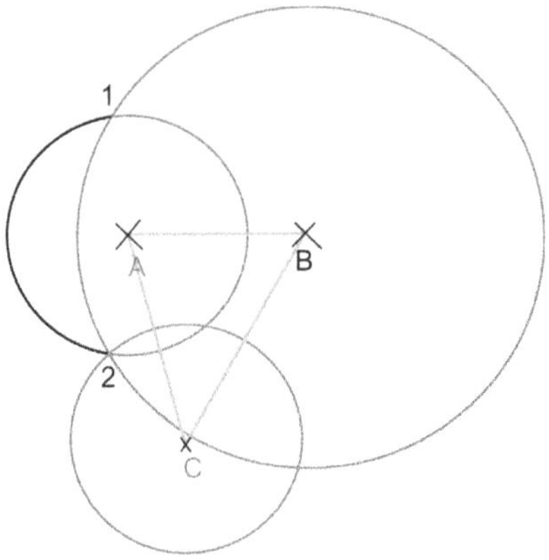

Figure 3.175 At a known distance from three fixed points A, B and C you must, in this case, be at position 2

The GPS receiver receives a signal from each GPS satellite. The satellites transmit the exact time at which the signals are sent. By subtracting the time when the signal was transmitted from the time it was received, the GPS can tell how far it is from each satellite. The GPS receiver also knows the exact position in the sky of the satellites, at the moment they sent their signals. So, given the travel time of the GPS signals from three satellites and their exact position in the sky, the GPS receiver can determine position in three dimensions: east/west, north/south and altitude (Fig. 3.176).

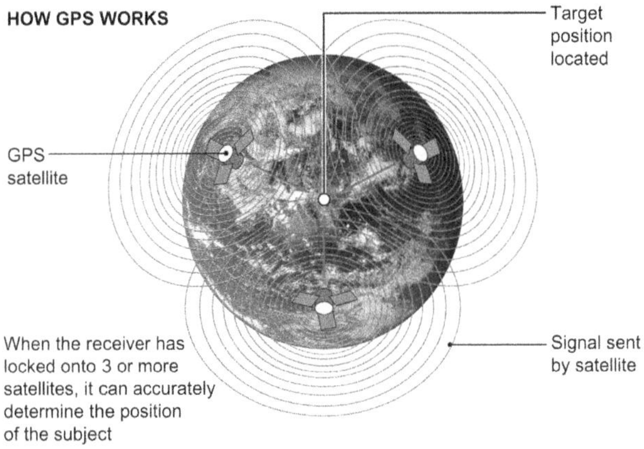

Figure 3.176 How GPS works

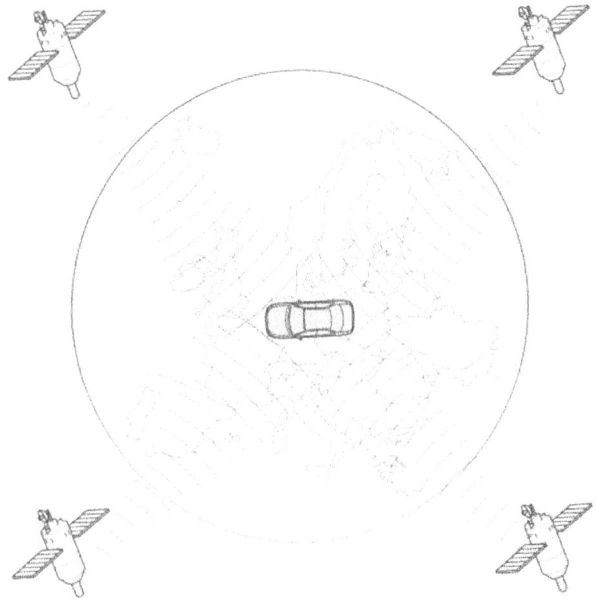

Figure 3.177 Four satellites determine vehicle position more accurately

To calculate the time the GPS signals took to arrive, the GPS receiver needs to know the time very accurately. The GPS satellites have atomic clocks that keep very precise time, but it is not feasible to equip a GPS receiver with such a device. However, if the GPS receiver uses the signal from a fourth satellite it can solve an equation that lets it determine the exact time, without needing an atomic clock (Fig. 3.177).

If the GPS receiver is only able to get signals from three satellites, position can still be calculated, but less accurately. If only three satellites are available, the GPS receiver can get an approximate position by making the assumption that you are at mean sea level. If you really are at sea level, the position will be reasonably accurate, but if you are driving in the

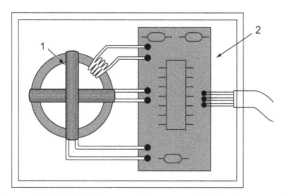

Figure 3.178 Field sensor: 1, crossed coils; 2, control circuit. (Source: Ford)

mountains, the two-dimensional fix could be several hundreds of metres out.

The magnetic field sensor in a GPS unit is a key component (Fig. 3.178). It determines direction of travel in relation to the Earth's magnetic field. It also senses the changes in direction when driving round a corner or a bend. The two crossed measuring coils sense changes in the Earth's magnetic field because it has a different effect in each of them. The direction of the Earth's field can be calculated from the polarity and voltage produced by these two coils. The smaller excitation coil produces a signal that causes the ferrite core to oscillate. The direction of the Earth's magnetic field causes the signals from the measuring coils to change depending on the direction of the vehicle.

To use most satellite navigation systems, the destination address is entered using a joystick control, cursor keys or something similar. The systems usually 'predict' the possible destination as letters are entered so it is not usually necessary to enter the complete address. Once the destination is set, the unit will calculate the journey. Options may be given for the shortest or quickest routes at this stage. Driving instructions, relating to the route to be followed, are given visually on the display and audibly through speakers.

Even though the satellite information provides a positional accuracy of only about 50 m, using dead-reckoning, intelligent software can still get the driver to their destination with an accuracy of about 5 m in some cases. Dead-reckoning means that the vehicle position is determined from speed and turn signals.

The computer can update the vehicle position from the GPS data by using the possible positions on the stored digital map (Fig. 3.179). This is because in many places on the map only one particular position is possible: it is assumed that short-cuts across fields are not taken! Dead-reckoning even allows navigation when satellite signals are disrupted.

Key fact

Dead-reckoning allows basic navigation when satellite signals are disrupted.

Vehicle global positioning systems use a combination of information from satellites and sensors to accurately determine the vehicle's position on a digital map. A route can then be calculated to a given destination. Like all vehicle systems, GPS continues to develop and will do for some time yet as more features are added to the software. Already it is possible to 'ask' many systems for the nearest fuel station or restaurant, for example.

3.5.4 Laser head-up displays (HUDs)

HUDs are now available as standard or optional equipment on many high-end cars. They are also being incorporated into a range of advanced driver assistance systems (ADAS). This is done to help drivers stay focused on the road ahead rather than looking down at their dashboard. The HUD places road speed, warning signals, and indicator arrows on the driver's windscreen directly in their line of sight.

One of the newest HUDs uses laser diode drivers to pulse high-intensity red, green, and blue (RGB) lasers. These project high-definition (HD) video onto the windscreen These are described as augmented-reality HUDs and can, for example, paint a transparent arrow directly onto the road in front of the car, to make GPS directions easy to follow. Another key feature is that they can highlight other objects, like pedestrians, animals or vehicles that might present a hazard.

The laser scanned projection system enables the use of lower cost optics. The main components are

Figure 3.179 As the driver follows the instructions for the first right turn here, the system will 'know' the location to within a metre or so if steering angle is used as an input

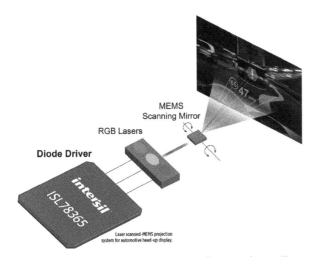

Figure 3.180 Laser HUD principle (Source: Intersil)

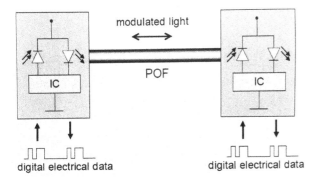

Figure 3.181 Laser driver circuit (Source: Intersil)

the laser diode driver, laser diodes, shaping optics, the oscillating mirror and control electronics. The RGB colour laser diodes are pulsed synchronously as the mirror is scanned across the display field. The image is then drawn pixel-by-pixel across the display field, which is shown on the screen. Of course, this happens very quickly so the driver just sees the completed images. Each pixel is pulsed very rapidly to create HD resolution.

Laser systems offer better electrical efficiency than earlier LCD frame-based projection systems. This is because the instrument information does not fill the entire HUD field display area as it only presents time-sensitive information for a short duration of time. Augmented reality information can have over 70% of the display pixels turned off. A laser scanned HUD only consumes electrical power when there are relevant pixels to be projected, thus reducing power consumption.

3.5.5 Connected vehicles

The car of the future will be connected. This is because using up-to-the-minute information from the

internet will get vehicle occupants to their destination even more safely, efficiently and conveniently. Integration into the internet of things also unlocks a host of vehicle-related services.

Information on traffic jams, black ice and wrong-way drivers is available in the cloud. When combined with infrastructure data from parking garages and charge spots, this provides a broader perspective known as the connected horizon. In the connected vehicle, the driver can 'see' over the top of the next hill, around the next bend and beyond. Because future cars will warn drivers in plenty of time about sudden fog or about a line of cars stopped behind the next bend, driving will be safer.

5G follows previous mobile network generations 2G, 3G, and 4G. Compared to current networks (mostly 4G and 3G technology), 5G is designed to be faster and more reliable. It also has greater capacity and lower response times. The key benefit of 5G is speed, which can be more than 1 Gb/s (1000 Mbit/s) and could reach ten times this figure. The speed we experience in the real world depends on many factors, such as how far we are from a base station and how many other people are using the network at the same time. However, a typical user-experienced data rate for downloads is expected to be a minimum of 100 Mbit/s (which is much faster than existing systems).

5G also has access to a wider spectrum and higher frequencies, in particular millimetre wave, which is the band between 30 GHz and 300 GHhz. This means that networks will be able to handle many high-demand applications all at once. Because of these improved speeds and short latency, automakers will be able to download more data and use the cellular network for some safety-related vehicle-to-everything (V2X) features.

Features like tele-operated driving and cooperative manoeuvres will become possible. Data collected from vehicle cameras, meanwhile, can be collated

Figure 3.182 Pedestrians can be 'enhanced' (Source: Nvidia)

259

Figure 3.183 Bosch technology puts the car online (Source: Bosch Media)

Figure 3.184 A connected car drives more proactively than a human! (Source: Bosch Media)

to create maps that include real-time updates such as accidents and temporary roadworks. Because of the higher bandwidth, large high-resolution map tiles could be uploaded and downloaded as needed. Cybersecurity will be a major issue, and most experts believe communications will have to be stored using techniques such as blockchain. This will ensure that information in messages cannot be tampered with, and that IDs are trusted.

Connectivity also enhances vehicle efficiency. For example, precise data about traffic jams and the road ahead makes it possible to optimize charging management in hybrid and electric vehicles along the selected route. And because the car thinks ahead, the diesel particulate filter can be regenerated just before the car exits the motorway, and not in the subsequent stop-and-go traffic. Connectivity

improves convenience as well, as it is a prerequisite for automated driving. It is the only way to provide unhurried braking in advance of construction zones, traffic jams and accident scenes.

Along with driving data and information on the vehicle's surroundings, the connected car also captures data on the operation of individual components. Running this data through sophisticated algorithms permits preventive diagnostics. For example, the data collected from an injection nozzle can be put through distributed algorithms in the cloud and in the vehicle to predict the part's remaining service life. The driver or fleet operator can be notified immediately, and an appointment made with the workshop in good time. In this way, it is often possible to avoid expensive repair and down times, especially for large commercial vehicles.

Buildings, hedges or a truck can quickly obscure drivers' view, especially at intersections. If a road user is driving carelessly, it is often a matter of milliseconds that decide whether there is a collision or not. However, vehicle connectivity can greatly reduce the number of resulting accidents by promptly providing information that is outside the driver's and the vehicle's field of vision.

Outside of cities, where vehicles travel at higher speeds, there is a definite advantage if data takes the short route via the local cloud. Compared to solutions that exchange information via a central cloud, local cloud approaches are at least three times faster and they have much lower variances in the case of vehicle-to-vehicle latencies under 2 ms. In some situations, this can make the difference as to whether the

The components of the "Drive with confidence" phase

▷ **Prevent exploits**	▷ **Detect suspicious behavior**	▷ **Respond by recovering gracefully and safely**
• Authenticating entities • Managing access • Encrypting data	• Uncovering anomalies • Applying security analytics and intelligence • Controlling versions	• Managing vulnerabilities • Implementing a Vehicle Security Operations Center • Improving continuously

Figure 3.185 Vehicle occupants will exchange lots of different types of data so security of the link will be essential (Source: IBM)

Figure 3.186 V2X can prevent accidents (Source: Continental)

Figure 3.187 Cloud connection (Source: Bosch Media)

261

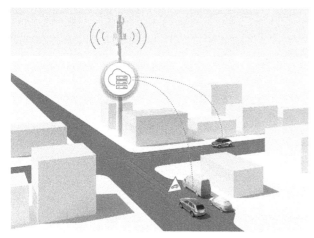

Figure 3.188 Local 'cell-based' cloud intersection assistant (Source: Bosch Media)

information reaches the car on time and whether the driver, or the safety function can react quickly enough.

To enable connected and automated driving in the future, vehicles must be able to easily communicate with one another as well as with their surroundings. There is currently no globally standardized technical basis for this exchange of data, which is known as vehicle-to-everything communication, or V2X. Vehicles will in future communicate using the wide variety of different standards implemented by countries and vehicle manufacturers around the world.

Cars can use the Wi-Fi networks available in cities, while elsewhere they can communicate using, for instance, cellular networks. The complex task of managing these diverse communication options

is handled by software. It continuously searches for the best transmission technology that suits the requirements and switches automatically between the available alternatives. The software therefore maintains continuous and seamless vehicle connectivity, ensuring cars can, for example, reliably alert one another to accidents, and passengers can enjoy uninterrupted music streaming.

Initially, most vehicles will connect directly to the cloud. Soon, increasing numbers of vehicles will also be able to communicate directly with one another (V2V) as well as with traffic signals, road construction sites, pedestrian crossings and buildings, etc. They will then be able to alert one another to potential hazards like the approaching tail end of a traffic jam, accidents and icy conditions.

As well as keeping an eye on which V2X communication technologies are currently available for use, the software also closely monitors the costs and data transmission latency of each alternative connection option, since not every technology is suitable in every situation.

For example, when it comes to alerting a driver to another vehicle that is about to pull out in front of them from a side street, every millisecond counts. This kind of critical information must be communicated in real time using highly reliable technology that is always ready for use, even if that means the resulting data transmission costs are greater. Large volumes of data can be transmitted via Wi-Fi in a short space of time, though a downside is that public or home Wi-Fi hotspots are not always available.

Figure 3.189 Communication with everything (Source: Bosch Media)

Figure 3.190 Connectivity unit (Source: Bosch media)

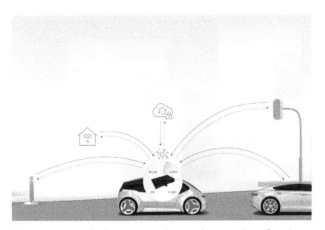

Figure 3.191 Software continuously searches for the best transmission technology (Source: Bosch media)

Figure 3.192 V2X connectivity

3.5.6 Hacking and cybersecurity

As more systems on vehicles are connected to the outside world by radio waves of some sort, or they scan the world outside of the car, then more opportunities are presented to hackers. Manufacturers are working very hard to reduce the chances of this happening and are helped in this process by what can be described as ethical hackers. There have been several interesting examples in the news recently; we look below at two areas that illustrate the new vulnerabilities that come with enhanced connectivity. In addition, there is currently a debate about the legality of hacking your own car as well as ways to stop other hackers.

Securing vehicles of the future is a cybersecurity challenge, to say the least. When something goes wrong with your home computer, 'crash' is only a metaphor. A recent survey found that nearly 100% of today's cars include wireless technologies that could be insecure, and most manufacturers may not be able to easily determine if their vehicles have been hacked. Physical attacks via onboard diagnostic devices have shown it could be possible to manipulate some systems, steering for example, even while cars are moving.

Security, and specifically cybersecurity, is an increasingly urgent issue for the automotive industry. Systems are becoming more complex, and the threat environment is also becoming more capable and sophisticated. This issue will only be made worse for automated driving vehicles (ADVs) because of vehicle to everything (V2X) communication. A range of best practices exist ranging from management focus down to technical measures, which can help to control the risk.

The shift from independent, closed vehicle systems to that of a connected environment is a massive change for the industry.

All vehicle systems must therefore have three mutually reinforcing properties:

- **Secure:** Prevention is better than cure and effective risk management begins by preventing system breaches in the first place.
- **Vigilant:** Hardware and software can degrade, and the nature and type of attacks can change. No level of security is perfect. Security must therefore be monitored to ensure it is still secure or to see if it has been compromised.
- **Resilient:** When a breach occurs, there must be a system in place to limit the damage and re-establish normal operations. The system should also neutralize threats and prevent further spread.

The importance of securing individual sensors is critical in connected cars. Keep in mind these vehicles are a kind of internet-connected data centre on wheels. A typical car can contain:

- About 70 computational systems running up to 100 million lines of code.
- GPS devices that aid navigation and report on real-time traffic.
- Diagnostic systems that check maintenance needs and send an alert in the event of an accident or breakdown.

As infrastructure evolves, cars will also be able to communicate with roadside devices such as traffic lights. Security must be part of the design and development, not bolted on at the end.

As vehicles get smarter, cybersecurity in the automotive industry is becoming an increasing concern. Whether we're turning cars into Wi-Fi-connected hotspots or equipping them with millions of lines of code to create fully autonomous vehicles, cars are more vulnerable than ever to hacking and data theft.

3.5.7 Automotive software updates

As smartphone and computer users we consider it quite normal to be told that a software update is available. We are then usually asked whether we want to install the update now or later, or we can choose to automatically download updates as they become available. Vehicle software updates are now becoming just as important, and the methods for delivering these updates are changing.

There are hundreds of millions of lines of code embedded in the software of a typical vehicle. The software is also controlling more functions. This could include calling for roadside assistance when an air bag is deployed, or braking automatically to prevent a collision. It is, therefore, essential that vehicle software is 100% reliable. A program crashing has far worse consequences than me losing a few lines of this document! Software updates are issued to correct bugs, improve performance, add new features, or protect against recently discovered security vulnerabilities.

> **Key fact**
> There are hundreds of millions of lines of code embedded in the software of a typical vehicle.

Updates have been delivered mostly by main dealers during regular servicing – or by recalls. However, it is also possible for independents to do updates using 'pass-through'.

A pass-through device is used in conjunction with a computer to reprogram vehicle control modules through the data link connector (DLC) and the CAN bus. The ECUs may require updating to improve functions or to recognize new components that have been fitted.

Each manufacturer has its own method, but SAE International standardized the J2534 universal requirements in 2004. This required all manufacturers of vehicles sold in the USA and Europe to accept

Figure 3.193 Data link connector (DLC).

powertrain reprogramming through specific universal parameters. In the USA and Europe, vehicle manufacturers must therefore provide ECU reprogramming functionality to all workshops, whether independent or franchised.

> **Key fact**
> SAE International standardized the J2534 universal requirements in 2004.

Because of the existence of pass-through, vehicle manufacturers have had to ensure that the reprogramming software applications (APIs) are compatible with standardized J2534 vehicle communication interfaces (VCIs). Because of this, independent workshops can access OEM applications by subscribing to the appropriate websites. By downloading the software to a PC and connecting it to the vehicle with a J-2534 VCI, you have the same level of access to the vehicle as a main dealer.

In summary, to reprogram a vehicle ECU you need:

1 computer equipped with a Windows operating system
2 J2534 vehicle communication interface (VCI)
3 OEM application programming interface (API)
4 knowledge of how to use the software!

> **Key fact**
> Reprogramming is sometimes described as 'flashing'.

The J2534 hardware works like a bridge between the vehicle's ECU and the PC. This pass-through device translates messages sent from the PC into messages with the same protocol being used in the vehicle ECU. J2534 supports a range of protocols.

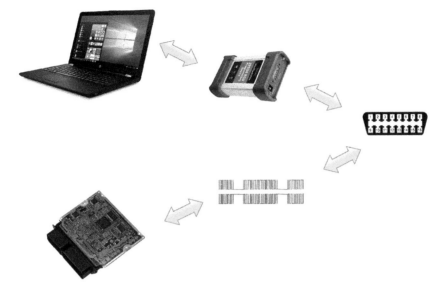

Figure 3.194 Pass-through principle and stages

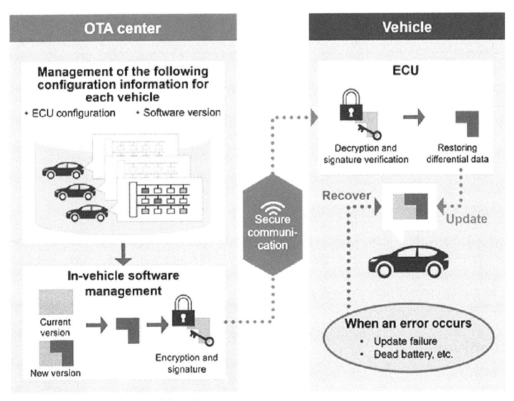

Figure 3.195 System outline (Source: Hitachi)

Software

Reprogramming an ECU using J2534 is done from a PC, preferably a laptop computer, with a Windows operating system. Each vehicle manufacturer has their own software application (API) used for analysing and programming their vehicles. The application will have complete information on the ECUs that are supported by it. The application also includes a user interface where choices can be made, depending on the ECU and what action is required. The APIs can be downloaded from the internet or installed from a DVD. How this API is provided depends on the manufacturer, but they do make a charge. The price differs a lot between

265

manufacturers, a one-year subscription, for example, currently costs between £75 and £2500 (2022).

It is now becoming more common for software to be updated 'over the air' as almost all new cars have a data connection to the cellular network.

Over the air (OTA)

Software updates over the air offer a tremendous potential for cost savings by avoiding recalls. Functional and security issues can be fixed remotely. OTA software updates also provide the opportunity to implement new business models with for instance new features installed on demand, after the vehicle has been delivered. There are a number of key factors to consider:

▷ Only authorized and verified data shall be flashed by authorized persons.
▷ Reprogramming has to be fail-safe; rollback strategies are needed in case of problems.
▷ Updates have to be dependable; the system must detect whether vehicles missed an update.

Hitachi's solution for wirelessly updating in-vehicle software, for example, applies standard security technologies such as encryption, digital signatures and mutual authentication. This enables secure updates that protect in-vehicle software from hacking and other attacks.

This solution uses differential updates, so instead of sending and updating in-vehicle software in its entirety, the solution sends and updates only parts of the software that have changed. This reduces the amount of data to be sent and is faster. This also improves availability and reduces the load on the vehicle battery. Depending on the state of the vehicle and the characteristics of its ECUs, the solution can also recover the software when an error occurs, such as an update failure or a dead battery (Fig. 3.195).

3.6 Air conditioning

3.6.1 Air conditioning fundamentals

A vehicle fitted with air conditioning allows the temperature of the cabin to be controlled to the ideal or most comfortable value. This is usually determined by the ambient conditions. Air conditioning can be manually controlled or, as is often the case, combined with some form of electronic control. The system as a whole can be thought of as a type of refrigerator or heat exchanger. Heat is removed from the car interior and dispersed to the outside air. To understand the principle of air conditioning or refrigeration, the terms and definitions described here will be useful:

▷ Heat is a form of energy.
▷ Temperature is the degree or intensity of heat of a body, and the condition that determines whether or not it will transfer heat to, or receive heat from, another body.
▷ Heat will only flow from a higher to a lower temperature.
▷ Change of state describes the changing of a solid to a liquid, a liquid to a gas, a gas to a liquid or a liquid to a solid.
▷ Evaporation describes the change of state from a liquid to a gas.
▷ Condensation describes the change of state from gas to liquid.
▷ Latent heat describes the energy required to evaporate a liquid without changing its temperature.

Latent heat, in the change of state of a refrigerant, is the key to air conditioning. As an example of this, if you put a liquid such as vodka on your hand it feels cold, particularly if you blow on it. This is because it evaporates and the change of state, from liquid to a gas, uses heat from your body (Fig. 3.196). This is why the process is often thought of as 'unheating' rather than cooling. The refrigerant used in many air conditioning systems changes state from liquid to gas at a much low temperature and works better than vodka!

> **Definition**
>
> **Latent heat:** Heat that is released or absorbed accompanying a change of state or of phase of a material.

The refrigerant used in older air conditioning systems is known as R134a. This substance changes state from liquid to gas at −26.3°C. R134a is HFC based. Earlier types were CFC based and caused problems

Figure 3.196 A liquid evaporating on your body uses heat so it feels cold

with atmospheric ozone depletion. While R134a has been one of the global standard automotive air conditioning refrigerants, it does not meet the EU F-gas legislation (517/2014), which came into force in January 2015. HFO gas known as R1234yf has a GWP of 4 and will therefore meet the EU regulation. It has physical properties like R134a so has the potential to be used in current R134a systems with limited system modifications. However, different types of refrigerant are NOT compatible and must NEVER be mixed.

Definition
CFC: Chlorofluorocarbon
HFC: Hydrofluorocarbons
HFO: Hydrofluoroolefin
GWP: Global warming potential

Figure 3.197 R134a refrigerant

Key fact
HFO R1234yf has a GWP rating one 335th that of HFC R-134a and only four times higher than carbon dioxide.

Key fact
R1234yf is flammable.

Key to understanding refrigeration is to remember that low-pressure refrigerant will have a low temperature, and high-pressure refrigerant will have a high temperature.

The layout of an air conditioning or refrigeration system is shown in Figure 3.198. The main components are the evaporator, the condenser and the pump or compressor. The evaporator is situated in the car and, the condenser outside the car, in the air stream, and the compressor is driven by the engine.

As the compressor operates it causes the pressure on its intake side to fall. This allows the refrigerant in the evaporator to evaporate and draw heat from the vehicle interior. The high-pressure or output side of the pump is connected to the condenser. The pressure causes the refrigerant to condense, in the condenser, and it thus gives off heat outside the vehicle as it changes state.

The compressor pumps low-pressure but heat-laden vapour from the evaporator, compresses it and pumps it as a super-heated vapour under high pressure to the condenser. The temperature of the refrigerant at this stage is much higher than the outside air temperature.

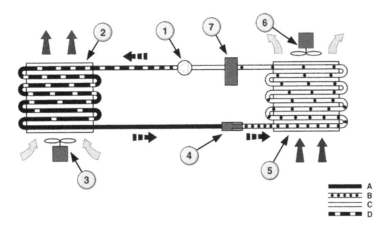

Figure 3.198 Air conditioning system layout: 1, compressor; 2, condenser; 3, auxiliary fan (depending on model); 4, fixed orifice tube; 5, evaporator; 6, heater/air conditioning blower; 7, suction accumulator/dryer; A, high-pressure warm liquid; B, low-pressure cool liquid; C, low pressure, gaseous and cool; D, high pressure, gaseous and hot. (Source: Ford Motor Company)

It therefore gives up its heat via the fins on the condenser, as it changes state back to a liquid. This high-pressure liquid is then passed to a receiver dryer, which stores any vapour that has not yet turned back to a liquid. Alternatively, a suction accumulator is used on the low-pressure side.

A desiccant, which is a drying agent, removes any moisture that is contaminating the refrigerant. Refrigerant, like brake fluid, is hygroscopic, which means it absorbs water. The high-pressure liquid is now passed through the thermostatic expansion valve, or a fixed orifice, and is converted back to a low-pressure liquid as it passes through a restriction into the evaporator.

Definition

Desiccant: A drying agent.

As the liquid changes state to a gas in the evaporator, it takes up heat from its surroundings, thus cooling or 'unheating' the air, which is forced over the fins. The low-pressure vapour leaves the evaporator, returning to the pump, thus completing the cycle.

If the temperature of the refrigerant increases beyond certain limits the condenser cooling fans can be switched on to supplement the ram air effect.

Changing a liquid into a gas uses energy. This energy, in the form of heat, is taken from inside the vehicle. When the gas is compressed, it gets hotter and the heat can be given off outside the vehicle. This turns the gas back into a liquid and the cycle starts again.

3.6.2 Air conditioning components

The main components of an air conditioning system are:

▶ a compressor
▶ a condenser
▶ an evaporator
▶ a control valve
▶ a dryer.

An air conditioning system compressor is shown in Figure 3.199. It is belt driven from the engine crankshaft and it causes refrigerant to circulate through the system. The compressor is controlled by an electromagnetic clutch, which may be under either manual control or electronic control, depending on the type of system.

Key fact

Some high voltage (HV) vehicles use HV compressors. The AC can be integrated into the HV to keep it cool.

Figure 3.199 Air conditioning system compressor in place

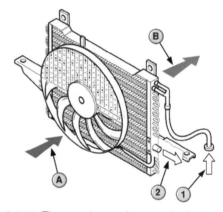
Figure 3.200 The condenser is very similar to the cooling system radiator: A, cooled air; B, heated air; 1, gaseous refrigerant; 2, liquid refrigerant

Figure 3.201 Receiver/dryer

The shaft is driven by the engine via a multi-V belt. Five double pistons are arranged around the driving shaft. The swash plate, which is mounted on the main shaft, causes the pistons to move backwards and forwards.

The condenser is fitted in front of the vehicle radiator. It is very similar in construction to the radiator and fulfils a similar role. The heat is conducted through the aluminium pipes and fins to the surrounding air and then by a process of radiation and convection is dispersed by the air movement (Fig. 3.200).

Figure 3.201 shows a typical receiver/dryer assembly. It is connected in the high-pressure line between the condenser and the thermostatic expansion valve. This component carries out four tasks:

▷ It holds refrigerant in a reservoir until a greater flow is required.
▷ It prevents contaminants circulating through the system by using a filter.
▷ It enables vapour retention until the vapour converts back to a liquid.
▷ It removes moisture from the system using a drying agent.

Systems that use a fixed orifice control system usually use a low-pressure accumulator instead of a receiver/dryer. This component carries out the same tasks as a receiver/dryer.

A sight glass is fitted to some receiver/dryers. This gives an indication of refrigerant condition and system operation. The refrigerant generally appears clear if all is in order.

The main function of the thermostatic expansion valve (Fig. 3.202) is to control the flow of refrigerant as demanded by the system. This, in turn, controls the temperature of the evaporator. A temperature sensor is fitted in the evaporator on some systems.

Some systems use a fixed orifice control valve. The operation is quite simple. A fixed orifice, which is a small hole, only allows a certain flow rate. Filters are included to prevent contamination. The fixed orifice is the connection between the low- and high-pressure systems.

The evaporator is similar in construction to the condenser, consisting of fins to maximize heat transfer. It is mounted in the car under the dash panel, forming part of the heating and ventilation system (Fig. 3.203). As well as cooling the air passed over it, the evaporator removes moisture from the air. This is because the moisture in the air condenses on the fins. The action is much like breathing on a cold pane of glass. A drain is fitted to remove water.

Figure 3.202 Thermostatic expansion valve

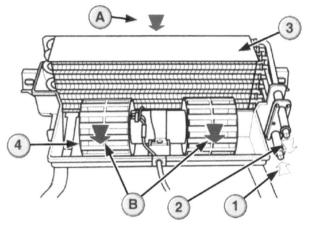

Figure 3.203 Evaporator housing and blower motor: A, hot air; B, cooled air; 1, liquid refrigerant; 2, gaseous refrigerant; 3, evaporator; 4, booster fan

Key fact

The main components of an air conditioning system are the condenser, evaporator and compressor.

The key components of an air conditioning system are the condenser, evaporator and compressor. Refrigerant takes heat from the car as it evaporates. It is then compressed, and condenses in the condenser. It gives off heat to the atmosphere.

3.6.3 Heat pumps

A heat pump is a device that transfers heat energy from a source of heat to a destination called a heat sink. Heat pumps are designed to move thermal energy in the opposite direction of natural heat flow by absorbing heat from a cold space and releasing it to a warmer one. A heat pump uses a small amount of external power to do this,

When a heat pump is used for heating, it uses the same basic refrigeration-type cycle used by an air

269

conditioner or a refrigerator but in the opposite direction. This therefore releases heat into the conditioned space rather than the surrounding environment. Imagine a car AC system working in reverse so it takes heat from outside and releases it in the cabin rather than the other way around. In heating mode, heat pumps are three to four times more efficient in their use of electric power than simple electrical resistance heaters.

Hot air naturally flows from warm to cold. A heat pump can absorb heat from a cold space and release it to a warmer one. Heat (energy) is not conserved in this process, which requires some level of external energy, such as electricity. Heat pumps are used to transfer heat because less high-grade energy is required for their operation than is released as heat. In electrically powered heat pumps, the heat transferred can be three or four times larger than the electrical power consumed, giving the system a coefficient of performance (COP) of 3 or 4, as opposed to a COP of 1 for a conventional electrical resistance heater, in which all heat is produced from input electrical energy.

Heat pumps use a refrigerant as an intermediate fluid to absorb heat where it vaporizes, in the evaporator, and then to release hot air where the refrigerant condenses, in the condenser. The refrigerant flows through insulated pipes between the evaporator and the condenser, allowing for efficient thermal energy transfer. Reversible heat pumps work in either direction to provide heating or cooling to the internal space. They use simple valves to reverse the flow of refrigerant from the compressor through the condenser and evaporator.

The refrigerant needs to reach a sufficiently high temperature, when compressed, to release heat through the condenser. Similarly, the fluid must reach a sufficiently low temperature when allowed to expand in the evaporator. The pressure difference must be great enough for the fluid to condense at the hot side and still evaporate in the lower pressure region at the cold side. The greater the temperature difference, the greater the required pressure difference and consequently, the more energy is needed to compress the fluid. Therefore, the efficiency decreases with increasing temperature difference.

The following three diagrams show the three modes of operation:

▶ heating
▶ cooling
▶ mixed.

In cooling mode, the system acts like normal air conditioning where excess heat is taken from the

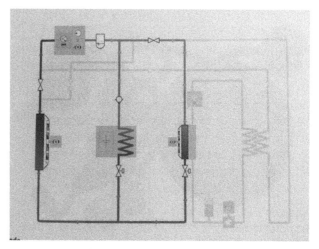

Figure 3.204 Cooling mode (Source: BMW)

cabin via the evaporator and is lost through the external condenser.

In heating mode, refrigerant shut-off valves are closed, and the heating ones opened. The hot (pressurized) refrigerant is now circulated through a heat pump heat exchanger, which causes the heat to be transferred to the coolant for the heater circuit. To ensure this transfer takes place, an expansion valve at the outlet of the pump is used to further increase refrigerant pressure and hence its heat. The normal expansion valve used for the AC is also activated, further increasing the temperature. The return circuit for the now cooled refrigerant is the opposite way through the condenser.

In mixed operating mode, the expansion valves are opened and closed as needed to achieve the desired temperature, and there is no reverse flow. The evaporator is cold to ensure dehumidification and the high voltage battery (in this example) can also be cooled. Refrigerant heat is used to ensure the desired cabin temperature is reached.

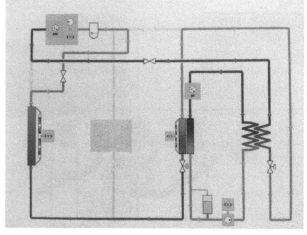

Figure 3.205 Heating mode (Source: BMW)

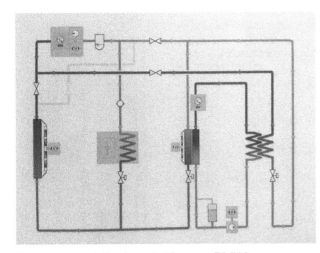

Figure 3.206 Mixed mode (Source: BMW)

Heat pumps are starting to be used on some electric vehicles because they improve overall efficiency. They can be thought of as an AC system working in reverse.

3.7 Electrical systems: new developments and updates

3.7.1 Touchscreen with haptic feedback

The idea behind haptic feedback is that users will be able to operate infotainment applications such as navigation, radio, or smartphone functions more interactively – and safely. The keys displayed on the touchscreen shown as Figure 3.207, have the feel of realistic buttons so that it is often possible for users to find their way around the keyboard without looking while operating the applications. They can therefore keep their eyes on the road for much longer periods, substantially enhancing safety while driving.

Figure 3.207 Touchscreen with feedback. (Source: Bosch Media)

Definition

Haptic: Relating to the sense of touch, in particular to the perception and manipulation of objects. Haptic feedback devices create the illusion of substance and force.

The new touchscreen offers a unique form of interaction. When touched, the display responds with haptic elements as well as visual and acoustic signals. Drivers can feel the keys on the touchscreen without looking thanks to variances in the surface structures, without immediately triggering an action. Rough, smooth or even patterned surfaces stand for different buttons and functions. The virtual button is not activated until the operator presses it more firmly. Users have the feeling that they are pressing a normal, mechanical button. In appearance, however, the touch screen with haptic elements does not differ from a conventional display.

Key fact

The touchscreen also recognizes the amount of pressure applied by the fingers.

The touch screen also recognizes the amount of pressure applied by the fingers and activates different functions accordingly. Light pressure, for example, initiates the Help function; by applying varying pressure, users can control how fast or slowly they scroll through a list. The touch screen is equipped with two sensors:

1 conventional touch sensor
2 pressure sensor.

Special software and suspension mechanics are employed to create the various surface structures.

3.7.2 Navigation and the new NDS data standard

The map display on built-in navigation systems is becoming even more engaging and relevant. Buildings extend skyward, enabling you to get your bearings more easily, and visible changes in terrain height combined with integrated satellite imagery produce an almost photorealistic look. This is made possible by Bosch's advanced navigation software, which takes data compliant with the new Navigation Data Standard (NDS) and processes it in a 3D rendering module to turn it into a visually stunning map.

The system will work offline but if an internet connection is available, the system can enhance the map display with dynamic data. In the future, this will

271

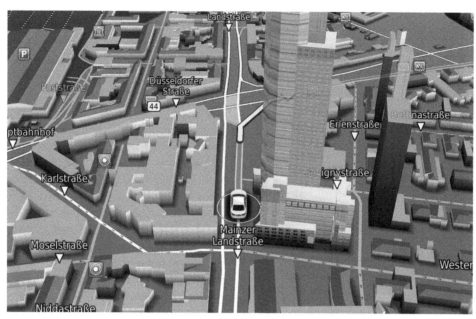

Figure 3.208 Powerful 3D map engine supports continuous zoom. (Source: Bosch Media)

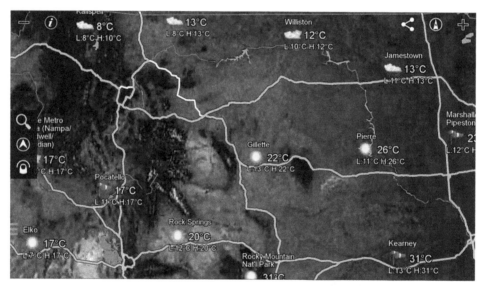

Figure 3.209 Dynamic data from the connected horizon – more than just traffi c information. (Source: Bosch Media)

allow integration of the latest weather information or prices at fuel stations along the route, for example.

The key component in the new navigation software is a 3D map engine based on OpenSceneGraph. It superimposes three-dimensional elements like buildings using an additional display layer and can also make them transparent, keeping the route visible to the driver when it goes behind structures. The driver can smoothly zoom the visible map area, from the highest level of detail to the world view. Using topographical information contained in the NDS data, the software displays differences in terrain height. It

will even be possible to artificially bend up the map in the direction of the horizon, thus maximizing the amount of screen area used to display the route.

Key fact

The key component in the new navigation software is a 3D map engine based on OpenSceneGraph. www.openscenegraph.org

For interacting with the system, the driver can choose between voice input, multi-touch and handwriting recognition. It is also possible to show different areas of the map on different screens at the same time, such as the displays in the centre console and instrument cluster. The level of display detail can be adapted to the infotainment system's computing power and memory. The navigation software can thus be configured to suit carmakers' requirements.

Traffic congestion can already be portrayed on the map in near real time. But if the infotainment system has internet access, it will be possible in the future to integrate even more information in the map display. The Bosch connected horizon, for example, gives real-time access to data on road conditions stored in the cloud. The 3D map engine can visualize this data, so that areas of the map appear in a different colour if there is particularly heavy rain or a risk of black ice.

In electric vehicles, the system uses a coloured, transparent overlay to indicate the current range on the map for the remaining battery charge.

3.7.4 **Alternator developments**

Bosch presented the world's first voltage regulating generator for motor vehicles about 100 years ago. Though initially designed as a power source for the innovation of electric headlights, it ultimately laid the foundations for the electrification of the automobile. Headlights were quickly followed by numerous other electrical consumers, such as the electric starter and the first electric drive for windscreen wipers (launched in 1926). The first generators provided a modest output of just 4 A. The latest Bosch generators (the Power Density Line for passenger cars) have four output categories ranging from 130 to 250 A. Without the continuous improvements made to generator (alternator) efficiency, today's machines would need to weigh as much as a small car to provide all the power needed by modern electrical systems with their numerous consumers!

The ratio between the power supplied to a machine and the power it produces is known as its efficiency. In an alternator, there are five key areas where power is lost as it is converted from mechanical (or kinetic) energy to electrical energy. Ongoing developments aim to optimize all these losses.

Definition

Efficiency: The ratio between the power supplied to a machine and the power it produces.

1 Copper losses: These are caused by the electrical resistance of the stator and rotor windings.

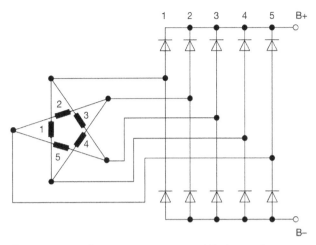

Figure 3.210 Pentagram connected 5-phase alternator with a bridge rectifier

2 Iron losses – stator: These occur in the iron of the stator core. They are caused by alternating magnetic fields, which results in eddy currents that produce heat. They can be reduced by laminating the stator core.
3 Iron losses – rotor: Eddy currents are produced on the surface of the rotor because of the fluctuations of magnetic fields caused by the slits in the stator core.
4 Rectifier losses: Rectifier diodes cause a voltage drop. This results in heat; hence the need for the rectifier heat sink. High-efficiency diodes (HEDs) can reduce, but not eliminate this loss.
5 Mechanical losses: These are caused by friction in the bearings and sliding contacts, but mostly by air resistance (known as windage). This is caused by the rotor and fan and is particularly noticeable at higher speeds.

Alternator noise is caused by the cutting of magnetic flux. Interestingly, on some alternators, it is possible to determine if they are working by listening for the 'whine' they produce. Of course, proper measurements are needed too! One area of research that may reduce the noise is the use of a 5-phase, pentagram-connected stator winding. Another method is to use two 3-phase windings offset by 30 degrees.

The 'Efficiency Line' (EL) series alternators are designed for vehicles that feature several safety and comfort functions, and therefore require a greater amount of electrical energy. By further improving the electrical design of this alternator and adding new diodes, its efficiency is as much as 77%. This means that the alternator needs less mechanical energy to generate the same electrical power. In

273

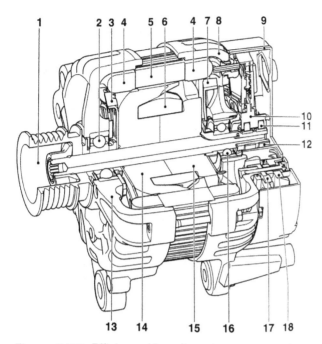

Figure 3.212 Alternator 28V regulator with a LIN interface

Figure 3.211 Efficiency Line alternator components: 1, Belt pulley; 2, A-side (drive-end) ball bearing; 3, A-side fan; 4, Stator winding heads; 5, Laminated stator core; 6, Rotor winding (excitation winding); 7, B-side fan; 8, B-side end shield; 9, Protective cap; 10, Brush holder; 11, Carbon brush; 12, Collector ring; 13, A-side end shield; 14, A-side claw pole; 15, B-side claw pole; 16, B-side ball bearing; 17, Negative heat sink of rectifier; 18, Positive heat sink of rectifier. (Source: Bosch)

real driving conditions, fuel consumption and CO_2 emissions can be reduced by as much as 2%. These alternators can also withstand temperatures of up to 125°C.

The EL alternators are available in three sizes, covering a range between 130 and 210 A. They ensure that the battery is well charged even when the engine is idling or running at low rpm. A well-charged battery is a condition for start-stop systems. This means that if this fuel-saving system is used, consumption is reduced even further. The combination of a modern alternator and a start-stop system can reduce fuel consumption (and therefore CO_2 emissions) in urban driving by as much as 10%.

More fuel can be saved if the electrical energy is generated as far as possible during overrun (when rolling or braking). This calls for intelligent alternator control, which is provided by an interface such as the local interconnected network (LIN). The variable control of charging voltage lowers fuel consumption and extends the battery's life cycle.

3.7.5 Electronic starter motor control and stop-start system

An electronically controlled starting system differs from a normal system in that the driver is not in direct control. The driver sends a 'start' request to an ECU, which then in turn can carry out several different checks. For example:

▶ Is the driver authorized to start the engine?
▶ Is the engine stationary?
▶ Is battery state of charge adequate at the current temperature?
▶ Is the clutch disengaged or is the automatic transmission in neutral?

The checks only take a fraction of a second after which the ECU initiates the start. Once a set engine speed has been reached the starter is switched off. This always ensures the shortest possible start time and reduces starter wear. A similar process can be used for start-stop operation.

Stop-start systems were mostly used on cars with manual transmission but it is now possible to achieve the same feature to work with automatic transmission. The benefit of this system is 5% or more reduction in CO_2 and fuel consumption in urban traffic. Almost all European automakers are now integrating stop-start technology into compact cars, premium sedans, light trucks and even in powerful sports cars. Interest in this fuel-saving technology is also on the rise in the USA and China.

To ensure reliable operation a high degree of control of other systems is necessary. For example, electrical energy management that incorporates battery charge detection. Measures such as a DC-DC converter may also be needed to stabilize the supply to other electrical systems during the voltage drop caused by the starter.

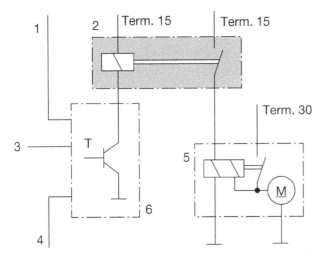

Figure 3.213 Automatic starting circuit: 1, Start signal from driver; 2, Relay; 3, Other input signals; 4, Park/neutral or clutch signal; 5, Starter; 6, ECU (Terminal 15 is ignition live and 30 is battery live)

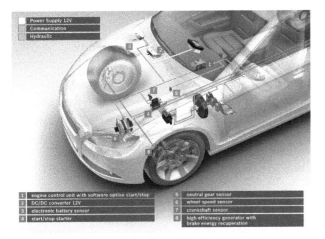

Figure 3.214 Stop-start systems can reduce fuel consumption and therefore CO_2 emissions by up to 5%. (Source: Bosch Media)

Engine systems must also be optimized to ensure a fast start and finally the starter and ring gear components must be enhanced to not wear out prematurely.

Stop-start systems, which largely use existing components such as the starter motor and engine control unit, can be quickly and cost-effectively adapted to different engines and vehicles. Therefore, this starter motor based method has now become very popular. Efficient generators that charge the battery as fast as possible are an ideal addition to their systems, since they allow the function to be used more often.

In vehicles fitted with automatic transmission, the stop-start function is very easy to use. The driver presses the brake pedal, and as soon as the car has stopped, the engine stops automatically. It starts again when the

brake pedal is released. This is very convenient for the driver, who uses the accelerator and brake as normal.

The system works completely in the background, and evaluates a wide array of sensor information before being activated. For instance, a battery sensor determines the battery's level of charge and the engine will only be switched off if a quick restart is guaranteed. Another example is the vehicle's cabin temperature. If the cabin temperature is too cold or too hot, the engine will continue to run until the occupants' desired temperature is reached. Finally, there is a DC-DC converter for stabilizing the voltage of the electrical system during starts to ensure that the radio, navigation system, or hands-free telephone operate without any interference or interruption.

To further reduce consumption and CO_2 emissions, engineers are continually working to extend engine shutdown times. This will apply initially to the time when the vehicle is coasting to a stop and then it will also apply to periods while driving, when the driver is not accelerating. The effectiveness of all safety and comfort functions will be guaranteed, even during these longer periods with the engine switched off. The enhanced stop-start systems could save up to an additional 10% of fuel.

Key fact
Enhanced stop-start systems could save up to 10% of fuel.

There are several enhancements to the starting system that are designed to improve driver experience and enhance component life but the main benefit is a reduction in consumption and CO_2 emissions.

3.7.6 Advanced driver assistance systems (ADAS)

Advanced driver assistance systems (ADAS) have been developed to automate/adapt/enhance vehicle systems for safety and better driving. Safety features are designed to avoid collisions and accidents by offering technologies that alert the driver to potential problems, or to avoid collisions by implementing safeguards and taking over control of the vehicle. Adaptive features may automate lighting, provide adaptive cruise control, automate braking, incorporate GPS traffic warnings, connect to smartphones, alert drivers to other cars or dangers, keep the car in the correct lane, or show what is in blind spots.

Definition
ADAS: Advanced driver assistance systems.

275

To not to be locked out of future service and repair work, workshops must understand these developments, as ADAS technology is an increasingly crucial component in vehicle design.

As one of the original equipment (OE) manufacturers that have pioneered ADAS technology, Hella is ideally placed to help the service and repair sector rise to the significant challenges it presents.

A perfect example of ADAS technology is adaptive lighting, such as Hella's Matrix LED system, which, through camera-controlled information processes, provides 'intelligent' intense, but glare-free lighting, which, it is claimed, has already helped to reduce road accidents at night by 30%.

From 2016, any new vehicle requires a minimum of two items of ADAS technology to achieve the current 'industry standard' 5-star Euro NCAP safety rating. This minimum requirement includes autonomous emergency braking (AEB) and lane departure warning (LDW), both of which

Figure 3.215 This unique, cost effective Camera System Calibration (CSC) tool is quick and simple-to-use and allows precise recalibration of the vehicle's ADAS components. (Source: Hella-Gutmann Solutions)

Figure 3.216 Wheel alignment is part of the system (Source: Hella Gutmann Solutions)

are complex systems that require specialist equipment to diagnose any faults and for their recalibration.

By its very nature, ADAS technology, which also includes hazard recognition, blind spot detection and fatigue warning, is of fundamental importance to driver and pedestrian safety and therefore needs to function precisely, efficiently and reliably. These systems, and the associated actions they perform, are all controlled by sensors, cameras and complex software, which means that incorrect inputs from any of these elements can have catastrophic consequences. Service and repair work on these vehicles, therefore, carries extra responsibility from those undertaking it.

ADAS technology will have the biggest impact on the bodyshop and windscreen replacement sectors, where accident damage and glass replacement logically requires the system to be reset. Traditional workshops concentrating on mechanical repair will also be confronted by it because even something as commonplace as adjustment to the vehicle's steering geometry will require the system's recalibration.

Ultimately therefore, every workshop will need to have access to the necessary equipment and training to be able to correctly reset the ADAS components following an accident, windscreen replacement or geometry-related repair. In fact, the vehicle manufacturers now specify that recalibration of ADAS must take place after work has been performed, otherwise the workshop could potentially face liability.

Naturally there could be serious repercussions if recalibration is not undertaken or is carried out incorrectly, with the potential deactivation of systems and corresponding ADAS function. There is also the possibility that the motorist could pursue legal action against the workshop if the calibration is not carried out properly.

Safety first

There could be serious repercussions if ADAS recalibration is not undertaken or is carried out incorrectly.

3.7.7 Adaptive headlamps

The term refers to technology that is designed to allow drivers to see further, around corners and past traffic using brighter and more efficient lights. Headlights that actively respond to changing conditions allow more time to react to conditions ahead. The term can include different strategies to achieve this and automate the use of lights on the motorist's behalf;

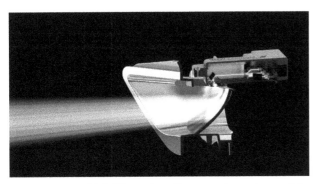

Figure 3.217 BMW laser headlamps where the rapid deflection of a laser creates whatever beam pattern is needed. (Source: BMW Media)

Figure 3.218 Headlamp patterns. (Source: BMW Media)

some systems are more technology dependent than others.

Cornering lights: This simple addition is the forerunner of adaptive light technology. Cornering lights sit next to the main lights of the vehicle, switching on and off when cornering. They are fixed in position and activate in response to the steering wheel turn or activation of an indicator. They switch off when the turn signal cancels, or the driver returns the steering wheel to a straight position. This system has been available, mainly for luxury vehicles, for decades.

Curve-adaptive lights: Just for fun, these are sometimes also referred to as cornering lights but are smarter than their earlier siblings. Curve-adaptive lights pivot the unit's light source, mounted on motors or servos, toward the direction of travel to better illuminate the path ahead. Some go further by incorporating subtle changes in angle in relation to the vehicle's speed and there are systems that can react to a detection of an upcoming curvature in the road. They can use the existing front fog lamps or be incorporated into the headlight unit.

Automatic high beams: The vehicle has the high beam set by default, switching to dipped beam when sensing another motorist by using sensors to detect the lights of other vehicles. The system can be manually overridden to allow headlamp controls

to be used in the conventional way if preferred or if environmental conditions do not switch from main to dipped fast enough and cause dazzling.

Adaptive driving beams (ADB): What makes ADB different is that the units are made up of many very bright, individual adjustable LED lights in an array. The precision control software reacts to sensors fitted to the vehicle, creating brightness around other vehicles but projecting a 'shadow' where dazzle to other motorists could occur. Instead of simply switching LEDs on or off, each one dims or brightens according to need, maximising brightness whilst reducing glare for oncoming traffic.

Evidence suggests that insurance claims are reduced in vehicles with adaptive headlights, and it is one less task for a motorist to concentrate on, leaving them to focus on the road whilst being considerate to other road users. Generally, the more complex the technology, the more expensive it is to design, install, maintain and repair – but like ABS, it is already filtering down the model ranges and due to the safety aspect, it may become a standard expectation in a short space of time.

Definition

Lux: A measurement of lighting intensity.

Chassis systems

4.1 Suspension

4.1.1 Overview of suspension

The suspension system is the link between the vehicle body and the wheels (Fig. 4.1). Its purpose is to:

▷ locate the wheels while allowing them to move up and down, and steer
▷ maintain the wheels in contact with the road and minimize road noise
▷ distribute the weight of the vehicle to the wheels
▷ reduce vehicle weight as much as possible, in particular the unsprung mass
▷ resist the effects of steering, braking and acceleration
▷ work in conjunction with the tyres and seat springs to give acceptable ride comfort.

This list is difficult to achieve completely, so some sort of compromise has to be reached. Because of this, many different methods have been tried, and many are still in use. Keeping these requirements in mind will help you to understand why some systems are constructed in different ways.

Unsprung mass (sometimes described as unsprung weight) is usually the mass of the suspension components, the wheels and the springs. However, only 50% of the spring mass and the moving suspension arms is included. This is because they form part of the link between the sprung and unsprung masses. It is beneficial to have the unsprung mass as small as possible in comparison with the sprung mass (main vehicle mass). This is so that when the vehicle hits a bump the movement of the suspension will have only a small effect on the main part of the vehicle. The overall result is therefore improved ride comfort.

Definition
Unsprung mass: The mass of the suspension components, wheels and springs.

Definition
Damper: A device that reduces vibration.

A vehicle needs a suspension system to cushion and damp out road shocks (Figs 4.2 and 4.3). This provides comfort to the passengers and prevents damage to the load and vehicle components. A spring between the wheel and the vehicle body allows the wheel to follow the road surface. The tyre plays an important role in absorbing small road shocks. It is often described as the primary form of suspension. The vehicle body is supported by springs located

Figure 4.1 Suspension plays a key role

Automobile Mechanical and Electrical Systems. 978-0-415-72578-1 © 2018 Tom Denton.
Published by Taylor & Francis. All rights reserved.

DOI: 10.1201/9781003299066-4

Figure 4.2 Spring just before a bump

Figure 4.3 Spring hitting a bump

Figure 4.4 Details of a coil spring

Figure 4.5 Coil spring in position

between the body and the wheel axles. Together with the damper (also described as a shock absorber), these components are referred to as the suspension system.

As a wheel hits a bump in the road, it is moved upwards with quite some force. An unsprung wheel is affected only by gravity, which will try to return the wheel to the road surface. However, most of the energy will be transferred to the body. When a spring is used between the wheel and the vehicle body, most of the energy in the bouncing wheel is stored in the spring and not passed to the vehicle body. The vehicle body only moves upwards through a very small distance compared with the movement of the wheel.

The springs in the suspension system take up the movement or shock from the road. The energy of the movement is stored in the spring. The actual spring can be in many different forms, ranging from a steel coil to a pressurized chamber of nitrogen. Soft springs provide the best comfort, but stiff springs can be better for high performance. Vehicle springs and suspension therefore are made to provide a compromise between good handling and comfort.

Modern vehicles use a number of different types of spring medium, but the most popular is the coil (or helical) spring. Coil or helical springs used in vehicle

suspension systems are made from round spring-steel bars. The heated bar is wound on a special former and then heat-treated to obtain the correct elasticity (springiness). The coil spring can withstand any compression load, but not side thrust. It is also difficult for a coil spring to resist braking or driving thrust. Suspension arms are used to resist these loads.

Key fact

Modern cars usually use a coil (or helical) spring.

Coil springs are generally used with independent suspension systems; the springs are usually fitted on each side of the vehicle, between the stub axle assembly and the body (Figs 4.4 and 4.5). The spring remains in the correct position because recesses are made in both the stub axle assembly and body.

The spring is always under compression owing to the weight of the vehicle and hence holds itself in place.

The leaf spring can provide all the control for the wheels during acceleration, braking, cornering and general movement caused by the road surface. Leaf springs are used with fixed axles and usually on larger vehicles. They can be described as:

▷ laminated or multileaf springs (Fig. 4.6)
▷ single-leaf or monoleaf springs.

Definition

Laminated: Consisting of many thin layers.

The multileaf spring was widely used at the rear of cars and light vehicles, and is still used in commercial vehicle suspension systems (Fig. 4.7). It consists of a number of steel strips or leaves placed on top of each other and then clamped together. The length, cross-section and number of leaves are determined by the loads carried.

The top leaf is called the main leaf and each end of this leaf is rolled to form an eye (Fig. 4.8). This is for attachment to the vehicle chassis or body. The leaves of the spring are clamped together by a bolt or pin known as the centre bolt.

The spring eye allows movement about a shackle and pin at the rear, allowing the spring to flex. The vehicle is pushed along by the rear axle through the front section of the spring, which is anchored firmly to the fixed shackle on the vehicle chassis or body. The curve of leaf springs straightens out when a load is applied to it, and its length changes.

Key fact

The curve of leaf springs straightens out when a load is applied to it, and its length changes.

Because of the change in length as the spring moves, the rear end of a leaf spring is fixed by a shackle bolt to a swinging shackle. As the road wheel passes over a bump, the spring is compressed and the leaves slide over each other. As it returns to its original shape, the spring forces the wheel back in contact with the road. The leaf spring is usually secured to the axle by means of U-bolts. As the leaves of the spring move, they rub together. This produces interleaf friction, which has a damping effect.

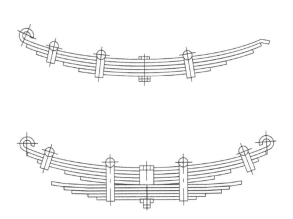

Figure 4.6 Laminated springs

Figure 4.7 Commercial vehicle leaf spring

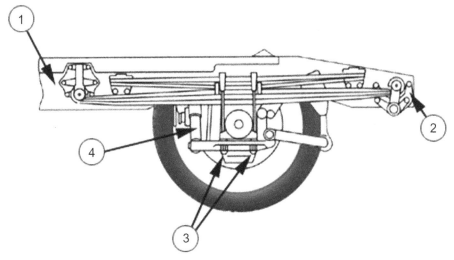

Figure 4.8 Details of a leaf spring: 1, fixed shackle; 2, swinging shackle; 3, U-bolts; 4, damper

281

A single-leaf spring, as the name implies, consists of one uniformly stressed leaf (Fig. 4.9). The spring varies in thickness from a maximum at the centre to a minimum at the spring eyes. This type of leaf spring is made to work in the same way as a multileaf spring. Advantages of this type of spring are:

▶ simplified construction
▶ constant performance over a period, because interleaf friction is eliminated
▶ reduction in unsprung mass.

Torsion bar suspension uses a metal bar, which provides the springing effect as it is twisted (Fig. 4.10). It has the advantage that the components do not take up too much room. The torsion bar can be round or square section, solid or hollow. The surface must be finished accurately to eliminate pressure points, which may cause cracking and fatigue failure. They can be fitted longitudinally or laterally.

> **Key fact**
>
> A torsion bar can be round or square section, solid or hollow.

Torsion bars are maintenance free but can sometimes be adjusted. They transmit longitudinal and lateral forces and have low mass. However, they have limited self-damping. Their spring rate is linear and their life may be limited by fatigue (Fig. 4.11).

Steel springs must be stiff enough to carry a vehicle's maximum load. However, this can result in the springs being too stiff to provide consistent ride control and comfort when the vehicle is empty. Pneumatic suspension can be made self-compensating. It is fitted to many heavy goods vehicles and buses, but is also becoming popular on some off-road light vehicles.

The pneumatic or air spring is a reinforced rubber bellow fitted between the axle and the chassis or vehicle body (Figs 4.12 and 4.13). An air compressor is used to increase or decrease the pressure depending on the load in the vehicle. This is done automatically, but some manual control can be retained for adjusting the height of the vehicle or stiffness of the suspension. Air springs can be thought of as being like a balloon or football on which the car is supported. The system involves compressors and air tanks. The system is not normally used on light vehicles.

> **Key fact**
>
> Air springs are like a (strong) balloon or football on which the car is supported.

Rubber is now a very old suspension method, but never say never, as old ideas often come back! The suspension medium, or spring, is simply a specially shaped piece of rubber (Figs 4.14 and 4.15). This technique was used on early Mini cars, for example. The rubber did not require damping in most cases. Nowadays, rubber springs are only used as a supplement to other forms of springs. They are, however, popular on trailers and caravans.

In a hydragas suspension system, each wheel has a sealed displacer unit (Figs 4.16 and 4.17). This contains nitrogen gas under very high pressure, which works in much the same way as the steel spring in a conventional system. A damper is also incorporated within the displacer unit. The lower part of the displacer unit is filled with a suspension fluid (usually

Figure 4.9 Tapered single leaf

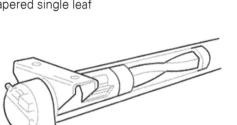
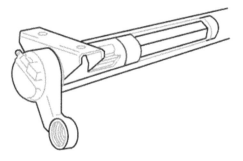

Figure 4.10 Torsion bar in a guide tube

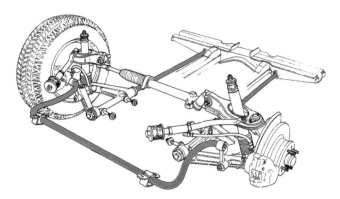

Figure 4.11 Torsion bar springs in blue and the anti-roll torsion bar in red

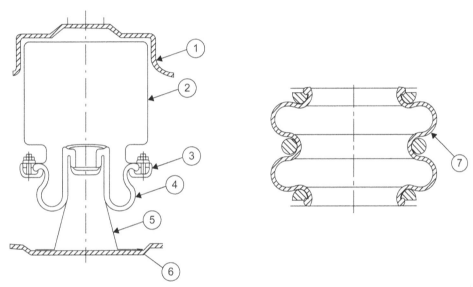

Figure 4.12 Air spring: 1, suspension mounting; 2, air chamber; 3, clamping ring; 4, rolling bellows; 5, piston; 6, suspension arm; 7, concertina bellows

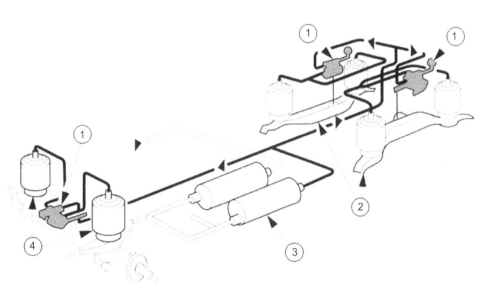

Figure 4.13 Air spring suspension system: 1, air spring valve; 2, axle support; 3, compressed air tank; 4, bellows

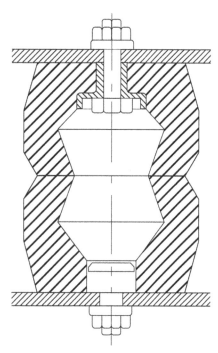

Figure 4.14 Hollow rubber spring

Figure 4.16 Hydragas suspension unit

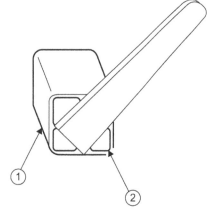

Figure 4.15 Torsion type rubber spring: 1, casing; 2, rubber insert

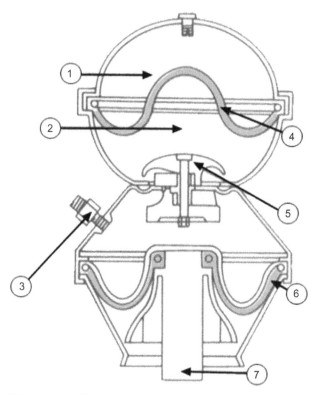

Figure 4.17 Suspension unit: 1, nitrogen gas spring; 2, fluid under high pressure; 3, connector for adding fluid; 4, flexible diaphragm; 5, damper valve; 6, flexible diaphragm; 7, strut connection to suspension

a type of wood alcohol). The units can be joined by pipes or used individually. Linking front to rear makes the rear unit rise as the front unit is compressed by a bump. This tends to keep the vehicle level and reduce pitch. Ride height control can be achieved by pumping oil into or out of the working chamber.

Key fact

A hydragas suspension unit contains nitrogen gas under very high pressure.

The energy stored in any type of spring after a bump has to be got rid of or else the spring would oscillate

(bounce up and down). A damper damps down these oscillations by converting the energy from the spring into heat. If working correctly, the spring should stop moving after just one bounce and rebound. Shock absorber is a term that is also used to describe a damper (Fig. 4.18).

Key fact

The energy stored in any type of spring after a bump has to be got rid of or else the spring would oscillate (bounce up and down).

Figure 4.18 Telescopic damper

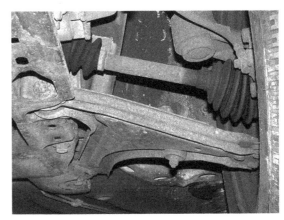

Figure 4.20 Front suspension wishbone

Figure 4.19 MacPherson strut

Figure 4.21 Rubber stop

The combination of a coil spring with a damper inside it, between the wheel stub axle and the inner wing, is often referred to as a strut or, after its inventor, a MacPherson strut (Fig. 4.19). This is a very popular type of suspension.

A wishbone is a triangular shaped component with two corners hinged in a straight line on the vehicle body (Fig. 4.20). The third corner is hinged to the moving part of the suspension.

When a vehicle hits a particularly large bump, or if it is carrying a heavy load, the suspension system may bottom out (reach the end of its travel). A bump stop, usually made of rubber, is used to prevent metal-to-metal contact, which would cause damage (Fig. 4.21).

A link is a very general term used to describe a bar or other similar component that holds or controls the position of another component. Other terms may be used, such as tie-bar or tie-rod.

A beam axle is a solid axle from one wheel to the other. It is not now used on the majority of light vehicles. However, as it makes a very strong

Figure 4.22 Heavy vehicle axle

construction, it is common on heavy vehicles (Fig. 4.22).

Independent front and rear suspension (IFS/IRS) was developed to meet the demand for improved ride quality and handling. The main advantages of independent suspension (i.e. not using a beam axle) are as follows:

▶ When one wheel is lifted or drops, it does not affect the opposite wheel.
▶ The unsprung mass is lower; therefore, the road wheel stays in better contact with the road.
▶ Problems with changing steering geometry are reduced.
▶ There is more space for the engine at the front.
▶ Softer springing with larger wheel movement is possible.

An anti-roll bar usually forms part of a suspension system (Fig. 4.23). The main purpose of an anti-roll bar is to reduce body roll on corners. The anti-roll bar can be thought of as a torsion bar. The centre is pivoted on the body and each end bends to make connection with the suspension/wheel assembly. When the suspension is compressed on both sides, the anti-roll bar has no effect because it pivots on its mountings. As the suspension is compressed on just one side, a twisting force is exerted on the anti-roll bar. Part of this load is transmitted to the opposite wheel, pulling it upwards. This reduces the amount of body roll on corners.

Key fact

An anti-roll bar is fitted to reduce body roll on corners.

Definition

Lateral: Sideways.

Longitudinal: Along the length of a vehicle.

The Panhard rod was named after a French engineer (Fig. 4.24). Its purpose is to link a rear axle to the body. The rod is pivoted at each end to allow movement. It takes up lateral forces between the axle and body, thus removing load from the radius arms. The radius arms have now to transmit only longitudinal forces.

A wide variety of suspension systems and components is used. Engineers strive to achieve optimum comfort and handling. However, these two main requirements are often at odds with each other. As is common with all vehicle systems, electronic control is one way that developments are now being made.

The requirements of the springs and suspension system can be summarized as follows. They:

▶ absorb road shocks from uneven surfaces
▶ control ground clearance and ride height
▶ ensure good tyre adhesion
▶ support the weight of the vehicle
▶ transmit gravity forces to the wheels.

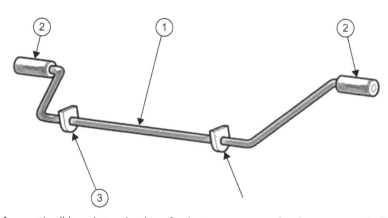

Figure 4.23 Shape of an anti-roll bar: 1, torsion bar; 2, pivots on suspension lower arms; 3, fixings to vehicle body

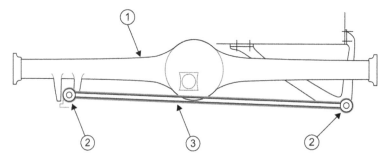

Figure 4.24 Rear axle with Panhard rod: 1, axle; 2, pivots; 3, Panhard rod

Figure 4.25 Modern front suspension system. (Source: Jaguar/Ford Media)

Suspension springs can be made from a variety of materials and in many different ways. However, the most common is the coil spring. This is because it has many advantages and is reasonably inexpensive. A modern front suspension system is shown in Figure 4.25.

4.1.2 Dampers/shock absorbers

As a spring is deflected, energy is stored in it. If the spring is free to move, the energy is released in the form of oscillations, for a short time, before it comes to rest. This principle can be demonstrated by flicking the end of a ruler placed on the edge of a desk. The function of the damper is to absorb the stored energy, which reduces the rebound oscillation. A spring without a damper would build up dangerous and uncomfortable bouncing of the vehicle.

Hydraulic dampers are the most common type used on modern vehicles (Fig. 4.26). In a hydraulic damper, the energy in the spring is converted into heat. This is caused as the fluid (a type of oil) is forced rapidly through small holes (orifices).

The oil temperature in a damper can reach over 150°C during normal operation. The functions of dampers can be summarized as follows. They:

▷ ensure directional stability
▷ ensure good contact between the tyres and the road
▷ prevent build-up of vertical movements
▷ reduce oscillations
▷ reduce wear on tyres and chassis components.

Figure 4.26 Dampers form part of the struts on this system. (Source: Ford Media)

Safety first

The oil temperature in a damper can reach over 150°C during normal operation.

The twin-tube telescopic damper is the most commonly used type of telescopic damper (Fig. 4.27). It consists of two tubes. An outer tube forms a reservoir space and contains the oil displaced

287

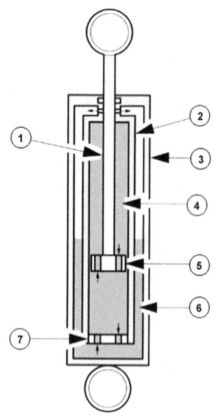

Figure 4.27 Twin-tube damper: 1, piston rod; 2, inner tube; 3, outer tube; 4, oil chamber; 5, piston with valves; 6, reservoir space; 7, inner tube bottom valves. (Source: Ford Motor Company)

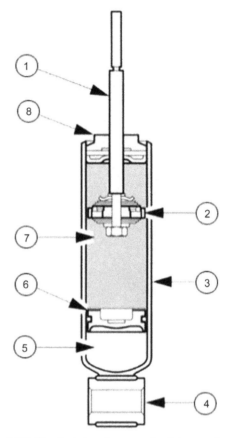

Figure 4.28 Single-tube damper: 1, piston rod; 2, working piston; 3, tube; 4, mounting bush; 5, gas chamber; 6, separator piston; 7, oil chamber; 8, seal. (Source: Ford Motor Company)

from an inner tube. Oil is forced through a valve by the action of a piston as the damper moves up or down. The reservoir space is essential to make up for the changes in volume as the piston rod moves in and out.

Key fact

The single-tube telescopic damper is often referred to as a gas damper.

The single-tube telescopic damper is often referred to as a gas damper (Fig. 4.28). However, the damping action is still achieved by forcing oil through a restriction. The gas space behind a separator piston compensates for the changes in cylinder volume, which are caused as the piston rod moves. The gas is at a pressure of about 25 bar.

A twin-tube gas damper is a combination of the standard twin-tube and the single tube gas system. The gas cushion is used in this case to prevent oil foaming. The gas pressure on the oil prevents foaming, which in turn ensures constant operation

under all operating conditions. Gas pressure is set at about 5 bar. If bypass grooves are machined in the upper half of the working chamber, the damping rate can be made variable. With light loads the damper works in this area with a soft damping effect. When the load is increased the piston moves lower down the working chamber away from the grooves, resulting in a full damping effect.

There are dampers where the damping rate can be controlled by solenoid valves inside the units. With suitable electronic control, the characteristics can be changed within milliseconds to react to driving and/ or load conditions. When it is activated, the solenoid allows some of the oil to be diverted (Fig. 4.29).

Dampers (or shock absorbers) are used to prevent the suspension springs oscillating. This improves handling, comfort and safety.

Key fact

Dampers (or shock absorbers) are used to prevent the springs oscillating.

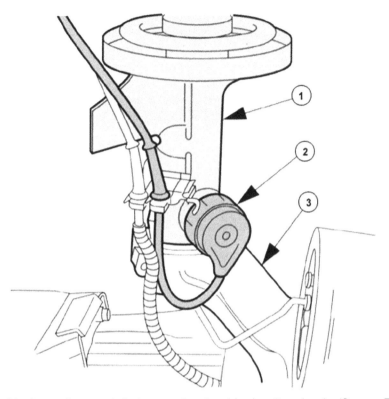

Figure 4.29 Damper with electronic control: 1, damper; 2, solenoid valve; 3, stub axle. (Source: Ford Motor Company)

4.1.3 Suspension layouts

As with most design aspects of the vehicle, compromise often has to be reached between performance, body styling and cost.

Twin, unequal-length wishbone suspension is widely used on light vehicles (Fig. 4.30). A coil spring is fitted between two suspension arms. The suspension arms are 'wishbone' shaped and the bottom end of the spring fits in a plate in the lower wishbone assembly. The top end of the spring is located in a section of the body. The top and bottom wishbones are attached to the chassis by rubber bushes. A damper is fitted inside the spring. The stub axle and swivel pins are connected to the outer ends of the upper and lower wishbones by ball or swivel joints.

This strut type of suspension system has been used now for many years. It is often referred to as the MacPherson strut system. With this system, the stub axle is combined with the bottom section of a telescopic tube, which incorporates a damper. The

Key fact

A strut type of suspension is often referred to as a MacPherson strut system.

Figure 4.30 Twin, unequal-length wishbone suspension system

bottom end of the strut is connected to the outer part of a transverse link by means of a ball joint. The inner part of the link is secured to the body by rubber bushes. The top of the strut is fixed to the vehicle body by a bearing, which allows the complete strut to swivel. A coil spring is located between the upper and lower sections of the strut.

289

A transverse arm with MacPherson strut is the most popular method in current use. This system uses a combination of the spring, damper, wheel hub, steering arm and axle joints in one unit. There are only slight changes in track and camber with suspension movement. Forces on the joints are reduced because of the long strut. However, the body must be strengthened around the upper mounting and a low bonnet line is difficult.

There are other variations of front suspension layouts. However, the two most popular are the MacPherson strut and the unequal length wishbone systems.

The systems used for the rear suspension of light vehicles vary depending on the requirements of the vehicle. In addition, the systems are different if the vehicle is front- or rear-wheel drive. Older and heavy vehicles use leaf-type springs. The two main types using independent rear suspension are:

▶ the strut type for front-wheel drive (Fig. 4.31)
▶ trailing and semi-trailing arms with coil springs for rear-wheel drive (Fig. 4.32).

Figure 4.31 Rear struts

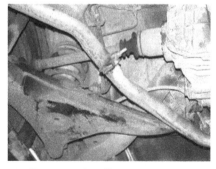

Figure 4.32 Rear semi-trailing arm

The strut system on the rear is much the same as used at the front of the vehicle (Fig. 4.33). Note that suitable links are used to allow up-and-down movement but to prevent the wheel moving in any other direction. Some change in the wheel geometry is designed in, to improve handling on corners.

Trailing arm suspension and semi-trailing arm suspension both use wishbone-shaped arms hinged on the body. Trailing arms are at right angles to the vehicle centre line and semi-trailing arms are at an angle (Figs 4.34 and 4.35). This changes the geometry of the wheels as the suspension moves. The final drive and differential unit is fixed with rubber mountings to the vehicle body. Drive shafts must therefore be used to allow drive to be passed from the fixed final drive to the movable wheels. The coil springs and dampers are mounted between the trailing arms and the vehicle body.

> **Key fact**
> Trailing arm and semi-trailing arm systems both use wishbone-shaped arms hinged on the body.

The final drive, differential and axle shafts are all one unit with a rigid axle and leaf spring layout. With this system, the rear track remains constant, reducing tyre wear. It has good directional stability because there is no camber change due to body roll on corners. This is a strong design for load carrying (Fig. 4.36). However, it has a high unsprung mass. The interaction of the wheels causes lateral movement, reducing tyre adhesion when the suspension is compressed on one side.

A rigid axle with an A-bracket system has a solid axle with coil springs and a central joint supports the axle on the body (Fig. 4.37). This tends to make the rear of the vehicle pull down on braking, which stabilizes the vehicle. It results in a high unsprung mass.

On a torsion beam trailing arm axle, two links are used, connected by a U-section that has low torsional stiffness but a high resistance to bending (Fig. 4.38). Track and camber do not change as the suspension moves. It has a low unsprung mass and is simple to produce. It is a space-saving design but torsion bar springing on this system can be more expensive than coil springs.

In the system shown in Figure 4.39 two links are welded to an axle tube or U-section. The lateral forces are taken by a Panhard rod. Track and camber do not change as the suspension moves and simple flexible joints connect it to the bodywork. Torsion bar springing on this system, however, can be more expensive than coil springs.

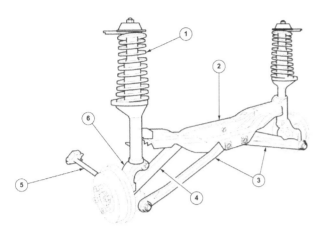

Figure 4.33 Rear suspension struts: 1, coil spring; 2, cross-member; 3, rear transverse arms; 4, front transverse arms; 5, radius arm; 6, stub axle

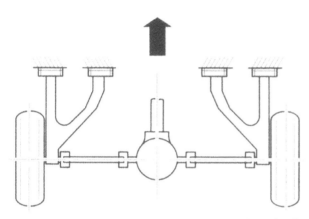

Figure 4.34 Trailing arms pivot at 90° to the length of the vehicle

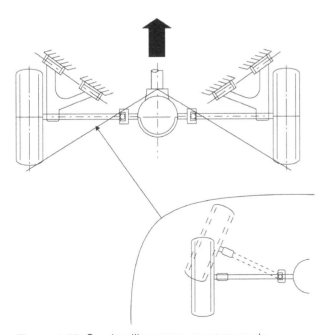

Figure 4.35 Semi-trailing arms are at an angle

Figure 4.36 This method is used on older and larger vehicles with live, fixed rear axles

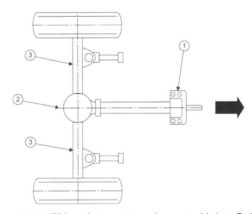

Figure 4.37 'A' bracket system: 1, central joint; 2, final drive; 3, half shafts

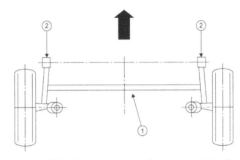

Figure 4.38 Torsion beams twist to provide the spring action: 1, U-section spring; 2, trailing arms mounted to the body

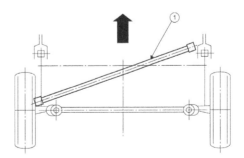

Figure 4.39 Panhard rod and torsion beam system: 1, Panhard rod

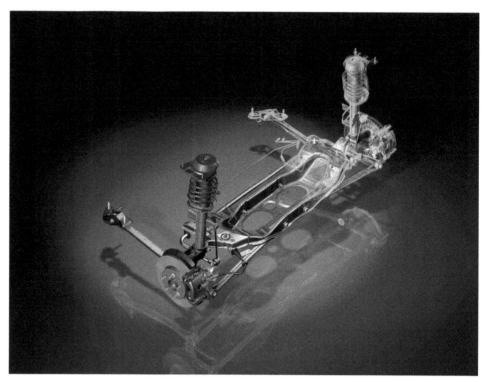

Figure 4.40 'Quadralink' suspension. (Source: Ford Media)

There are various rear suspension systems, each with its advantages and disadvantages, and engineers strive to achieve the optimum design. The system shown in Figure 4.40 is known as 'quadralink'. It is very similar to front strut systems.

4.1.4 Active suspension

Electronic control of suspension or active suspension, like many other innovations, was born in the world of Formula 1 (F1). It is now slowly becoming more popular on production vehicles. Conventional suspension systems are always a compromise between soft springs for comfort and harder springing for better cornering capability. Active systems have the ability to switch between the two extremes.

A traditional or a conventional suspension system, consisting of springs and dampers, is passive. In other words, once it has been installed in the car, its characteristics do not change.

The main advantage of a conventional suspension system is its predictability. Over time, the driver will become familiar with a car's suspension and

understand its capabilities and limitations. The disadvantage is that the system has no way of compensating for situations beyond its original design. The main benefits of active suspension are:

▶ improvements in ride comfort, handling and safety
▶ predictable control of the vehicle under different conditions
▶ no change in handling between laden and unladen.

The benefits are considerable and as component prices fall, the system will become available on more vehicles. It is expected that more off-road vehicles may be fitted with active suspension in the near future.

An active suspension system (also known as computerized ride control) has the ability to adjust itself continuously (Fig. 4.42). It monitors and adjusts its characteristics to suit the current road conditions. As with all electronic control systems, sensors supply information to an electronic control unit (ECU) which in turn outputs to actuators. By changing its characteristics in response to changing road conditions, active suspension offers improved handling, comfort, responsiveness and safety.

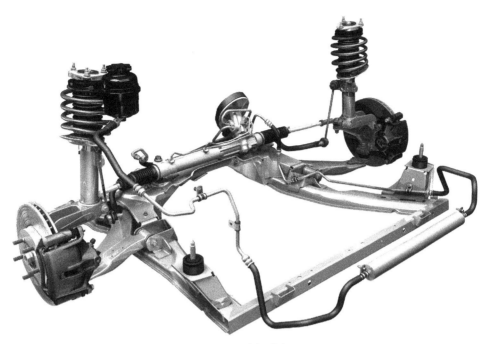

Figure 4.41 Jaguar suspension system. (Source: Jaguar Media)

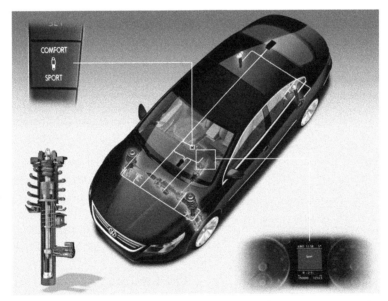

Figure 4.42 Active suspension also allows adjustments, in this case, between sport and comfort settings. (Source: Volkswagen Media)

Active suspension systems usually consist of the following components:

▷ ECU
▷ adjustable dampers and springs
▷ sensors at each wheel and throughout the car
▷ levelling compressor (some systems).

Key fact

Active suspension works by sensing changes in conditions and feeding that information to the ECU.

Active suspension works by constantly sensing changes in the road surface and feeding that information to the ECU, which in turn controls the suspension springs and dampers. These components then act upon the system to modify the overall suspension characteristics by adjusting damper stiffness, ride height (in some cases) and spring rate.

The suspension can be controlled in a number of ways. However, in most cases it is done by controlling the oil restriction in the damper. On some systems, ride height is controlled by opening a valve and

293

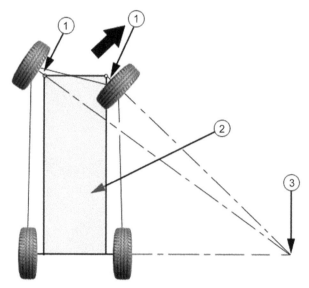

Figure 4.43 Suspension strut and actuator connection. (Source: Delphi Media)

Figure 4.44 Ackermann steering method: 1, stub axles; 2, vehicle support area; 3, imaginary common centre

supplying pressurized fluid from an engine-driven compressor. Later systems are starting to use special fluid in the dampers that reacts to a magnetic field, which is applied from a simple electromagnetic coil.

The improvements in ride comfort are considerable, which is why active suspension technology is becoming more popular. In simple terms, sensors provide the input to a control system that in turn actuates the suspension dampers in a way that improves stability and comfort.

4.2 Steering

4.2.1 Introduction to steering

The development of steering systems began before cars were invented. On early cars, the entire front axle was steered by way of a pivot (fifth wheel) situated in the centre of the vehicle. The steering accuracy was not very good, there was a serious risk of overturning and the tyre wear was significant.

> **Key fact**
>
> On early cars, the entire front axle was steered by way of a pivot (fifth wheel) – like a horse and cart!

In 1817, Rudolph Ackermann patented the first stub axle steering system in which each front wheel was fixed to the front axle by a joint. This made it possible to cover a larger curve radius with the wheel on the outside of the curve than with the front wheel on the inside of the curve (Fig. 4.44).

Rack and pinion steering was developed at an early age in the history of the car. However, this became popular when front-wheel drive was used more, since it requires little space and production costs are lower (Fig. 4.45). The first hydraulic power steering was produced in 1928. However, since there was no real demand for this until the 1950s, the development of power steering systems stagnated.

Increasing standards of comfort stimulated the demand for power steering systems. Speed-sensitive or variable-assistant power steering (VAPS) systems were developed using electronic controls. The demand for safety and comfort will lead to further improvements in steering systems.

> **Definition**
>
> **Ackermann angle:** The angle of the inside wheel in relation to the outside wheel when the front wheels are steered. The inside front wheel turns at a sharper angle than the outside wheel.

Motor vehicles are generally steered via the front wheels, the rear wheels following the front wheels on a smaller radius. With motor vehicles, two factors have to be taken into account:

▶ dead weight or axle loading
▶ the contact area of the steered wheels.

To overcome the friction forces more easily, many different types of steering gear have been developed. Power steering, in particular, reduces the effort

Figure 4.45 Steering rack in position. (Source: Ford Media)

required and increases safety and comfort. Steering systems must be capable of:

- automatically returning the steered front wheels to the straight-ahead position after cornering (self-centring action)
- translating the steering wheel rotation so that only about two rotations of the steering wheel are necessary for a steering angle of about 40°.

Steering and suspension must always be regarded as a unit (Fig. 4.46). If the suspension system is not working correctly, it will have a considerable influence on the vehicle's steering characteristics. For example, defective shock absorbers or dampers reduce the wheel contact with the road, limiting the ability to steer the vehicle. The driving safety of a motor vehicle depends largely on the steering. Reliable steering at high speeds is required, together with easy manoeuvrability.

Key fact

Suspension faults will influence the vehicle's steering characteristics.

Crucial to the manoeuvrability of a motor vehicle is the turning circle, which in turn is directly dependent on the track circle. Designers strive for the smallest possible track and turning circle. The wheel housing should enclose the wheels as tightly as possible;

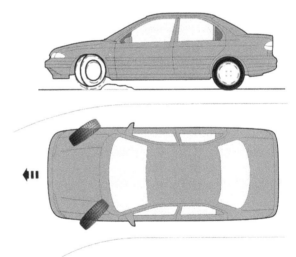

Figure 4.46 Suspension and steering interact

however, sufficient clearance must be left so that the tyres do not rub when the wheels are turned.

With the stub axle, Ackermann type of steering, the stub axle of the steered front wheel is swivelled about the steering axis (Fig. 4.47). When steering, the wheelbase remains constant. The space between the steered wheels can be used for the installation of deep-seated components such as the engine. The low centre of gravity contributes to road handling characteristics. Even at large steering angles, the stability of the vehicle is maintained since the area of support is only slightly reduced.

295

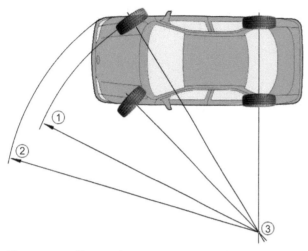

Figure 4.47 Front axle geometry: 1, track circle; 2, turning circle; 3, common centre

4.2.2 Steering racks and boxes

In order to transmit the steering movements of the driver to the wheels, several components are required (Fig. 4.48). The steering movement is transmitted by way of the steering wheel, shaft, gear and linkage to the front wheels. The rotational movement of the steering wheel is transmitted via the steering shaft to the steering pinion in the steering gear. The steering shaft is supported in the steering column tube, which is fixed to the vehicle body.

The steering gear translates (reduces) the steering force applied by the driver. It also converts the rotational movement of the steering wheel into push or pull movements of the track rods. The converted movement is transmitted to the linkage, which in turn moves the wheels in the desired steering direction. Track rods are required to transmit the steering movement from the steering gear to the front wheels. Different track rods are used depending on the type of front axle.

> **Key fact**
> The steering gear translates (reduces) the steering force applied by the driver.

The steering rack is the simplest design of steering linkage, needing only three joints. One-piece track rods are found only with rigid axles, since the distance of the steering swivel pins or joints cannot vary (Fig. 4.49).

Two-piece track rods may be split centrally or to one side (Fig. 4.50). They are necessary on vehicles with independent suspension, since the suspensions of the

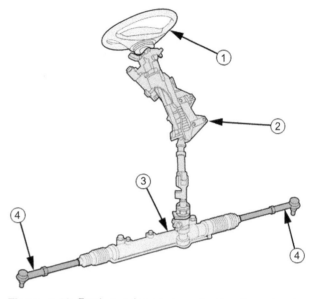

Figure 4.48 Ford steering system: 1, steering wheel; 2, column; 3, rack; 4, track rod ends

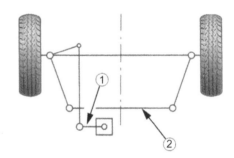

Figure 4.49 One-piece track rod: 1, steering box drop arm; 2, solid track rod

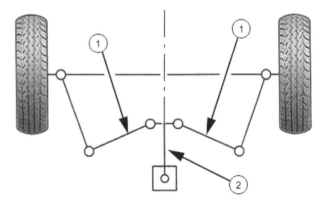

Figure 4.50 Two-piece track rod: 1, track rod; 2, drop arm

steered wheels are compressed independently of one another. The split reduces the effect of bump steering.

In the type of steering that is now almost universal on light vehicles, the steering linkage is operated by a rack and pinion. Two designs are encountered as

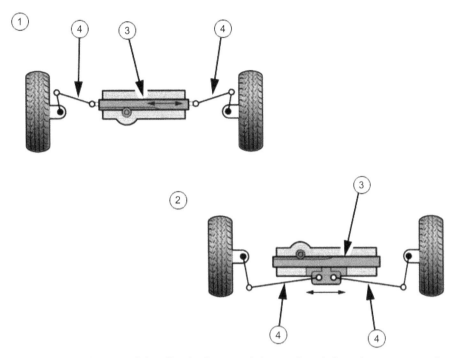

Figure 4.51 Basic steering racks: 1, rack is effectively part of the track rod; 2, rack acts on a split track rod; 3, rack; 4, track rod

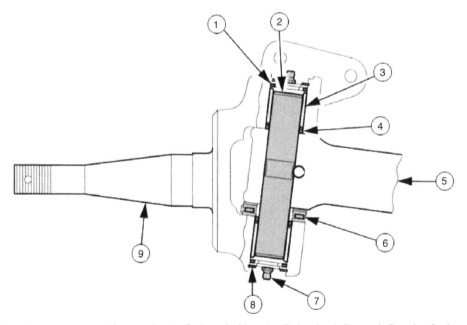

Figure 4.52 Kingpin on a steered front axle: 1, O-ring; 2, kingpin; 3, bush; 4, lip seal; 5, axle; 6, thrust bearing; 7, grease nipple; 8, circlip/snap ring; 9, stub axle

shown in Figure 4.51, number 1 being by far the more common. The rack either forms part of the track rod or acts directly on the split track rod.

Swivels of some sort are necessary for steering movement. The kingpin is the predecessor of the ball joint (Fig. 4.52). It is only fitted in commercial vehicles

and a few off-road vehicles, because these generally have rigid front axles in which the distances of the track rods do not vary. The kingpin is not maintenance free – it must be supplied with grease via a grease nipple.

Ball joints allow parts of the steering linkage to rotate about the longitudinal axis of the ball joint

297

(Fig. 4.53). They also allow limited swivel movements transversely to the longitudinal axis. The lubricated ball pivot is supported in steel cups or between preloaded plastic cups. A gaiter prevents lubricant losses. Ball joints are generally maintenance free and must always be renewed if the gaiter is damaged.

There are several different steering box designs. The first, shown in Figure 4.54, consists of a steering screw on

which the steering nut is displaced axially as the steering wheel is moved. Slide rings on the circumference of the steering nut transmit the movement to the steering fork and thereby to the drop arm. The drop arm performs a movement of up to 90°. In this type of steering the wear is relatively high. The steering nut play cannot be adjusted, and this is a disadvantage. With this type of steering gear, the steering is linear.

> **Key fact**
> There are a number of different steering box designs: a steering rack is a type of steering box; it just doesn't look like one.

Owing to the high friction in the screw and nut steering gear, gears with roller friction have become more common. In the recirculating ball steering gear, the steering screw and steering nut have ball groove threads (Fig. 4.55). The threads do not touch one another because they form channels for the balls. When the steering screw is turned, the balls roll in the ball groove thread in two closed recirculating ball races. The balls are returned by two tubes. The drop arm is moved by means of a gear sector. The advantage of the recirculating ball steering gear is that it functions virtually free of wear.

> **Key fact**
> The recirculating ball steering box has very low friction and is therefore virtually free of wear.

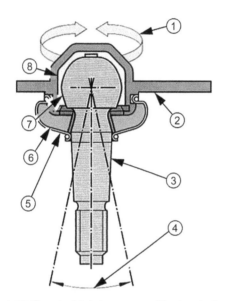

Figure 4.53 Two ball joints are used instead of a kingpin: 1, rotation; 2, connecting flange; 3, taper; 4, possible swivel movement; 5, gaiter; 6, lubricating grease; 7, ball pivot; 8, plastic cup

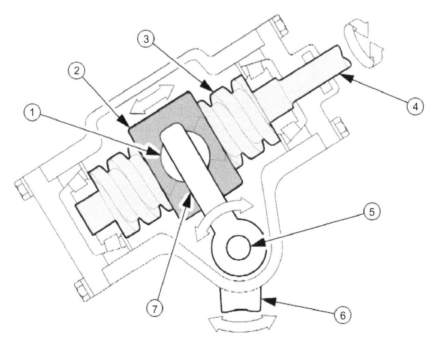

Figure 4.54 Screw and nut steering box: 1, slide rings; 2, steering nut; 3, steering screw; 4, steering shaft; 5, shaft to drop arm; 6, drop arm; 7, steering fork. (Source: Ford Motor Company)

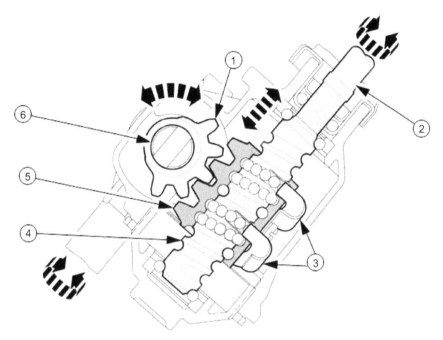

Figure 4.55 Recirculating ball steering box: 1, gear sector; 2, steering shaft; 3, ball return tubes; 4, steering screw; 5, steering nut; 6, shaft to drop arm. (Source: Ford Motor Company)

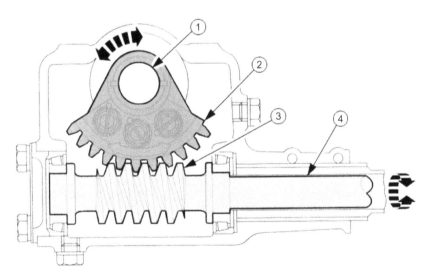

Figure 4.56 Worm and sector steering box: 1, shaft to drop arm; 2, sector; 3, cylindrical worm gear; 4, steering shaft. (Source: Ford Motor Company)

The worm and sector steering gear has a cylindrical worm, which, owing to its screw motion, turns a steering sector back and forth (Fig. 4.56). The drop arm is fixed to the sector. It can perform a swivel movement of up to about 70°. Worm steering gears are characterized by high transmission ratios, for example 22:1. One disadvantage is the high wear due to the sliding friction between the sector and the cylindrical worm. In addition, large steering forces are required. In this type of steering gear, the steering is linear.

The worm and roller steering gear has a roller instead of the sector (Fig. 4.57).

The steering worm is not cylindrical but tapers towards the middle like an hourglass.

The roller, driven by the worm, can thus perform a steering movement about its centre when the steering wheel is turned. The drop arm can perform a swivel movement of up to 90°. Advantages include low wear, ease of steering and that a small space is required. The steering play can be adjusted and the steering is

299

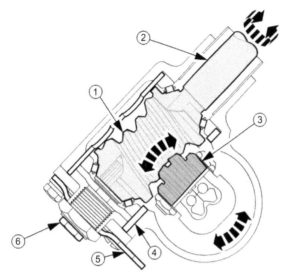

Figure 4.57 Worm and roller steering box: 1, hourglass work gear; 2, steering shaft; 3, roller; 4, eccentric bush; 5, backlash adjustment lever; 6, steering shaft adjustment screw. (Source: Ford Motor Company)

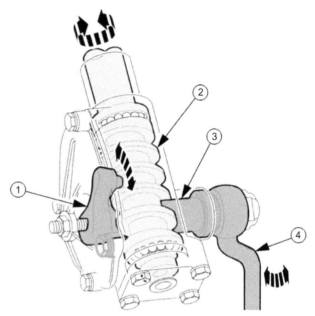

Figure 4.58 Worm and rolling finger steering box: 1, rolling finger; 2, work with progressive pitch; 3, shaft; 4, drop arm. (Source: Ford Motor Company)

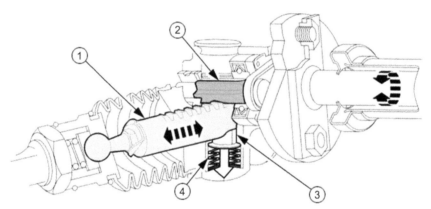

Figure 4.59 Standard linear steering rack: 1, rack; 2, pinion; 3, pressure pad; 4, vibration damper. (Source: Ford Motor Company)

free of play when running in a straight line. With this type of steering gear, the steering is linear.

The worm and rolling finger steering gear has a cylindrical screw with an uneven thread pitch (Fig. 4.58). When the worm is rotated, the tapered rolling finger rolls on the flanks of the worm. The rolling finger is displaced. This movement is converted by the shaft into a swivel movement of the drop arm. This system has low wear and ease of steering. The longitudinal play of the worm and the shaft, and the play between rolling finger and worm thread, are adjustable. In this type of steering gear, the steering is progressive owing to the uneven thread pitch on the worm.

Now back to the most common method used on cars (Fig. 4.59); but remember, old ideas often come back.

The steering rack housing contains a toothed pinion, which meshes with the rack. By turning the steering wheel and hence the pinion, the rack is displaced transversely to the direction of travel. A spring-loaded pressure pad presses the rack against the pinion. For this reason, the steering gear always functions without backlash. At the same time, the sliding friction between pressure pad and rack acts as a damper to absorb road shocks. Advantages of rack and pinion steering include the shallow construction, very direct steering, good steering return and the low cost of manufacture.

It is also possible to make a non-linear or progressive pitch steering rack (Fig. 4.60). The basic construction and the advantages are similar to those of a rack

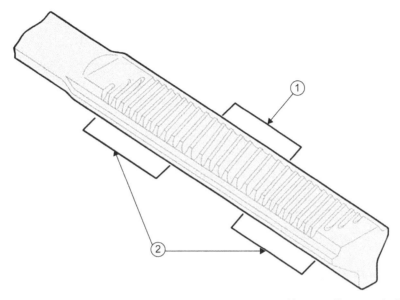

Figure 4.60 Progressive pitch steering rack: 1, large rack movement and larger effort needed; 2, smaller movement and less effort required

and pinion steering gear with constant pitch. In a rack and pinion steering gear with variable pitch, a rack is used which has teeth that diminish in size towards the ends. This makes it possible to increase the transmission ratio constantly. This means, in practice, that more steering wheel turns but less effort is required to turn the wheels. As a result, the steering moves more easily when applying lock than when moving in a straight line. This makes parking considerably easier.

There is a wide range of steering boxes and steering layouts. On light vehicles, the most common by far is the steering rack. This is because it has a shallow construction, is very direct, has good steering return and the cost of manufacture is low.

> **Key fact**
>
> A steering rack is described as being 'direct' because there are fewer joints than in other types of steering box, and hence there is reduced free play.

4.2.3 Steering geometry

The wheels of a vehicle cover different distances when cornering. At low speed, optimum rolling of the wheels is only possible if the centre lines of the stub axles, with the front wheels turned, meet the extended centre line of the rear axle. In this case, the paths covered by the front and rear wheels have a common centre.

> **Key fact**
>
> In an Ackermann steering system, the name 'trapezium' comes from the geometrical shape that the two steering arms and the track rod form with the front axle.

The inside front wheel must, therefore, be turned more than the outside front wheel. This is achieved by the Ackermann trapezium system (Figs 4.61 and 4.62).

The 'trapezium' name is derived from the geometrical shape that the two steering arms and the track rod form with the front axle. The stub axle and steering arm are rigidly connected to one another. The stub axles are swivel mounted on the kingpins or in ball

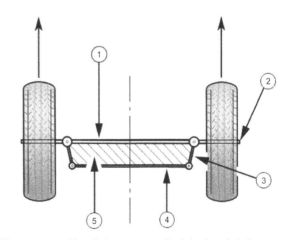

Figure 4.61 The Ackermann principle (straight): 1, front axle; 2, stub axle; 3, steering arm; 4, track rod; 5, trapezium

301

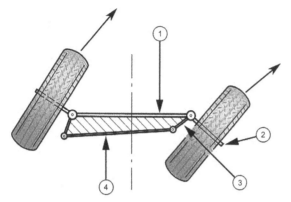

Figure 4.62 The Ackermann principle (cornering):
1, front axle; 2, stub axle; 3, steering arm; 4, track rod

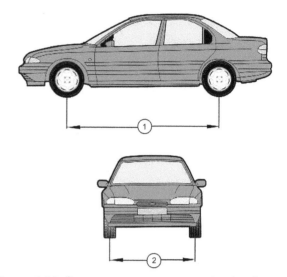

Figure 4.63 Common measurements: 1, wheelbase; 2, track

joints. Track rod and steering arms are movably connected to one another. When in the straight-ahead position, track rod and front axle are parallel. When cornering, the stub axles are swivelled, thereby turning the front wheels. With the front wheels turned, the track rod is no longer parallel to the front axle; this results in the inside front wheel being turned more than the outside front wheel.

There are numerous terms and phrases associated with steering geometry. Wheelbase is the distance between the wheel centres of the front and rear wheels. The track is the distance between the wheels, measured from tyre centre to tyre centre on the wheel contact plane. A greater track and wheelbase results in improved safety, especially when cornering.

The wheel toe is the difference in the distance between the rim flanges in front of and behind the axle in the straight-ahead position. If the distance in front of and behind the axle is the same, the vehicle has zero wheel toe. There is generally some toe-in or toe-out (Fig. 4.64). The wheel toe is given in millimetres or angular degrees and minutes. It is often referred to as tracking. Toe-in occurs when the distance between the rim flanges in the direction of travel is smaller in front of the axle than behind the axle. Toe-out occurs when the distance between the rim flanges in the direction of travel is greater in front of the axle than behind the axle.

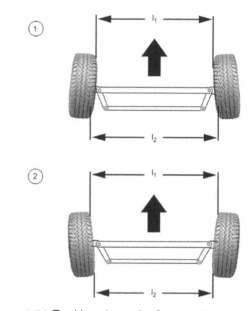

Figure 4.64 Tracking: 1, toe-in; 2, toe-out

Key fact

The wheel toe or tracking setting is given in millimetres, fractions of an inch, or angular degrees and minutes.

The ideal running direction of the wheels is parallel to the vehicle's longitudinal axis. Owing to deformations in the suspension elements, however, the front wheels are

diverted from their ideal line. In the case of front-wheel drive, they are forced inwards in the toe-in direction and in the case of rear-wheel drive, outwards in the toe-out direction. Undesirable toe-out is counteracted by toe-in and undesirable toe-in by toe-out.

Camber is the angle between the wheel plane and a line perpendicular to the road (Figs 4.65–4.67). The wheels must be straight ahead. Camber is described as positive when the wheel is inclined out at the

Definition

Camber: The angle between the wheel plane and a line perpendicular to the road.

Figure 4.65 Positive camber (r = scrub radius)

Figure 4.67 F1 cars use negative camber

Figure 4.66 Negative camber (r = scrub radius)

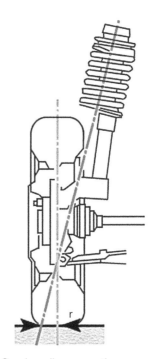

Figure 4.68 Scrub radius: negative

top. It has the effect of reducing scrub radius and influences the wheel forces when cornering. Camber is described as negative when the wheel is inclined in at the top. It produces a good cornering force and allows a low vehicle centre of gravity.

The scrub radius is the distance between the contact point of the steering axis with the road surface plane and the wheel centre contact point. The functions of the scrub radius are to reduce the steering force, prevent shimmy and stabilize the straight-ahead position.

Definition

Scrub radius: The distance between the contact point of the steering axis with the road surface plane and the wheel centre contact point.

When the point of contact of the steering axis with the road surface is between the wheel centre and the outside of the wheel, it is termed negative scrub radius (Fig. 4.68). The result of negative scrub radius is that the brake forces acting on the wheel produce a torque, which tends to turn the wheel inwards. As a result, the wheel with the greater braking action is turned inwards, i.e. steered away from the more heavily braked side. This produces automatic counter-steer, stabilizing the vehicle.

When the contact point of the steering axis with the road surface is between the wheel centre and the inside of the wheel, this is termed positive scrub radius (Fig. 4.69). The greater the positive scrub

303

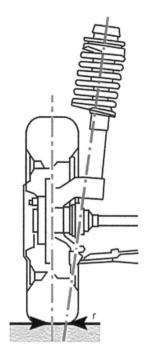

Figure 4.69 Scrub radius: positive

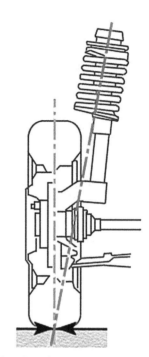

Figure 4.70 Scrub radius: zero

radius, the more easily the wheels can be turned. The result of positive scrub radius is that the brake forces acting on the wheel produce a torque, which tends to turn the wheel outwards. With a large positive scrub radius, the vehicle can be steered very easily. Disturbing forces, such as different road surfaces, act on a long lever and can produce an unwanted steering angle.

When the point of contact of the steering axis with the road surface is in the wheel centre, the scrub radius is zero. With a zero scrub radius, the wheel swivels on the spot (Fig. 4.70). Steering is heavy when the vehicle is stationary, since the wheel cannot roll at the steering angle. In this case no separate torques occur.

Kingpin angle is the angle between the steering axis and the perpendicular to the road surface, viewed in the direction of travel. Scrub radius, wheel camber and kingpin inclination all influence one another. The kingpin inclination mainly affects the aligning torque, which brings the wheels back into the straight-ahead position.

Definition

Kingpin angle: The angle between the steering axis and a line perpendicular to the road surface, viewed from the front.

Owing to the inclination of the steering axes, the vehicle is raised slightly at the front when the steering

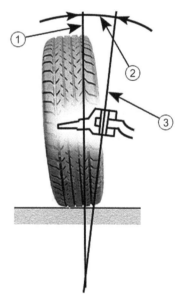

Figure 4.71 Kingpin or swivel axis inclination: 1, perpendicular to road surface; 2, kingpin angle; 3, steering axis

is turned. The weight of the vehicle, therefore, forces the wheels back into the straight-ahead position (Fig. 4.71).

If the wheels are turned, they move downwards on the inclined plane. On the road, the wheels obviously cannot penetrate the road surface. Therefore, for the load condition on the road, this means that the vehicle is raised. This is counteracted by the weight of the vehicle. As a result, the steered wheels attempt to

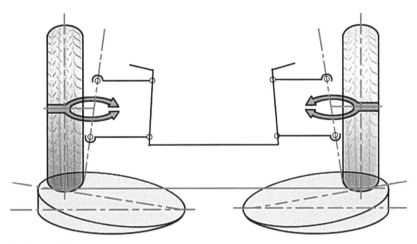

Figure 4.72 Inclined plane effect

return to the straight-ahead position. This self-centring action increases with a greater kingpin angle. After cornering, the vehicle is steered back into the straight-ahead position as a result of this effect and the axle stabilizes itself (Fig. 4.72).

Turning out of the straight-ahead position is made more difficult. This is an advantage when travelling over uneven road surfaces. The restoring forces of the kingpin inclination counteract the disturbing forces. They help the driver to hold a course without heavy counter-steering. The steering therefore becomes smoother. This principle cannot operate if the scrub radius is equal to zero. In this case, the wheel turns on its contact point and does not raise the vehicle body. No steering return forces of any kind are generated. In such a case, the steering return forces are obtained by a positive castor.

Castor angle is the angle in the vehicle's longitudinal direction between the steering axis and the perpendicular through the wheel centre. The castor trail is the distance between the point of intersection of the steering axis with the road surface plane and the perpendicular through the wheel centre. If the wheel contact point is situated between the point of intersection of the steering axis with the road surface in the direction of travel, the castor angle and castor trail is positive. Positive castor causes the wheels to return to the straight-ahead position. It influences the steering torque when cornering and the straight-ahead stability (Fig. 4.73).

Definition

Castor angle: The angle in the vehicle's longitudinal direction between the steering axis and the perpendicular through the wheel centre.

If the wheel contact point is situated in front of the point of intersection of the steering axis with the road

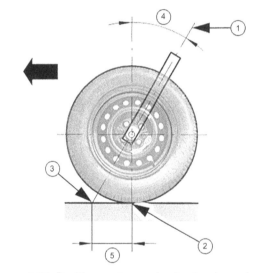

Figure 4.73 Positive castor angle: 1, steering axis; 2, wheel contact point; 3, positive castor point of intersection of steering axis with the road surface; 4, castor angle; 5, trail

surface in the direction of travel, the castor angle and castor trail are negative. Negative castor, or at least only slight positive castor, is frequently present in front-wheel drive vehicles. This is used to reduce the return forces when cornering (Fig. 4.74).

Tyres are a key part of the steering and suspension system. To allow the specially tuned steering and suspension systems in today's cars to operate, there must be good contact between vehicle and road surface. The tyres are therefore designed to:

▷ support the weight of the vehicle
▷ ensure good road adhesion
▷ transmit the drive, braking, steering and cornering forces
▷ improve the ride comfort through good suspension
▷ achieve a high mileage.

305

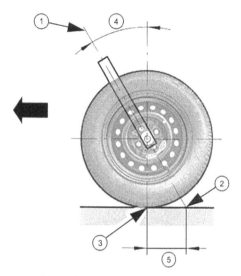

Figure 4.74 Negative castor angle: 1, steering axis; 2, wheel contact point; 3, negative castor point of intersection of steering axis with the road surface; 4, castor angle; 5, trail

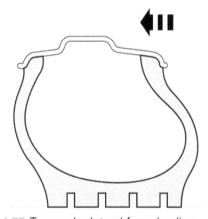

Figure 4.75 Tyre under lateral force loading

When drive force is transmitted at the contact area of a rolling wheel, a relative movement occurs between the tyre and road surface. In this case, the distance covered by the vehicle is shorter than that corresponding to the rolling circumference; in other words, slip occurs. The slip percentage represents the difference between the distance covered by a wheel rolling without power transmission and the distance actually covered with power transmission. When braking with locked wheels the slip is 100%. The slip varies as a function of drive, braking and cornering forces, as well as the friction of tyre and road surface.

A tyre can only transmit lateral forces when it is rolling at an angle to the direction of travel. For this reason, the tyre does not roll straight ahead when cornering, but flexes laterally (Fig. 4.75). As a result of the flexing, the tyre develops a resistance, or a side force, which keeps the vehicle on course. The side deflection of the tyre is introduced by the camber and the toe-in of the wheels. It is necessary to transmit the lateral forces in order to absorb disturbing forces such as side winds or negative lift force. When cornering, the centrifugal force represents an additional disturbing force.

Key fact

A tyre can only transmit lateral forces when it is rolling at an angle to the direction of travel.

At higher cornering speeds, the centrifugal force drives the vehicle mass towards the outside of the curve. So that the vehicle can be kept on track, the tyres must transmit cornering forces, which counteract the centrifugal force. This is only possible, however, if the tyre flexes laterally. In so doing, the wheels no longer move in their turned direction, but drift off at a certain angle from this direction. This means that the tyre is running at an angle to the direction of travel. This angle, which occurs between the tyre's longitudinal axis and the actual direction, is the slip angle (Fig. 4.76).

Definition

Slip angle: The angle between the tyre's longitudinal axis and the actual direction.

Tyres corner best at a slip angle of 15–20°. The lateral adhesion depends on the slip angle, the wheel load and the type of road surface. Steering systems are

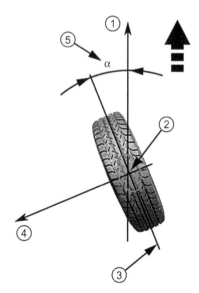

Figure 4.76 Representation of slip angle: 1, direction of movement; 2, contact patch; 3, tyre longitudinal axis; 4, cornering force; 5, slip angle

generally designed so that on bends with radii of more than 20 m, the two steered front wheels lie virtually parallel (that is, not in accordance with the Ackermann principle). On bends with smaller radii, the angles of the stub axles differ significantly from one another, in accordance with the Ackermann principle. In high-speed cornering, this adjustment leads to improved cornering of the wheel due to the greater turning of the outside front wheel.

The centrifugal force acting at the vehicle's centre of gravity is distributed to the front and rear wheels, according to the position of the centre of gravity. This may result in a direction of travel that deviates from the desired direction. A vehicle oversteers when the rear of the vehicle tends to swing outward more than the front during cornering. The slip angle is significantly greater on the rear axle than on the front axle. The vehicle therefore travels in a tighter circle. If the steering angle is not reduced, the vehicle may break away.

A vehicle understeers when the front of the vehicle tends to swing outward more than the rear during cornering. The slip angle is greater on the front axle than on the rear axle. The vehicle therefore travels in a greater circle. It must be forced into the bend with a greater steering angle. Vehicles with understeer can be carried out of the bend. Front-engined vehicles have a tendency to understeer, since the centre of gravity is situated in front of the vehicle centre.

4.2.4 Power steering

The effort required to steer the front wheels depends primarily on the axle load. This is particularly apparent in the following situations:

▷ low speed
▷ low tyre pressures
▷ large tyre contact area
▷ tight cornering.

> **Definition**
>
> **Oversteer:** This occurs when a car turns by more than the amount commanded by the driver.
>
> **Understeer:** This occurs when a car steers less than the amount commanded by the driver.

Steering ratio cannot be increased too much, because a large number of steering wheel turns would be necessary for the steering movement. In general, a steering force of 250 N (184 ft lbs) should not be exceeded. Therefore, the need arises for power steering in heavier cars, trucks and buses. The power assistance is generally produced by hydraulic pressures. However, electric systems are now becoming popular. The requirements of a power steering system are:

▷ precise onset of power assistance
▷ maintenance of driver feel
▷ continued ability to steer should the power system fail.

> **Key fact**
>
> The effort required to steer the front wheels depends primarily on the axle load.

4.2.4.1 Hydraulic power steering

Hydraulic power-assisted steering (PAS) systems use an engine-driven pump to supply pressurized fluid (Fig. 4.77). A control valve directs the fluid to a ram that assists with movement of the steering. If the fluid supply or ram cylinder fails, the steering works like a manual system.

In PAS systems of modular design, parts of the steering gear take the form of a hydraulic piston and cylinder (Fig. 4.78). This gives a compact construction.

Most control valves incorporate a torsion bar. This is designed to twist by a small amount as steering force is exerted. As the torsion bar twists, it allows valves to open and close. These valves supply fluid under pressure to the appropriate side or the ram cylinder. Splines limit the amount of torsion bar twist. In the event of a failure of the hydraulic power assistance, the driver can steer the vehicle by purely mechanical means.

> **Key fact**
>
> In most hydraulic power-assisted steering systems, parts of the steering gear take the form of a hydraulic piston and cylinder.

When the steering wheel is turned, the control valve is actuated, admitting hydraulic fluid into the ram cylinder (Fig. 4.79). Hydraulic fluid under pressure in the ram cylinder assists with the steering force exerted by the driver. Return hydraulic fluid flows through the outlet at the other end of the ram cylinder into the reservoir.

When the steering movement is interrupted, the control piston assumes a neutral position. In this neutral position, the pressure in the ram cylinder is reduced.

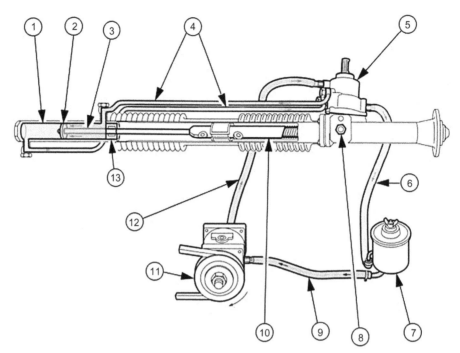

Figure 4.77 Power-assisted steering (PAS) hydraulic system: 1, ram cylinder; 2, piston; 3, piston rod; 4, pressure lines; 5, steering gear; 6, return line; 7, reservoir; 8, pressure pad adjusting screw; 9, suction line; 10, rack; 11, high-pressure pump; 12, high-pressure expansion hose; 13, seal. (Source: Ford Motor Company)

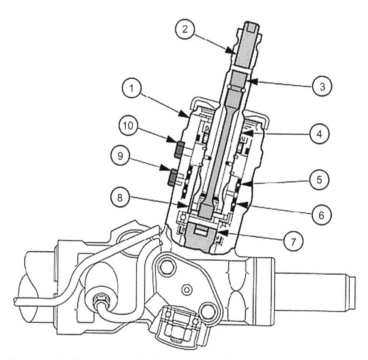

Figure 4.78 Steering rack control unit: 1, control valve housing; 2, input steering shaft; 3, torsion bar; 4, control valve sleeve; 5, upper radial groove feed to right side of piston; 6, lower radial groove feed to left side of piston; 7, pinion shaft; 8, control valve sleeve return groove; 9, pump line connection; 10, return line connection. (Source: Ford Motor Company)

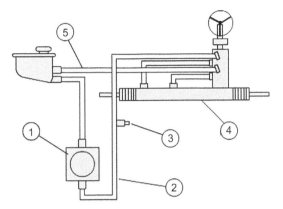

Figure 4.79 Hydraulic circuit: 1, pump; 2, pressure line; 3, pressure switch; 4, rack and ram; 5, return line

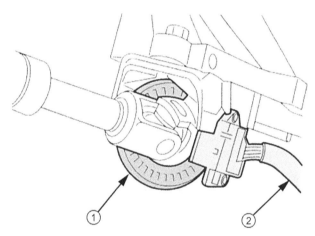

Figure 4.80 This sensor monitors steering position: 1, rotating disc; 2, sensor

Variable assist power steering (VAPS) systems are controlled electronically. Variable power steering (sometimes called progressive power steering) makes steering easier at low speeds but provides good driver feel at higher speeds. An electronic control unit monitors the signals from the vehicle speed sensor and the steering position sensor (Fig. 4.80). From this data, it can work out the power assistance required. The solenoid valve controls the amount of assistance, because the valve in turn controls fluid pressure. Maximum power assistance occurs at speeds less than 10 km/h (6 mph) or when the steering wheel is rotated more than 45°.

Hydraulic PAS systems use an engine-driven pump to supply pressurized fluid. A control valve directs the fluid to a ram that assists with the movement of the steering. Variable assistance systems are used. These

usually involve some electronic control. Progressive PAS is controlled in this way or by a restrictor valve that changes with road speed. Pressure switches, when used, often inform engine management systems that PAS is in use. This allows idle speed to be increased if necessary.

4.2.4.2 Electric power steering

There are two main ways of using electric power for steering assistance (the second is now the more common):

▷ replacing the conventional system pump with an electric motor while the ram remains much the same
▷ using a drive motor, which directly assists with the steering and has no hydraulic components.

With a direct-acting type, an electric motor works directly on the steering via an epicyclic gear train (Fig. 4.81). This completely replaces the hydraulic pump and servo cylinder. This eliminates the fuel penalty of the conventional pump and greatly simplifies the drive arrangements. Engine stall when the power steering is operated at idle speed is also eliminated.

On many systems, an optical torque sensor is used to measure driver effort on the steering wheel (all systems use a sensor of some sort). The sensor works by measuring light from a light-emitting diode (LED), which shines through some holes. These are aligned in discs at either end of a torsion bar, fitted into the steering column. An optical sensor element identifies the twist of two discs on the steering axis with respect to each other, each disc being provided with

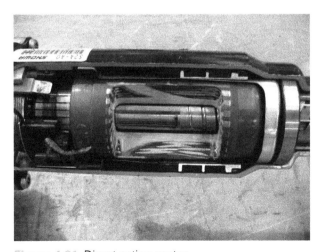

Figure 4.81 Direct acting motor

309

Figure 4.82 Steering sensor. (Source: Bosch Media)

Figure 4.83 Electric power-assisted steering (PAS). (Source: Ford Media)

appropriate codes. From this sensor information the system calculates the torque as well as the absolute steering angle (Fig. 4.82).

Key fact

Electrical PAS occupies little under-bonnet space and the motor only averages about 2 A current draw under urban driving conditions.

Electric PAS (Fig. 4.83) occupies little under-bonnet space, something that is at a premium these days, and the 400 W motor only averages about 2 A under urban driving conditions. The cost benefits over conventional hydraulic methods are considerable.

Electric power for steering assistance can be applied in a number of ways. However, using a drive motor, which directly assists with the steering, is the most common. This method uses less power and takes up less space than other methods.

4.3 Brakes

4.3.1 Disc, drum and parking brakes

The main purpose of the braking system is simple: to slow down or stop a vehicle. To do this, the energy in the vehicle movement must be taken away or converted.

This is achieved by creating friction. The resulting heat takes energy away from the movement. In other words, kinetic energy is converted into heat energy.

The main braking system of a car works by hydraulics. This means that when the driver presses the brake pedal, liquid pressure forces pistons to apply brakes on each wheel. Disc brakes are used on all wheels of most cars nowadays (Fig. 4.84). Braking pressure forces brake pads against both sides of a steel disc.

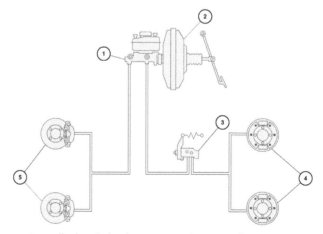

Figure 4.84 Brake system: 1, master cylinder; 2, brake servo or booster; 3, pressure regulator; 4, brake shoes; 5, brake discs (rotors) and pads

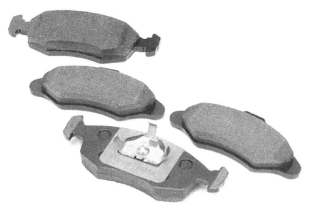

Figure 4.85 Brake pads

Figure 4.86 Brake shoes

Drum brakes are fitted on the rear wheels of some cars and on all wheels of older vehicles. Braking pressure forces shoes to expand outwards into contact with a drum. The important part of brake pads and shoes is the friction lining.

Key fact

To slow down a vehicle its movement energy (kinetic) must be converted to another form.

Brake pads (Fig. 4.85) are steel-backed blocks of friction material, which are pressed onto both sides of the disc (also known as a brake rotor). Older types were asbestos based, so you must not inhale the dust. Pads should be changed when the friction material wears down to 2 or 3 mm. The circular steel disc rotates with the wheel. Some are solid but many have ventilation holes. Modern brake pad and shoe lining material is made from mineral fibres, cellulose, aramid, polyacrylonitrile (a resinous, fibrous or rubbery organic polymer), ceramics, chopped glass, steel, and copper fibres, carbon fibres or a similar material. There is no need to remember all these; just choose the correct replacement types as recommended by the manufacturer.

Key fact

Brake pads should be changed when the friction material wears down to 2 or 3 mm, or as specified by the manufacturer.

Brake shoes are steel crescent shapes with a friction material lining (Fig. 4.86). They are pressed inside a steel drum, which rotates with the wheel. The rotating action of the brake drum tends to pull one brake shoe

harder into contact. This is known as self-servo action. It occurs on the brake shoe after the wheel cylinder in the direction of wheel rotation. This brake shoe is described as the leading shoe. The brake shoe before the wheel cylinder in the direction of wheel rotation is described as the trailing shoe.

The master cylinder piston is moved by the brake pedal. In its basic form, it is like a pump that forces brake fluid through the pipes. Pressure in the pipes causes a small movement to operate either brake shoes or pads. The wheel cylinders and callipers work like a pump only in reverse (Figs 4.87 and 4.88).

Key fact

Vacuum, from the engine inlet manifold, is used to work most brake servos.

Figure 4.87 Brake master cylinder on a vehicle

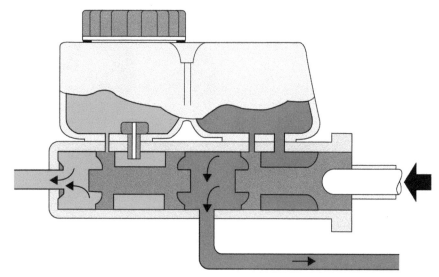

Figure 4.88 Master cylinder operation

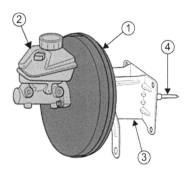

Figure 4.89 Servo unit: 1, brake servo (booster);
2, master cylinder and fluid reservoir; 3, bracket;
4, diaphragm rod connection to brake pedal

Figure 4.90 Flexible and metal brake pipes

The brake servo (booster) increases the force applied by the driver on the pedal. It makes the brakes more effective. Vacuum, from the engine inlet manifold, is used to work most brake servos (Fig. 4.89).

Strong, high-quality pipes are used to connect the master cylinder to the wheel cylinders. Fluid connection, from the vehicle body to the wheels, has to be through flexible pipes to allow suspension and steering movement (Fig. 4.90). As a safety precaution (because brakes are important!), brake systems are split into two sections. If one section fails, say by a pipe breaking, the other will continue to operate.

If the brakes cause the wheels to lock and make them skid, steering control is lost. In addition, the brakes will not stop the car as quickly. ABS (anti-lock braking system) uses electronic control to prevent this happening.

On most car braking systems, about 70% (or more) of the braking force is directed to the front

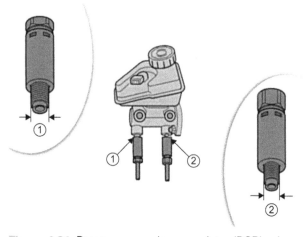

Figure 4.91 Pressure-conscious regulator (PCR) valves:
1, PCR valve for circuit one with larger thread diameter;
2, PCR valve for circuit two with smaller thread diameter

wheels. This is because, under braking, the weight of the vehicle transfers to the front wheels. Load compensation, however, allows the braking pressure to the rear wheels to increase as load in the vehicle increases.

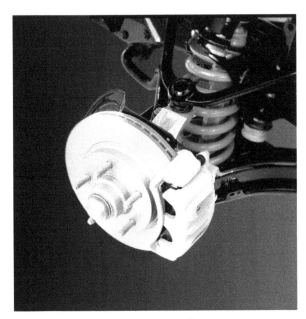

Figure 4.92 Sliding disc brake calliper

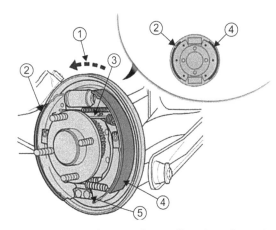

Figure 4.93 Rear drum brake: 1, direction of rotation; 2, leading shoe; 3, self-adjusting device; 4, trailing shoe; 5, lower bracket

Devices such as pressure-conscious regulator (PCR) valves (Fig. 4.91) to may be used to control the braking force.

If brakes become so hot that they cannot convert energy rapidly enough, they become much less efficient, or in other words, fade away! This is described as brake fade. A more serious form of brake fade can also be caused if the heat generated is enough to melt the bonding resin in the friction material. This reduces the frictional value of the linings or pads.

Safety first

If brakes become so hot that they cannot convert energy rapidly enough, they become much less efficient. This is known as brake fade.

All components of the braking system must be in good working order, in line with most other vehicle systems. Braking efficiency means the braking force compared to the weight of the vehicle. For example, the brakes on a vehicle with a weight of 10 kN (1000 kg × 10 ms^{-2} [g]) will provide a braking force of, say, 7 kN. This is said to be 70% efficiency. During an annual test this is measured on brake rollers. The current efficiency requirements in the UK are as follows: service brake efficiency 50%, second line brake efficiency 25% and parking brake efficiency 16%.

Key fact

Service brake minimum efficiency: 50%; second line brake efficiency: 25%; parking brake efficiency: 16%.

Disc brakes are less prone to brake fade than drum brakes (Fig. 4.92). This is because they are more exposed and can get rid of heat more easily. They also throw off water better than drum brakes. Brake fade occurs when the brakes become so hot they cannot transfer any more energy, and they stop working. Disc brakes are self-adjusting. When the pedal is depressed, the rubber seal is preloaded. When the pedal is released, the piston is pulled back owing to the elasticity of the rubber sealing ring.

Brake shoes, when used, are mounted inside a cast iron drum (Fig. 4.93). They are mounted on a steel backplate, which is rigidly fixed to a stationary part of the axle. The two curved shoes have friction material on their outer faces. One end of each shoe bears on a pivot point. The other end of each shoe is pushed out by the action of a wheel cylinder when the brake pedal is pressed. This puts the brake linings in contact with the drum inner surface. When the brake pedal is released, the return spring pulls the shoes back to their rest position.

Definition

DOR: Direction of rotation.

Drum brakes are more adversely affected by wet and heat than disc brakes, because both water and heat are trapped inside the drum. However, they are easier to fit with a mechanical handbrake linkage (Fig. 4.94).

Brakes must be adjusted so that the minimum movement of the pedal starts to apply the brakes. The adjustment in question is the gap between the pads and disc and the shoes and drum. Disc brakes are self-adjusting because as pressure is released it moves the

Figure 4.94 Brake drum being removed

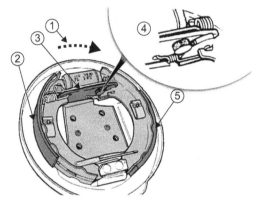

Figure 4.95 Self-adjusting device: 1, direction of rotation; 2, trailing shoe; 3, thrust link; 4, toothed cam and pawl mechanism; 5, leading shoe

pads just away from the disc (Fig. 4.95). Drum brakes are different because the shoes are moved away from the drum to a set position by a pull-off spring. Self-adjusting drum brakes are almost universal now on light vehicles. A common type uses an offset ratchet, which clicks to a wider position if the shoes move beyond a certain amount when operated.

If an automatic adjuster is not used, manual adjustment through a hole in the back plate is often employed (Fig. 4.96). This involves moving a type of nut on a threaded bar, which pushes the shoes out as it is screwed along the thread. This method is similar to the automatic adjusters. An adjustment screw on the back plate is now quite an old method. A screw or square head protruding from the back plate moves the shoes by a snail cam. As a guide, tighten the adjuster until the wheels lock, and then move it back until the wheel is just released. You must ensure that the brakes are not rubbing as this would build up heat and wear the friction material very quickly.

Safety first

Brake shoes must not be rubbing (binding) as this builds up heat and wears the friction material very quickly, and prevents the brakes working properly.

The precise way in which the shoes move into contact with the drum affects the power of the brakes. If the shoes are both hinged at the same point then the system is said to have one leading and one trailing shoe. As the shoes are pushed into contact with the drum, the leading shoe is dragged by the drum rotation harder into contact, whereas the rotation tends to push the trailing shoe away. This 'self-servo' action on the leading shoe can be used to increase

Figure 4.96 Drum brake adjustment hole

the power of drum brakes. This is required on the front wheels of all-round drum brake vehicles.

The shoes are arranged so that they both experience the self-servo action. The shoes are pivoted at opposite points on the backplate and two wheel cylinders are used. The arrangement is known as twin leading shoe brakes (Fig. 4.97). It is not suitable for use on the rear brakes because if the car is travelling in reverse then it would become a twin trailing shoe arrangement, which means the efficiency of the brakes would be seriously reduced. The leading and trailing layout is therefore used on rear brakes, as one shoe will always be leading no matter in what direction the vehicle is moving.

The standard layout of drum brake systems on a vehicle is normally:

▶ twin leading shoe brakes on the front wheels
▶ leading and trailing shoe brakes on the rear wheels.

Disc brakes are now used on the front wheels of all light vehicles and the rear wheels of most. However,

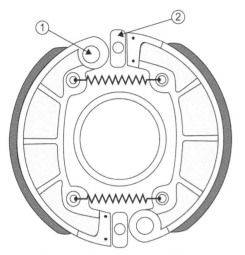

Figure 4.97 Twin leading shoe principle: an expander at the top and the bottom: 1, pivot; 2, actuator

Figure 4.98 Leading and trailing system

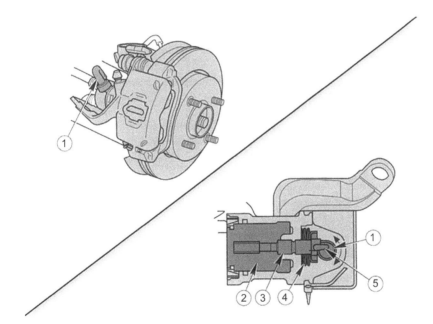

Figure 4.99 Sliding calliper parking brake: 1, moving lever; 2, piston; 3, adjusting screw; 4, spring washers; 5, cam

some retain leading and trailing shoe brakes on the rear (Fig. 4.98). In most cases, it is easier to attach a handbrake linkage to the system with shoes on the rear. This method will also provide the braking performance required when the vehicle is reversing.

Inside a brake drum, the handbrake linkage is usually a lever mechanism as shown in Figure 4.99. This lever pushes the shoes against the drum and locks the wheel. The handbrake lever pulls on one or more cables and has a ratchet to allow it to be locked in the 'on' position. There are several ways in which the handbrake linkage can be laid out to provide equal force, or compensation, for both wheels:

▷ two cables, one to each wheel
▷ equalizer on a single cable pulling a U-section to balance effort through the rear cable (as shown in Fig. 4.99)
▷ single cable to a small linkage on the rear axle.

Key fact

The handbrake is usually on the rear axle, but it can appear on the front.

Some sliding calliper disc brakes incorporate a handbrake mechanism. The footbrake operates as normal. Handbrake operation is by a moving lever. The lever acts through a shaft and cam, which works on the adjusting screw of the piston. The piston presses one pad against the disc and, because of the sliding action, the other pad also moves.

Some manufacturers use a set of small brake shoes inside a small drum, which is built into the brake disc. The calliper is operated as normal by the footbrake. The small shoes are moved by a cable and lever.

There are some cars that use two sets of calipers on the same axle, one to operate the service brake, and another smaller set for the handbrake.

In summary, remember that the purpose of the braking system is to slow down or stop a vehicle. This is achieved by converting the vehicle's movement energy into heat using friction.

4.3.2 Hydraulic components

A complete braking system includes a master cylinder, which operates several wheel cylinders (Fig. 4.100). The system is designed to give the power amplification

needed for braking the particular vehicle. On any vehicle when braking, a lot of the weight is transferred to the front wheels. Most braking effort is therefore designed to work on the front brakes. Some cars have special hydraulic valves to limit rear wheel braking. This reduces the chance of the rear wheels locking and skidding.

> **Key fact**
> Most braking effort is directed to the front wheels.

Brake shoes can be moved by double- (Fig. 4.101) or single-acting wheel cylinders. A common layout is to use one double-acting cylinder and brake shoes on each rear wheel of the vehicle and disc brakes on the front wheels. A double-acting cylinder simply means that as fluid pressure acts through a centre inlet, pistons are forced out of both ends.

Disc brake callipers are known as fixed, floating or sliding types (Figs 4.102– 4.104). The pistons are moved by hydraulic pressure created in the master

Figure 4.100 Master cylinder and fluid reservoir fitted on to the servo

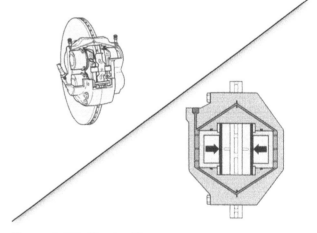

Figure 4.102 Fixed calliper

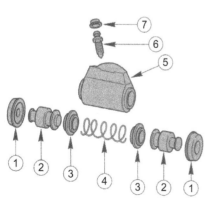

Figure 4.101 Slave cylinder components: 1, dust seal; 2, piston; 3, piston seal; 4, spring; 5, cylinder; 6, bleed nipple; 7, dust cap

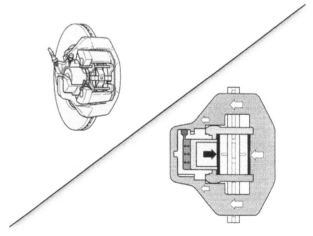

Figure 4.103 Floating calliper

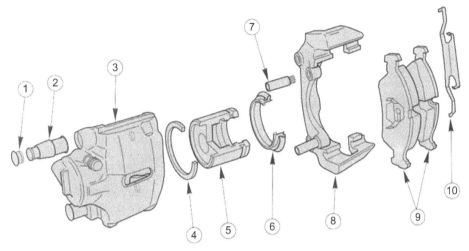

Figure 4.104 Sliding disc brake calliper components: 1, dust cap; 2, bleed nipple; 3, piston housing (calliper assembly); 4, piston seal; 5, piston; 6, dust seal; 7, locating stud; 8, housing bracket; 9, pads; 10, retaining clip

cylinder. A number of different callipers are used. Some high-performance callipers include up to four pistons. However, the operating principle remains the same. The sliding calliper tends to be used most.

Key fact

Some high-performance callipers include up to four pistons.

Always use new and approved brake fluid when topping up or refilling the system. Manufacturers' recommendations must always be followed. Brake fluid is hygroscopic, which means that, over time, it absorbs water. This increases the risk of the fluid boiling owing to the heat from the brakes. Pockets of steam in the system would not allow full braking pressure to be applied. Many manufacturers recommend that the fluid be changed at regular intervals. Make sure the correct grade of fluid is used. The common recommended types are known as DOT4 and DOT5.

Safety first

Many manufacturers recommend that the brake fluid is changed at regular intervals.

Key fact

A tandem master cylinder is like two separate cylinders but inside one housing.

Safety is built into braking systems by using a double-acting master cylinder. This is often described as tandem and can be thought of as two master cylinders inside one housing. The pressure from the pedal acts on both cylinders but fluid cannot pass from one to the other. Each cylinder is then connected to a separate circuit. These split lines can be connected in a number of ways. Under normal operating conditions, the pressure developed in the first part of the master cylinder is transmitted to the second. This is because the fluid in the first chamber acts directly on the second piston (Fig. 4.105).

If one line fails, the first piston meets no restriction and closes up to the second piston. Further movement will now provide pressure for the second circuit. The driver will notice that pedal travel increases, but some braking performance will remain. If the fluid leak is from the second circuit, then the second piston will meet no restriction and close up the gap. Braking will now be just from the first circuit. Diagonal split brakes are the most common and are used on vehicles with

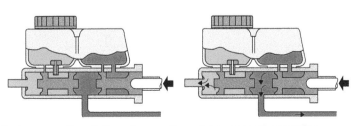

Figure 4.105 Master cylinder operation: left, not operating; right, normal operation

317

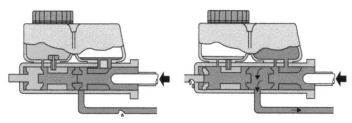

Figure 4.106 Tandem master cylinder showing the operation when there is a failure in different circuits

a negative scrub radius. Steering control is maintained under brake failure conditions (Fig. 4.106).

Three common 'splits' or multicircuit systems are used on braking systems. The first two types listed are the most common:

▷ diagonal split type, where, if a fault occurs, the driver loses half of the front and half of the rear brakes (Fig. 4.107)
▷ separate front and rear, where, if a fault occurs, the driver loses all of the front or all of the rear brakes (Fig. 4.108)
▷ duplicated front, where, if a fault occurs, the driver loses the rear and part of the front brakes only. Special front callipers are required when using this method.

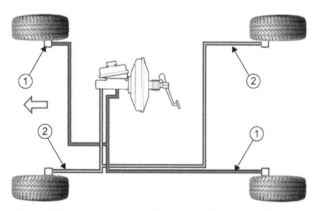

Figure 4.107 Diagonal split: 1, circuit 1; 2, circuit 2

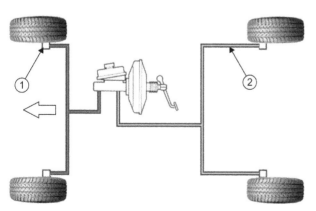

Figure 4.108 Front/rear split: 1, circuit 1; 2, circuit 2

4.3.3 Brake servo operation

The brakes of a vehicle must perform well, while keeping the effort required by the driver to a reasonable level. This is achieved by the use of a brake servo. It is also called a brake booster. Vacuum-operated systems are commonly used on light vehicles (Fig. 4.109).

Hydraulic power brakes use the pressure from an engine-driven pump. The pump will often be the same as the one used to supply the PAS. Pressure from the pump is made to act on a plunger in line with the normal master cylinder. As the driver applies force to the pedal, a servo valve opens in proportion to the force applied by the driver. The hydraulic assisting force is therefore also proportional. This maintains the all-important 'driver feel'.

A hydraulic accumulator (a reservoir for fluid under pressure) is incorporated into the system (Fig. 4.110). This is because the pressure supplied by the pump varies with engine speed. The pressure in the accumulator is kept between set pressures in the region of 70 bar. A warning, therefore: if you have to disconnect any components from the braking system on a vehicle fitted with an accumulator, you must follow the manufacturer's recommendations on releasing the pressure first.

Figure 4.109 Vacuum servo and master cylinder

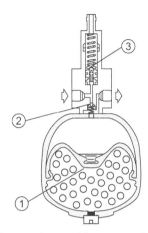

Figure 4.110 Accumulator with arrows showing the inlet and outlet: 1, diaphragm; 2, check valve; 3, pressure-relief valve

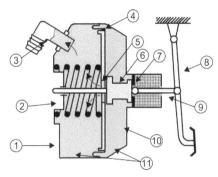

Figure 4.111 Servo construction: 1, vacuum chamber; 2, compression spring; 3, vacuum line; 4, diaphragm; 5, vacuum passage; 6, poppet valve; 7, exterior air passage; 8, brake pedal; 9, filter; 10, working chamber; 11, chambers

The more common servo system uses low pressure (vacuum) from the manifold on one side, and the higher atmospheric pressure on the other side of a diaphragm (Fig. 4.111). The low pressure is taken via a non-return safety valve from the engine inlet manifold. A pump is often used on diesel-engined vehicles as most do not have a throttle butterfly and hence do not develop any significant manifold vacuum. The pressure difference, however created, causes a force, which is made to act on the master cylinder.

The vacuum servo is fitted in between the brake pedal and the master cylinder. The main part of the servo is the diaphragm. The larger this diaphragm, the greater the servo assistance provided. A vacuum is allowed to act on both sides of the diaphragm when the brake pedal is in its rest position. When pedal force is applied to the piston a valve cuts the vacuum connection to the rear chamber and allows air at atmospheric pressure to enter. This causes a force to act on the diaphragm, so assisting with the application of the brakes.

Once the master cylinder piston moves, the valve closes again to hold the applied pressure. Further effort by the driver on the brake pedal will open the valve again and apply further vacuum assistance. In this way, the driver can 'feel' the amount of braking effort being applied. The cycle continues until the driver effort reaches a point where the servo assistance remains fully on.

A non-return valve is fitted in the line to keep a vacuum in the servo chamber. This means that it is possible to carry out three or four braking operations, with servo assistance, without the engine running. The valve also prevents fuel vapours getting in the servo and damaging the diaphragm (Fig. 4.112).

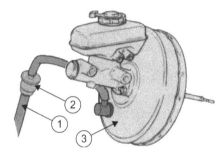

Figure 4.112 A check valve is fitted in the vacuum supply: 1, vacuum connection to inlet manifold; 2, check (non-return valve); 3, chamber

If the vacuum servo (booster) stops working the brakes will still operate, but extra force will be required from the driver. The connection to the inlet manifold will normally be via a check valve as an extra safety feature.

Safety first

If the vacuum servo (booster) stops working the brakes will still operate, but extra force will be required from the driver.

A brake servo assists the driver when the brakes are applied. The 'feel' must be maintained during operation. Most servos are vacuum operated.

4.3.4 Braking force control

Some cars use devices to control the braking force. There are three main types:

▷ load-apportioning valve (Fig. 4.113)
▷ pressure-conscious regulator
▷ deceleration-sensing brake pressure reducer.

The purpose of these devices is to ensure braking force is distributed so that most of the force goes to the front brakes. This improves performance and stability.

319

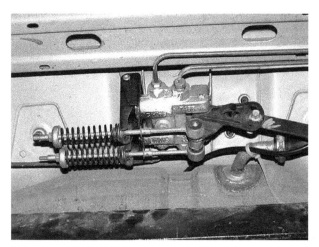

Figure 4.113 Load apportioning valve

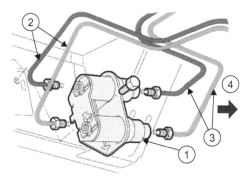

Figure 4.114 Pressure reducer location: 1, brake pressure reducers (2); 2, pipes from master cylinder; 3, pipes to rear axle; 4, direction of travel

Load-apportioning valves are fitted between the rear axle and vehicle floor assembly. A single valve is used for vehicles with front to rear split lines, and two valves are used when the split is diagonal. A lever and tension spring changes the force necessary to make a plunger move. The lever and spring adjust position depending on the vehicle load. Fluid pressure moves the plunger; however, the position of the lever limits the movement. Load in the vehicle sets the valve position. Pressure, and therefore braking force, is controlled by the valve.

The pressure-conscious regulator is simply fitted in the line, or lines, to the rear brakes. It reduces braking pressure by a fixed amount. An internal control spring is used to set the operating pressure.

Where used, one deceleration sensor is used in each brake circuit (Fig. 4.114). The sensors are mounted on the vehicle floor at a set angle to the horizontal.

When the deceleration is greater than about 0.5 g, the valves allow the pressure to the rear brakes to rise more slowly than the front. During deceleration (usually above 0.5 g), a ball moves against a spring and a valve, which reduces the pressure to the rear brakes. When deceleration reduces, the ball rolls back against the washer.

Controlling brake pressure ensures that braking force is distributed so that most of the force goes to the front brakes. As a guide, more than 70% of the braking takes place on the front wheels. This improves performance, control and stability.

Key fact

About 70% of the braking takes place on the front wheels.

4.3.5 Anti-lock brake systems

The reason for the development of the anti-lock braking system (ABS) (Fig. 4.115) is simple. Under braking conditions, if one or more of the vehicle's wheels locks and begins to skid, then this has serious consequences:

▷ Braking distance increases.
▷ Steering control is lost.
▷ Tyre wear is abnormal.

The maximum deceleration of a vehicle is achieved when maximum energy conversion is taking place in the brake system. This is the conversion of kinetic energy to heat energy at the discs and brake drums. The potential for this conversion process between a tyre skidding and the road, even a dry road, is far less. A good driver can pump the brakes on and off to prevent locking (cadence braking) but electronic control can achieve even better results (Fig. 4.116).

Key fact

The maximum deceleration of a vehicle is achieved when maximum energy conversion is taking place in the brake system, rather than between skidding tyres and the road.

ABS is now common on lower price vehicles, which should be a contribution to safety. It is important to remember, however, that for normal use, the system is not intended to allow faster driving and shorter braking distances. It should be viewed as operating in an emergency only. Good steering and road holding must continue when ABS is operating. This is arguably the key issue, as being able to swerve round a hazard, while still braking hard, is often the best course of action.

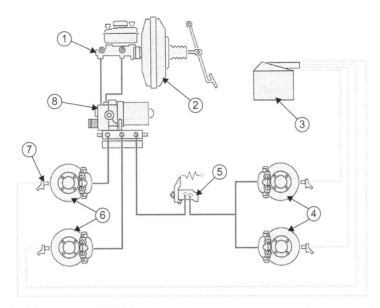

Figure 4.115 Anti-lock braking system (ABS) layout: 1, master cylinder; 2, brake servo or booster; 3, electronic control unit; 4, rear disc brakes; 5, load apportioning valve (if used); 6, front disc brakes; 7, wheel speed sensor; 8, hydraulic modulator

Figure 4.116 Electronic control unit control of the brakes

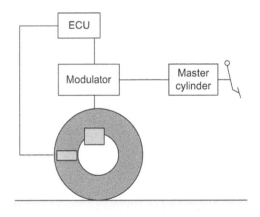

Figure 4.117 Representation of an anti-lock braking system (ABS)

In the event of the ABS failing, then conventional brakes must still operate to their full potential. In addition, a warning must be given to the driver. This is normally in the form of a simple warning light. The system must operate under all speed conditions above walking pace. At this very slow speed, even when the wheels lock, the vehicle will come to rest quickly. If the wheels did not ever lock then in theory the vehicle would never stop!

ABS must be able to recognize aquaplaning and react accordingly. It must also operate on an uneven road surface. The one operating condition that has still not been perfected is braking from slow speed on snow. The ABS can actually increase stopping distances in snow. However, steering control will be improved and this is considered a suitable trade-off.

An ABS is represented by the closed loop system block diagram shown in Figure 4.117. The most important of the inputs are the wheel speed sensors. The main output is some form of brake system pressure control (the modulator). The task of the ECU is to compare signals from each wheel sensor. From these signals, it can determine the acceleration or deceleration of an individual wheel. Brake pressure can be reduced, held constant or allowed to increase. The maximum pressure is determined by the driver's pressure on the brake pedal.

321

A vehicle reference speed is determined from the combination of two diagonal wheel sensor signals (Fig. 4.118). After the start of braking, the ECU uses this value as its reference. The acceleration and deceleration values are live measurements, which are constantly changing.

Although brake slip cannot be measured directly, a value can be calculated from the vehicle reference speed. This figure is then used to determine whether and when ABS should take control of the brake pressure. There are variations between manufacturers involving a number of different components. However, for the majority of systems, there are three main components:

▷ wheel speed sensors (Fig. 4.119)
▷ electronic control unit (Fig. 4.120)
▷ hydraulic modulator (Fig. 4.121).

Most wheel speed sensors are inductance sensors and work in conjunction with a toothed wheel. They consist of a permanent magnet and a soft iron rod around which is wound a coil of wire. As the toothed wheel rotates, the changes in inductance of the magnetic circuit generates a signal. The frequency and voltage of the signal are proportional to wheel speed. The frequency is the signal used by the ECU. Some systems now use Hall effect sensors, which are more accurate at lower speed. The main parts of the sensor are a magnet and an integrated circuit containing the sensing element.

The ECU takes in information from the wheel sensors and calculates the best course of action for the hydraulic modulator. At the heart of an ABS ECU are two microprocessors, which run the same program independently of each other. This ensures greater security against any fault that could adversely affect braking performance. If a fault is detected, the ABS disconnects itself and operates a warning light. Both processors have non-volatile memory into which fault codes can be written for later service and diagnostic access. The ECU performs a self-test after the ignition is switched on. A failure results in disconnection of the system.

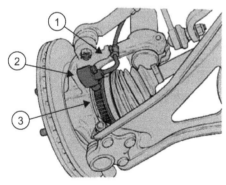

Figure 4.118 Front wheel sensor: 1, carrier; 2, wheel speed sensor; 3, pulse rotor (toothed wheel)

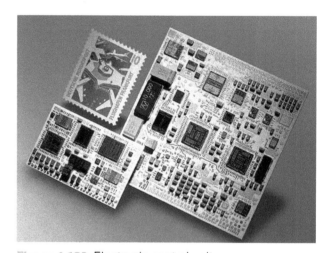

Figure 4.120 Electronic control unit

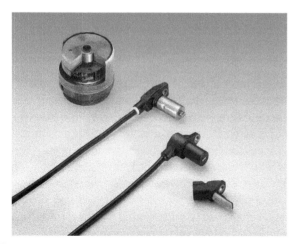

Figure 4.119 Wheel speed sensors

Figure 4.121 Hydraulic modulators

A hydraulic modulator has three operating positions:

▷ pressure release, where the brake line is open to the reservoir

▷ pressure holding, where the brake line is closed

▷ pressure build-up, where the brake line is open to the pump.

The start of ABS engagement is known as 'first control cycle smoothing'. This smoothing stage is necessary in order not to react to minor disturbances such as an uneven road surface, which can cause changes in the wheel sensor signals. The threshold of engagement is critical. If it started too soon, it would be distracting to the driver and cause unnecessary component wear. If too late, steering and stability could be lost on the first control cycle.

> **Key fact**
>
> A hydraulic modulator has three operating positions:
> ▶ pressure release
> ▶ pressure holding
> ▶ pressure build-up.

When braking on a road surface with different adhesion under the left and right wheels, the vehicle will yaw or start to twist. The driver can control this with the steering if time is available. This can be achieved if, when the front wheel with poor adhesion becomes unstable, the pressure to the other front wheel is reduced. This acts to reduce the vehicle yaw, which is particularly important when the vehicle is cornering.

Wheel speed instability occurs frequently and at random because of axle vibration on rough roads. Because of this instability, brake pressure tends to be reduced more than it is increased during ABS operation. This could lead to loss of braking under certain conditions. A slight delay in the reaction of the ABS due to delay in signal smoothing, the time taken to move control valves and a time lag in the brake lines, helps to reduce the effect of axle vibration.

The control strategy of the anti-lock brake system can be summarized as follows:

▷ a rapid brake pressure reduction during wheel speed instability; the wheel will therefore reaccelerate without too much pressure reduction and avoid underbraking

▷ a rapid rise in brake pressure during and after a reacceleration to a value just less than the instability pressure

▷ a discrete increase in brake pressure in the event of increased adhesion

▷ sensitivity suited to the prevalent conditions

▷ anti-lock braking not initiated during axle vibration.

The application of these five main requirements leads to the need for compromise. Optimum programming and prototype testing can reduce the level of compromise but some disadvantages have to be accepted. The best example of this is braking on uneven ground in deep snow, because deceleration is less effective unless the wheels are locked up. In this example, priority is given to stability rather than stopping distance, as directional control is favoured under these circumstances.

An interesting further development in braking systems is the electrohydraulic brake (EHB) (Figs 4.122–4.124). In this system, when the brakes are activated, the EHB control unit calculates the desired target brake pressures at the individual wheels. Braking pressure for each of the four wheels is regulated individually via a wheel pressure modulator, which consists of one inlet and one outlet valve controlled electronically. Normally, the brake master cylinder is detached from the brake circuit, with a pedal travel simulator creating normal pedal feedback. If the electronic stability program (ESP) intervenes, the high-pressure reservoir supplies the required brake pressure quickly and precisely to the wheel brakes.

> **Definition**
>
> **ESP:** Electronic stability program.

With EHB, the driver's brake pedal input is translated into an electrical signal that is calculated by the EHB control unit and transferred to a central hydraulic unit. The central hydraulic unit develops a hydraulic braking pressure in line with the driver's input and transfers the pressure to the wheel cylinders.

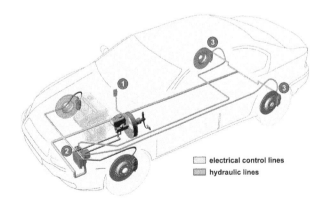

electrical control lines
hydraulic lines

Figure 4.122 Electrohydraulic brake (EHB) system: 1, push button for automatic brake; 2, ESP hydraulic unit with ABS software; 3, callipers with locking device. (Source: Bosch Media)

323

Typical braking manoeuvre at 50 kph (30 mph)

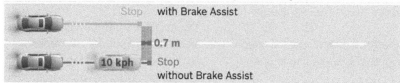

Typical braking manoeuvre at 100 kph (60 mph)

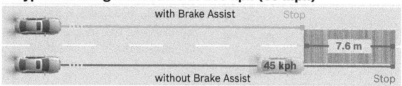

Figure 4.123 Shorter braking distance using EHB. (Source: Bosch Media)

Figure 4.124 Brake by wire: components of Bosch's electrohydraulic brake (EHB) with electronic stability program (ESP). (Source: Bosch Media)

In summary, the principle of operation of all anti-lock brake systems is the same. Three discrete operating phases have to be achieved: pressure reduction, pressure holding and pressure decrease. There are further developments such as EHB and stability control, but the actual brakes still work in the same way, with pads and discs.

Key fact

Even in complex systems, the brakes still use pads and discs to cause friction, which creates heat, therefore taking energy away from the movement.

4.3.6 Traction control

The 'steerability' of a vehicle is lost if the wheels spin when driving off under severe acceleration. Electronic traction control has been developed as a

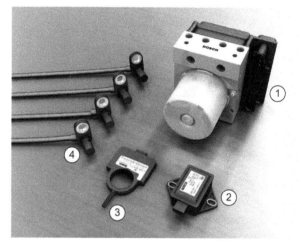

Figure 4.125 Components of an electronic stability program (ESP): 1, anti-lock braking system and traction control modulator with built-in electronic control; 2, position sensor; 3, steering sensor; 4, wheel speed sensors

supplement to ABS (Fig. 4.125). This control system prevents the wheels from spinning when moving off or when accelerating sharply while on the move. In this way, an individual wheel that is spinning is braked in a controlled manner. If both or all of the wheels are spinning, the drive torque is reduced by means of an engine control function. The traction control system (TCS) is also known as anti-slip regulation (ASR).

Key fact

Traction control is not normally available as an independent system, but is available in combination with ABS.

Traction control is not normally available as an independent system, but is available in combination

Figure 4.126 Steel wheels are often covered with attractive trims

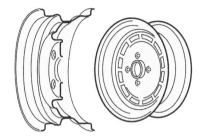

Figure 4.127 Standard wheel design

with ABS. This is because many of the components required are the same for each. A traction control system is linked to the ABS and the engine control system. Traction control will intervene to achieve the following:

- driving stability
- reduction of yawing moment reactions
- optimum propulsion at all speeds
- reduced driver workload.

An automatic control system can intervene more quickly and precisely than the driver of the vehicle. This allows stability to be maintained at a time when the driver might not be able to cope with the situation.

Tractive force can be controlled by a number of methods:

- throttle control
- ignition control
- braking effect.

When wheel spin is detected, the throttle position and ignition timing are adjusted. However, better results are gained when the brakes are applied to the spinning wheel. When the brakes are applied, a valve in the hydraulic modulator assembly moves over to allow traction control operation. This allows pressure from the pump to be applied to the brakes on the offending wheel. The valves, in the same way as for ABS, can provide pressure build-up, pressure hold and pressure reduction. This all takes place without the driver touching the brake pedal.

Traction control is designed to prevent wheel spin when a vehicle is accelerating. This improves traction and ensures vehicle stability. Anti-lock brakes and

traction control have now developed into complex stability control systems.

ESPs intervene to ensure stability under a wide range of situations. Sensors supply an ECU with information on vehicle movement such as rotation about a vertical axis (yaw). By controlling the driving force from the engine and the braking force to individual wheels, the vehicle can be kept in a stable condition.

4.4 Wheels and tyres

4.4.1 Wheels and rims

Together with the tyre, a road wheel must support the weight of the vehicle. It must also withstand side thrusts when cornering, and torsional forces when driving. Road wheels must be strong, but lightweight. They must be cheap to produce, easy to clean, and simple to remove and refit.

The centre of this type of wheel is made by pressing a disc into a dish shape, to give it greater strength. The rim is a rolled section, which is circled and welded. The rim is normally welded to the flange of the centre disc. The centre disc has several slots under the rim. This is to allow ventilation for the brakes as well as the wheel itself.

The manufacture of this type of wheel makes it cheap to produce and strong. The bead of a tyre is made from wire, which cannot be stretched for fitting or removal. The wheel rim therefore, must be designed to allow the tyre to be held in place, but also allow for easy removal.

To facilitate fitting and removal a 'well-base' is manufactured into the rim. For tyre removal, one bead must be forced into the well. This then allows the other bead to be levered over the edge of the rim. The bead seats are made with a taper so that as the tyre is inflated the bead is forced up the taper by the air pressure. This locks the tyre on to the rim making a good seal.

Taper seat | Taper seat | Flat ledge on inside

Old type taper rim | Flat edge reversed rim

Figure 4.128 All car rims are well base. Early ones were intended for cross-ply, tubed type small section tyres, which were usually less than 5 inch rim width and a low flange height

Figure 4.129 From left to right: Flat hump rim (top left), contre pente rim (bottom middle), double hump rim (top right)

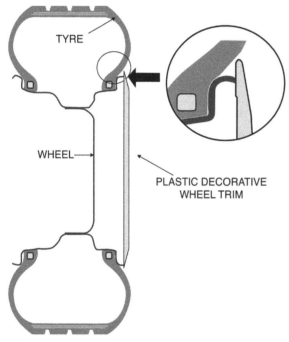

TYRE

WHEEL

PLASTIC DECORATIVE WHEEL TRIM

Figure 4.130 Wheel trim

Figure 4.131 Alloy wheel, note dual calliper system

Key fact

To facilitate fitting and removal of a tyre, a 'well-base' is manufactured into the rim.

Most wheels in current use have safety humps. The rim base width will be different on one side compared with the other. The safety hump provides some handling control characteristics when a tyre is in a fully deflated condition as it ensures the bead doesn't totally dislodge into the well-base.

Steel wheels are a very popular design. They are very strong and cheap to produce. Steel wheels are usually covered with plastic wheel trims. Trims are available in many different styles. To improve the appearance of the wheel trim some designs protrude above the wheel rim flange, i.e. their outer diameter is greater than the diameter of the wheel rim at the tip of the flange.

Wheel trims can abrade the tyre under load and, if not a tight fit rotationally, they can rotate around the wheel rim and foul the tyre inflation valve. A severe case could lead to valve damage and air leakage.

Alloy wheels, or 'alloys', are good, attractive-looking wheels. They tend to be fitted to higher specification vehicles. Many designs are used. They are lightweight but can be difficult to clean. Wheels of this type are generally produced from aluminium alloy castings, which are then machine finished. Alloy wheels can be easily damaged by 'kerbing'!

Many commercial vehicles use split rims, either of a two- or three-piece construction. The tyre is held in place by what could be described as a very large

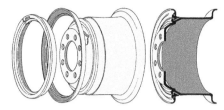

Figure 4.132 Do not work on split rims unless trained

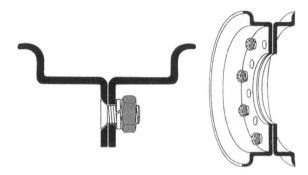

Figure 4.133 Do NOT work on divided rims unless trained

circlip. Do NOT remove or fit tyres on this type of wheel unless you have received proper instruction.

Safety first

Do not remove or fit tyres on split rim wheels unless you have received proper instruction.

On a few specialist vehicles, the rims are divided into two halves which are bolted together. The nuts and bolts holding them together should be specially marked. Undoing them with the tyre inflated would be very dangerous.

Light vehicle road wheels are usually held in place by four nuts or bolts. The fixing holes in the wheels are stamped or machined to form a cone-shaped seat. The wheel nut or bolt heads fit into this seat. This ensures that the wheel fits in exactly the right position. In the case of the steel pressed wheels, it also strengthens the wheel centre around the stud holes.

When fitting a wheel, the nuts or bolts must be tightened evenly in a diagonal sequence. It is also vital that they are set to the correct torque. Ensure the cone-shaped end of the wheel nuts is fitted towards the wheel.

Markings on the wheel rims are used to indicate the width and diameter. A code letter indicates the flange height and a letter and number code indicate the safety hump feature.

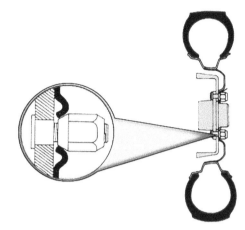

Figure 4.134 Wheel fixing

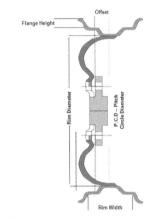

Figure 4.135 Rim sizes

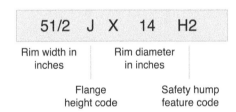

Figure 4.136 Diameter is the distance between bead seats, width is measured between the vertical faces of the flanges, and the flange height is denoted by a code such as J 17.3 mm, JK 18 mm, K 19.6 mm, L 21.6 mm etc.

The rim offset is measured in millimetres and is the distance between the centre of the rim and inner surface of the wheel disc (the hub contact face). This measurement can be positive or negative. The rim offset is normally specified by the car manufacturer and determines aspects such as wheel track. Different cars may have different offsets for the same tyre size. Rims used for upsizing may have a smaller rim offset

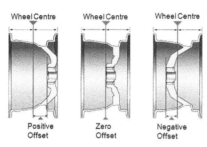

Figure 4.137 From left to right: positive, zero and negative offset

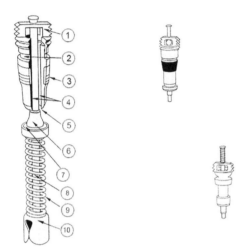

Figure 4.138 Details of the valve construction: 1, strong bridge; 2, sturdy swivel connection; 3, wedge-fit; 4, clear passage; 5, knife edge seat; 6, special bell-shaped guide; 7, seat washer; 8, full length plunger pin; 9, special spring; 10, spring cup

Figure 4.139 Various 'tubeless' valves

Figure 4.140 Overview of tyre materials. (Source: Continental)

Definition
PCD: Pitch circle diameter.

to increase the wheel track to ensure a fit into the wheelhouse.

On the front axle, it is extremely important to keep offset within the manufacturer's specs. The consequences of incorrect offset are: increased steering effort, increased steering wheel kick-back during acceleration, and increased load on wheel bearings and suspensions.

On the rear axle, negative offset can increase wheel track and improve stability and handling, but will increase the loading on the vehicle's suspension.

Always ensure that the correct wheel PCD dimensions are used and consider the tyre dimensions carefully. Increasing the wheel diameter will require adjustment to the tyre's aspect ratio to ensure the overall rolling circumference remains the same. Always remember to check the rear overall clearance dimensions carefully in a loaded and unloaded condition. Try jacking each wheel on the car as far as it will go; check with the suspension at full height.

The valve is to allow the tyre to be inflated with air under pressure, prevent air from escaping after inflation, and to allow the release of air for adjustment of pressure. The valve assembly is contained in a brass tube, which is bonded into a rubber sleeve and mounting section.

The valve core consists of a centre pin, which has metal and rubber disc valves. When the tyre is inflated, the centre pin is depressed, the disc valve moves away from the bottom of the seal tube and allows air to enter the tyre. To release air, or for pressure checking, the centre pin is depressed. During normal operation, the disc valve is held onto its seat by a spring and by the pressure of air. If all the air needs to be released, the valve core assembly can be removed. The upper part of the valve tube is threaded to accept a valve cap. This prevents dirt and grit from entering and acts as a secondary seal.

The tubeless valve core is as described previously. However, the valve body must be made so that when fitted into the wheel, an airtight seal is formed. Wheel

Figure 4.141 Steel and fabric reinforcing materials. (Source: Continental)

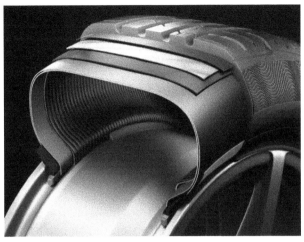

Figure 4.142 Tyre construction. (Source: Continental)

rims used for tubeless tyres must be sealed and airtight. Most wheels and tyres in use are of the tubeless design.

4.4.2 Tyre basics

The tyre performs four functions. These are:

1 **Support –** The tyre supports the vehicle when it is stationary. However, when it is in motion the tyre must resist considerable load shifts during acceleration and braking. At times, a car tyre must carry over 50 times its own weight.
2 **Traction and braking –** The tyre must be able to transmit forces. These are the engine's power output as acceleration and the braking forces when the brakes are applied. How well these forces are transmitted depends on the quality of just a few square centimetres of tyre in contact with the ground.
3 **Steering –** The tyre should steer the vehicle accurately, regardless of road and weather conditions. The car's ability to keep a straight path depends on the tyre's ability to maintain its course. The tyre must absorb transverse forces without deviating from its trajectory. Each vehicle has a set inflation pressure for the tyres of each axle. By respecting the differences in pressure between the front and rear axles, the best driving accuracy can be obtained.
4 **Road shocks –** The tyre absorbs road shocks to make life more comfortable for the driver and passengers. This also helps the vehicle components to last longer. The main characteristic of the tyre is its flexibility, especially in a vertical direction. The elasticity of the air in the tyre enables it to withstand successive deformations, caused by obstacles and uneven road surfaces. Correct pressures ensure a reasonable degree of comfort and maintain the correct steering capacity.

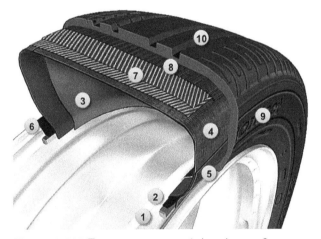

Figure 4.143 Tyre components: 1, bead core; 2, apex; 3, inner liner; 4, textile carcass; 5, bead reinforcement; 6, flange cushion or rim strip; 7, steel cord belt; 8, cap ply; 9, sidewall; 10, tread. (Source: Continental)

The tyre must continue at its best performance level for millions of revolutions of the wheel. Wear patterns depend on how the tyre is used, but especially on the quality of the ground contact. Tyre pressure plays a major role because it affects the size and shape of the contact area. It also affects the distribution of forces, to the different parts of the tyre, in contact with the ground. Several different materials are used in the construction of the tyre as shown in Figures 4.140 to 4.143.

Steel bead core wires hold the tyre firmly on the rim and steel belt cords provide driving characteristics and improve mileage. Carcass fabric cords keep the tyre in shape and cap ply fabric cords improve high-speed performance.

The tyre is flexible casing, which contains air. Tyres are manufactured from reinforced synthetic rubber. The tyre is made from an inner layer of fabric plies, which are wrapped around bead wires at the inner edges.

329

The bead wires hold the tyre in position on the wheel rim. The fabric plies are coated with rubber, which is moulded to form the side walls and the tread of the tyre. Behind the tread is a reinforcing band, usually made of steel, rayon, or glass fibre. Modern tyres are mostly tubeless, so they also have a thin layer of rubber coating the inside to act as a seal.

> **Key fact**
>
> Tyres are manufactured from reinforced synthetic rubber.

Beads clamp the radial tyre firmly against the wheel rim and can withstand very high forces. The tyre has supple rubber walls, which protect the tyre against the impacts (with kerbs, etc.) that might otherwise damage the carcass. There is also a hard rubber link between the tyre and the rim. Crown plies consist of oblique overlapping layers of rubber reinforced with very thin, but very strong, steel wires. The overlap between these wires, and the carcass cables, forms a series of non-deformable triangles. This arrangement lends great rigidity to the tyre structure.

As the sidewalls of the radial tyre are very flexible, they stretch in proportion to the increase in force. The sidewall acts like a moving hinge between the wheel and the crown. This allows the crown to remain flat against the ground. The path of the tyre therefore remains constant, even when subject to lateral forces.

There are various forms of tyre size markings and these differ to differentiate between tyre types. The size markings should be treated the same as a part number on a vehicle, so the driver should ensure that the tyres on their vehicle carry the precise markings indicated in the vehicle handbook or are an approved alternative fitment.

> **Safety first**
>
> Drivers should ensure that the tyres on their vehicle carry the precise markings indicated in the vehicle handbook or are an approved alternative fitment.

Most size markings indicate the dimensions, the type of structure and the speed capacity of the tyre.

Table 4.1 shows the weight that each index specification can carry.

Table 4.1

Load index	Load in kg	Load index	Load in kg	Load index	Load in kg
62	265	84	500	106	950
63	272	85	515	107	975
64	280	86	530	108	1000
65	290	87	545	109	1030
66	300	88	560	110	1060
67	307	89	580	111	1090
68	315	90	600	112	1120
69	325	91	615	113	1150
70	335	92	630	114	1180
71	345	93	650	115	1215
72	355	94	670	116	1250
73	365	95	690	117	1285
74	375	96	710	118	1320
75	387	97	730	119	1360
76	400	98	750	120	1400
77	412	99	775	121	1450
78	425	100	800	122	1500
79	437	101	825	123	1550
80	450	102	850	124	1600
81	462	103	875	125	1650
82	475	104	900	126	1700
83	487	105	925		

Table 4.2 shows the speed ratings in mph and kph.

Table 4.2

Speed rating	Mile/hour	Kilometres/hour	Speed rating	Miles/hour	Kilometres/hour
N	87	140	U	124	200
P	93	150	H	130	210
Q	99	160	V	149	240
R	106	170	Z	150+	240+
S	112	180	W	168	270
T	118	190	Y	186	300

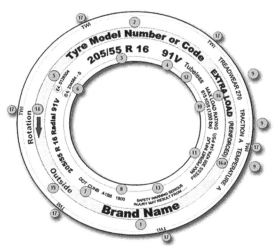

Figure 4.144 Tyre markings: 1, tyre brand name; 2, pattern or model code; 3, tyre size; 4, service description (load index and speed symbol); 5, ECE or EEC type approval number (C & U regulations stipulate all tyres used in the UK must carry an E number); 6, ECE noise approval; 7, North American recall code; 8, date of manufacture code; 9, UTQG quality rating; 10, maximum permitted load (USA); 11, maximum permitted inflation pressure (USA); 12, denotes tubeless construction; 13, safety warning; 14, direction of rotation (directional or composite tyres); 15, rim fitment instruction (asymmetrical or composite tyres); 16, denotes extra load tyre construction; 16a, denotes reinforced tyre construction; 17, position of tread wear indicators

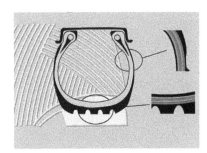

Figure 4.145 Plies embedded in layers of rubber

The tyre labelling regulation introduces labelling requirements regarding the display of information on the fuel efficiency, wet grip and external rolling noise of tyres. Its aim is to increase the safety and the environmental and economic efficiency of road transport by promoting fuel-efficient and safe tyres with low noise levels. This regulation allows end-users to make more informed choices when purchasing tyres by considering this information.

The rubber compound of a tyre provides grip – not the tread. However, without the tread to disperse water, very little grip can be achieved. Formula 1, and other racing, tyres do not have tread. These are known as 'slicks'. A version of these high-performance tyres has grooves to disperse water in wet conditions!

> **Key fact**
> The rubber compound of a tyre provides grip – not the tread.

Cross-ply tyres are not now used on any mass-produced modern car. However, the construction details are useful to show how tyre technology has developed. The cut-away picture (Fig. 4.145) shows the construction of a cross-ply tyre. Several textile plies are laid across each other, running from bead to bead in alternate directions. The same number of plies is used on the crown and the sidewalls. There is no difference between the sidewalls and crown, because each has the same plies.

Another tyre not now in common use is the bias belted tyre. This starts with two or more bias-plies to which stabilizer belts are bonded directly beneath the tread. This construction provides a smoother ride that is like the cross-ply tyre, while lessening

331

rolling resistance because the belts increase tread stiffness.

The functions of security, comfort and economy must continue for the whole lifetime of the tyre. The tyre needs air to function correctly and be long lasting. Its inflation pressure must be checked regularly. Tread depth and general condition should also be checked regularly.

4.4.3 Tread patterns

All tyre designs and tread patterns are, to some extent, a compromise between wet grip, dry grip and rolling resistance. There are four basic tread pattern concepts:

1 symmetrical
2 directional
3 composite
4 asymmetrical.

Symmetrical or multi-directional tyres have a pattern design that may be fitted either way round on the rim and to any position on the vehicle. The advantages are that they can be fitted on any position on the vehicle, without considering rotation and sidewall 'outside' marking, and they have good overall performance

characteristics. The disadvantage is that aquaplaning and pass-by noise may be compromised.

There are several different directional tyre tread patterns. They have a pattern that is designed for one direction of rotation only. The pattern grooves generally follow the shape of an arrow head. They offer very good protection against aquaplaning, have good directional stability and are attractive. However, there is an increased complexity at fitting and limitations when rotating after usage.

Composite tyres are both directional and asymmetric and are for specialist applications. These tyres can provide a combination of advantages of both asymmetric and directional tyres. They are therefore optimized for handling in both wet and dry conditions, offer good aquaplaning protection, have good directional stability and low noise emissions. They are,

Figure 4.146 Symmetrical. (Source: Continental)

Figure 4.148 Composite. (Source: Continental)

Figure 4.147 Directional. (Source: Continental)

Figure 4.149 Asymmetrical. (Source: Continental)

Figure 4.150 Water dispersal. (Source: Continental)

Figure 4.151 ContiSportContact™ 5. (Source: Continental)

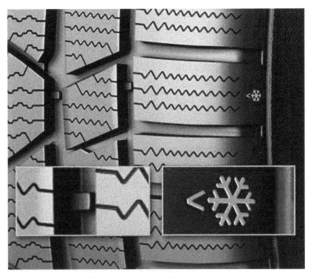

Figure 4.152 Once the tread is worn down to 4 mm, the Winter Indicator becomes level with the surface of the tread pattern, indicating that the tyre has reached its limit of winter suitability. (Source: Continental)

however, more complex to fit and rotation after usage is limited.

Asymmetrical pattern tyres have a tread pattern that differs between inner and outer shoulder. The outer shoulder has more rubber contact area to the road and the inside contact area is more open. They offer optimal adaptation to the road and driving conditions in both wet and dry conditions. They can be mounted at any position on the vehicle (if the 'outside' mark is correctly positioned). Pass-by noise is reduced and they have an improved wear potential. The asymmetrical design may have a slight compromise in aquaplaning vs. directional concept but this depends on the particular design.

The specially formulated natural rubber compound of winter tyres ensures rubber flexibility at a wide range of low temperature contrary to summer tyre tread compound. The winter tyre tread design offers its benefits especially on roads covered with snow or mud. The snow gets pressed into the wider and deeper tread grooves and by this it utilizes the effect of shear forces on snow for additional grip. The wider grooves can also absorb more slush compared with

summer tyres and therefore have more contact with the road surface.

This section has outlined the four main tread patterns and the winter (or M+S) tyre concept. All aspects of tyre operation are a compromise to some extent, but a key point to remember is that you generally get what you pay for!

4.4.4 Tyre pressure monitoring

A tyre pressure monitoring system (TPMS) is a safety feature that continually monitors a vehicles' tyres and alerts the driver to changes in tyre pressure. The changes in pressure can be detected by either direct or indirect means. Both methods will, as a minimum, illuminate a warning light on the vehicle dashboard display and sound an audible alert when 25% deflation has occurred. Early TPMS were introduced as an option on high-end luxury vehicles as early as the 1980s, although it wasn't until the year 2000 that it was first fitted as a standard feature.

Definition

TPMS: Tyre pressure monitoring system.

Key fact

TPMS is a safety feature that continually monitors a vehicle's tyres and alerts the driver to changes in tyre pressure.

333

It is estimated that 50% of all passenger car tyres have the wrong inflation pressure. This has several serious consequences:

▷ If the inflation pressure is not adapted to the axle loads, tyre flexing increases, resulting in energy loss and therefore higher fuel consumption.
▷ Increased flexing leads to higher temperatures, which cause damage to the tyre structure and can even lead to tyre failure.
▷ Hidden tyre damage will not be eliminated by later adjustment of the tyre pressure.
▷ The influence on a tyre's life expectancy is disadvantageous.
▷ Driving safety is influenced because in case of wrong inflation, the risk of tyre damage increases significantly.

Direct TPMS means each wheel of the vehicle has a sensor fixed to monitor the changes in pressure from the tyre. They also measure temperature. Each sensor sends its signal to the receiver inside the vehicle using a wireless connection. In Europe, the transmission frequency is 433 MHz.

▷ If low pressure or a leak is detected (generally 25% less than normal operating pressure), the driver is alerted by the in-car system and generally the deflated tyre is identified.
▷ Direct TPMS is very accurate, measuring to 1 or 2 psi.
▷ A puncture after parking is immediately identified.
▷ Sensors send their signals approximately every 30 seconds; when parked they transmit every 20 to 30 minutes. At 25 km/h the sensor switches back on to transmit every 30 seconds.
▷ Sensors have an approximate life of 5 years or 160,000 km.

Indirect TPMS is generally fitted to a vehicle that has had fitted or can be fitted with run flat tyres. This is because it is difficult to see or feel deflation in this type of tyre. Indirect tyre pressure monitoring systems do not use pressure sensors to monitor tyre pressure; they work from the ABS or speed sensors on the vehicle. Indirect systems monitor tyre pressure by assessing the rotational speeds of each tyre, and work on the premise that an under-inflated tyre has a slightly different diameter than a fully inflated tyre. An algorithm is used to assess the differences in wheel speeds. The under-inflated tyre would therefore rotate at a different speed than the correctly inflated one, causing a tyre pressure warning. The deflated tyre is not identified so the driver must check all four.

Figure 4.153 This sensor measures pressure and temperature and transmits the data to a control unit

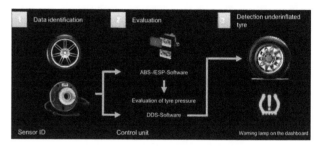

Figure 4.154 Indirect TPMS operation (Source: Continental)

Negative aspects of indirect TPMS

▷ The system is not very accurate.
▷ When tyres are re-inflated, the system needs to be re-calibrated.
▷ When tyre positions are changed, the system needs to be re-calibrated.
▷ When the tyres are replaced, the system needs to be re-calibrated.
▷ The system can be re-calibrated by the driver without first ensuring that the pressure is correct in all tyres.
▷ A puncture after parking is not immediately identified.

4.4.5 Tyre pressure monitoring and the law in Europe

▷ From November 2012, all new type vehicles in the M1 category (vehicles under 3.5 tonnes with fewer than eight seats) were required by law to have TPMS installed. This applies to the road wheels not the spare.
▷ From November 2014, all new passenger vehicles were required to have TPMS installed by the manufacturer.

The law is not currently retrospective, and does not apply to older vehicles. This law applies to passenger vehicles only, with no more than seven seats.

When a tyre is replaced or a puncture repaired on a vehicle without TPMS, it is normal practice to replace the rubber valve. With a TPMS sensor it is also important to replace the rubber sealing grommet, which deteriorates just like the rubber valve. At the same time as replacing the rubber grommet, it is good practice to replace other parts too.

▷ The metal sealing washer (if fitted) often becomes distorted, replacing this ensures a good seal when re-assembling the sensor.
▷ The outer collar is replaced because it becomes metal fatigued, and sometimes badly corroded. This metal fatigue can often make the outer securing collar crack and sometimes this will fall off completely.
▷ The core is replaced to create a good airtight seal in the sensor internally. Only nickel-plated cores should be used in metal valves.
▷ The cap is replaced with a new pressure cap. The rubber washer in the old cap deteriorates, replacing the cap helps keep the sensor airtight and free from debris, or fluids that might affect the operation of the core.

TPMS is now part of the annual vehicle test (in the UK), and applies to all newly registered cars from 1 January 2012. This means that a car with a faulty TPMS will fail the vehicle test from January 2015. Until 2015, a faulty TPMS was only an advisory and be listed on the testing certificate (UK). Different European and other countries may interpret the EU legislation differently, so please refer to your own country legislation.

There are hundreds of thousands of cars on European roads with sensors removed, because the cost of replacement is high. Rubber tyre valves are used to replace the sensor and the warning light disabled; however, this practice was prohibited on new-type 2012 cars. Insurance implications will be the same as for disabling an airbag or ABS. Disabling a safety system on a vehicle may become an offence.

4.4.6 Spare wheels and extended mobility

At first view, it would seem obvious that a vehicle should carry a spare wheel. However, modern tyre technology and associated extended mobility solutions mean that this is not necessarily the case. The information presented here is supplied by Continental Tyres but work by other manufacturers on similar systems has produced similar results. There are four main ways to deal with a roadside puncture:

1 spare wheel (full size or space-saver)
2 ContiKit (or similar kit from other manufacturers)
3 ContiSeal (or other self-sealing system)
4 self-supporting run-flat (SSR) systems.

> **Definition**
>
> **Extended mobility:** Measures taken to get a vehicle moving again after a flat tyre!

Most manufacturers prefer to eliminate the spare wheel to reduce weight, gain space and save costs. The choice of alternatives now allows this to happen. Other than a spare wheel, the three main options are outlined here.

A repair kit is designed to be simple for the driver to use. It is compact and versatile whilst being cost effective. The pump is adequate for all car tyre sizes up to 225/55 R16 with an inflation pressure up to 3 bar. A hole up to 6 mm in diameter can be sealed followed by a driving distance of about 200 km. Various sizes of the sealant bottles are available to suit the tyres fitted to the car. Note: the lifetime of the sealant is limited. Always check the latest data but a four-year life is typical. The system will work in the temperature range: −30°C to 70°C.

The ContiSeal system shown in Figure 4.156 has an extra air-proof sealant layer in the tyre tread area. It can automatically seal punctures up to 5 mm

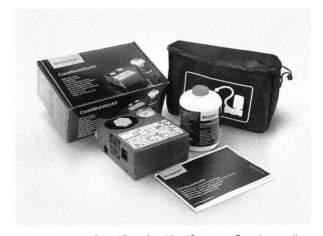

Figure 4.155 ContiComfortKit. (Source: Continental)

Figure 4.156 ContiSeal system. (Source: Continental)

diameter in the tread area. There is no influence on ride comfort or rolling resistance and it can be mounted on standard rims. It can seal about 80% of all punctures and works on all tyre types and dimensions. It has no effect on vehicle dynamics and distance and speed are not affected due to the instant seal. It is, however, recommended that this system is used in conjunction with a tyre pressure monitoring system (TPMS).

Run-flat tyres are designed to keep going in the event of full deflation. Stiff rubber inserts are used inside the tyre walls to carry the temporary additional load in the case of pressure loss. The system is compatible with current standard rims. They will keep working for up to 150 km at 80 km/h. There is an additional weight of about 20% associated with run-flat tyres. It is essential that a pressure warning system is used and ESP/DSC is also recommended.

Figure 4.157 Normal usage with full air pressure. (Source: Continental)

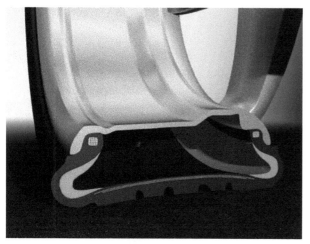

Figure 4.158 Pressure loss. (Source: Continental)

Definition

ESP: Electronic stability program.

DSC: Dynamic stability control.

Key fact

Run-flat tyres are designed to keep going in the event of full deflation.

There are several different extended mobility systems (EMS) on the market; none of them offers a perfect solution as even a full-size spare has only a 70% success rate! For this reason, conventional spares, mini-spares, sealant kits, sealant tyres and run-flat tyres co-exist in the market.

4.5 Chassis systems: new developments and updates

4.5.1 Electric power steering

Honda introduced the first 'all electric' power-steering system on the 1993 Acura NSX. Since then development has continued and electric power assisted steering (EPAS), also called electric power steering (EPS) is now fitted to most new cars and light commercial vehicles.

Definition

EPAS: Electric power assisted steering.

EPS: Electric power steering.

The basic principle is that as the driver turns the steering wheel, sensors register the corresponding steering torque and steering speed with absolute precision. The signals are used by an ECU to calculate the required steering assistance and, based on the calculated results, to control the servomotor. In the final step, the motor transmits the optimum servo torque, via a worm gear or recirculating ball gear. This is applied to the steering column, to a second pinion or directly to the steering rack.

If the vehicle power supply fails, the driver can continue to steer the vehicle, due to the mechanical

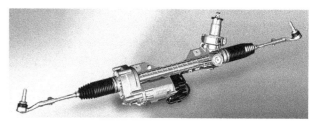

Figure 4.159 Electric power steering. (Source: ZF)

connection between the steering wheel and the steered wheels.

Motor currents in the region of 80 A are typical and therefore suitable wiring is needed as well as powerful control electronics.

> **Key fact**
>
> If the vehicle power supply fails, the driver can continue to steer the vehicle, due to the mechanical connection between the steering wheel and the steered wheels.

On many systems, an optical torque sensor is used to measure driver effort on the steering wheel (all systems use a sensor of some sort). This sensor works by measuring light from an LED, which is shining through holes. These are aligned in discs at either end of a torsion bar, fitted into the steering column. An optical sensor element identifies the twist of two discs on the steering axis with respect to each other. From this information, the system calculates the torque as well as the absolute steering angle.

Processing of the sensor signals is then done by the electronic control unit, which calculates the position, rotation direction and rotation speed of the steering

wheel. The control unit also validates the sensor output signals and detects short-circuits.

This sensor shown in Figure 4.161 measures the torque applied by the driver at the steering wheel. Based on this, the control unit calculates the steering assistance for the motor. The torque sensor sits on the steering pinion (5). A pole wheel (9) is fitted on the input shaft (2) which is connected to the steering pinion by means of the torsion bar (1). The measuring range covered by the sensor is between +/– 8 and +/– 10 Nm.

If the steering torques are very high, a mechanical angle limiter prevents overload of the torsion bar. When the driver applies a torque on the steering wheel, the torsion bar is rotated, as is the magnet relative to the sensor. The sensor consists of elements that change their resistance when the field direction changes.

The high safety demands on electric steering systems require detection of all faults occurring on the sensor and creation of a safe condition of the steering system. The sensor data are transmitted to the electronic control unit via a very rugged digital interface. Optionally, the torque sensor can also accommodate an index magnet (7) and an index sensor (8). The index sensor delivers a signal to the ECU for each full steering wheel turn. In combination with the data from the rotor position sensor and the wheel speeds, the electronic control unit can calculate the steering angle with a resolution < 0.05°.

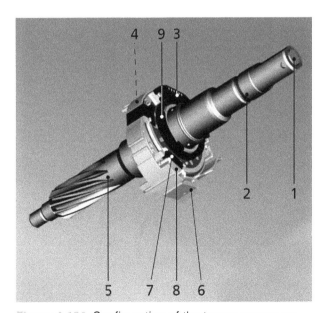

Figure 4.161 Configuration of the torque sensor on the steering pinion: 1, torsion bar; 2, input shaft; 3, sensor module; 4, clock spring; 5, steering pinion; 6, plug; 7, index magnet (optional); 8, index sensor (optional); 9, pole wheel. (Source: ZF Servoelectric)

Figure 4.160 Electric PAS. (Source: Ford Media)

The rotational movement of an electric motor must be changed into a linear movement. The ZF Servolectric drive system uses a drive concept consisting of a toothed belt drive and recirculating ball gear. The recirculating ball gear is a system like the non-power steering box technology that has been in use for many

Figure 4.162 Electric power steering motor. (Source: Bosch)

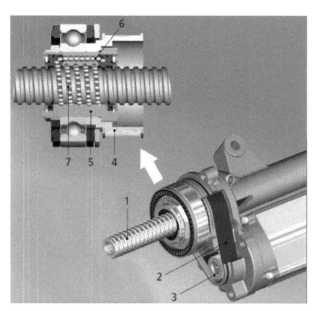

Figure 4.163 Belt drive system: 1, steering rack; 2, toothed belt; 3, toothed disc, small; 4, toothed disc, big; 5, ball recirculation nut; 6, ball return channel; 7, balls/ball chain. (Source: ZF Servoelectric)

years. The noise performance of this system is very good and as such allows the recirculating ball gear to be rigidly connected to the sub-frame of the vehicle. This in turn gives a very direct steering feel. Despite its ability to transmit high torques safely, the slip-free toothed belt (like a timing belt in construction) is very quiet.

4.5.2 Electronic parking brake

The electric (or electronic) parking brake (EPB), also known as an automatic parking brake (APB), is a function offering the driver increased comfort and convenience. In addition, as the hand lever is not used, car manufacturers have more freedom of choice as to where they site the operating parts within the car. Features such as hill- or auto-hold are also possible with an EPB. Hill-hold stops the car rolling away accidentally when standing still or setting off. Auto-hold keeps the brake pressure applied after the driver releases the pedal. If the ABS sensors detect any movement the pressure is increased. If the accelerator is pressed (or the clutch released on a manual) the brakes are let off.

There are two main systems in use, namely the cable-pull and electric-hydraulic type. Both methods include a visual warning light on the dashboard. The second of these is now the most common. A third full-electric type may start to be used.

The cable pull system is simply a development of the traditional lever and cable method. As the switch is operated, a motor, or motors, pull the cable by either rolling it on a drum or using an internally threaded gear on a spiral attached to the cable. The electronic

Figure 4.164 Electric parking brake switch

parking brake module shown here, also known as the EPB actuator, is fitted to some Range Rover and Land Rover models. The parking brake can be released manually on most vehicles. After removing a plastic cover or similar, pulling a wire cable loop will let off the brake.

Electric-hydraulic calliper systems are usually employed as part of a larger control system such as an electronic stability program (ESP). When the driver presses the switch to activate the parking brake, the ESP unit automatically generates pressure in the braking system and presses the brake pads against the disc. The callipers are then locked in position by an electrically controlled solenoid valve. The calliper remains locked without any need for hydraulic pressure. To release the brake, the ESP briefly generates pressure again, slightly more than was needed to lock the calliper, and the valve is released.

Testing the parking brake on rollers is possible on both systems. Cable pull types can be tested much like ordinary hand or parking brakes. But there is a risk of

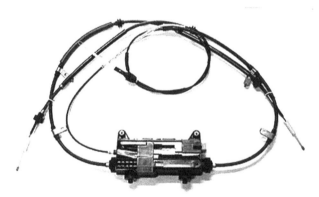

Figure 4.165 Cable pull system. (Source: Land Rover)

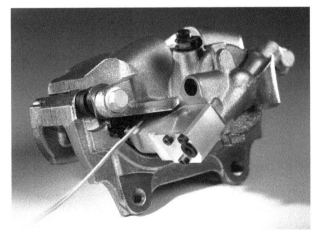

Figure 4.166 Electric-hydraulic parking brake calliper. (Source: Bosch Press)

Figure 4.167 Brake roller testing. (Source: Tecalemit)

locking the wheels. Some manufacturers have test modes – so double check!

The types with calliper motors can also be tested on rollers but the procedure is slightly different. Most of the calliper-motor systems have special software incorporated in the ECU for brake testing. When the car is put on a rolling road and the rear wheels are driven by the test equipment, the ECU detects this as a test situation because the rear wheels are moving and the fronts are not. It therefore puts the system into test mode. If a multi-function display is used on the vehicle dashboard, it will display an appropriate message. The technician can then operate the handbrake-switch. The ECU applies the electric parking brake with enough force to obtain a reading on the roller brake tester. The wheels should not lock. After the test is complete the rollers are stopped and the switch is released. When the switch is activated again, the brakes are re-applied in the normal way and the wheels will be locked.

The key drivers for enhanced systems such as the EPB are increased functions, comfort and safety as well as greater freedom for vehicle designers. The addition of electrical functions to the normal hydraulic calliper is currently the most common. However, brake-by-wire is just around the corner and development is, erm, unlikely to stop…

Safety first

Always check and follow specific instructions and specifications provided by the vehicle manufacturer.

4.5.3 Carbon ceramic brakes

The Brembo Group has been manufacturing carbon ceramic discs for automotive applications since 2002, when it first supplied these components for the Ferrari Enzo. This high-performance material, made from a special mixture of powders, resins and fibres in a complex manufacturing process, has been used since the 1970s in braking systems for aerospace applications and since the 1980s in motorsport.

Carbon ceramic materials are ceramics greatly strengthened by the inclusion of carbon fibres. They are therefore very light, extremely strong and capable of operating at temperatures well in excess of 1500°C (2732°F).

> **Key fact**
>
> Carbon ceramic materials are strengthened by carbon fibres, they are very light, extremely strong, and capable of operating at more than 1500°C.

Carbon ceramic offers substantial benefits in terms of performance, in wet and dry conditions, weight, comfort, corrosion resistance, durability and high-tech appeal. Brembo SGL Carbon Ceramic Brakes is currently the leading producer of carbon ceramic discs, and supplies these components for the most exclusive models of prestigious brands such as Ferrari, Maserati, Alfa Romeo, Aston Martin, Corvette, Nissan, Lexus, McLaren, Volkswagen, Porsche, Audi, Bentley, Lamborghini, Bugatti and AMG.

Figure 4.168 Supercars favour carbon ceramic discs; these are made by Brembo

Figure 4.169 Lamborghini brake callipers

Brembo designed six-piston front and four-piston rear callipers for the Lamborghini Aventador. The design of the Brembo callipers integrates seamlessly with the overall aesthetics of the most exclusive models.

4.5.4 Antilock braking system (ABS) update

In an emergency braking situation, the braking force applied by the driver may be greater than the tyre can handle, resulting in a locked wheel. In this case, the tyre can now no longer transfer any lateral traction forces. The vehicle becomes unstable and uncontrollable, since the vehicle no longer reacts to the steering input of the driver. In a vehicle equipped with an antilock braking system, wheel-speed sensors measure the speed of rotation of the wheels and pass this information to the ABS control unit.

If the ABS control unit detects that one or more wheels tend to lock, it intervenes within milliseconds by modulating the braking pressure at each individual wheel. In doing so, ABS prevents the wheels from locking and ensures safe braking because the vehicle remains steerable and stable. Generally, the stopping distance is also reduced.

Generation 9 of the Bosch ABS is part of the continuous development of this active safety system. The scalable product concept with modular software architecture characterizes this latest technical evolution. Compared with the previous generation, the

Figure 4.170 Driving tracks covered on one side with ice or suddenly emerging animals are typical traffic risks. Brake control systems such as the anti-locking system ABS and the electronic stability program increase the safety in such critical situations

Figure 4.171 Generation 9 ABS and ESP hydraulic modulator and valve block with integrated ECU. It is also the basis for many high-performance safety and assistance functions, such as the ACC adaptive cruise control with stop-and-go function

Figure 4.172 ABS 8 exploded view showing the ECU and eight 2/2-way solenoid valves (two hydraulic positions and two hydraulic connections)

weight and size have been reduced by up to 30%. The most compact ABS unit now weighs only 1.1 kg.

The ECU uses the signals from the wheel-speed sensors to compute the speeds of the wheels. Two different operating principles are used: passive (inductive) and active (Hall) speed sensors. Active sensors are becoming more and more widespread. They use a magnetic field for the contactless detection of the wheel speed and can recognize the direction of rotation as well as a standstill. ABS ECUs contain two microcontrollers for safety and they operate at 20 MHz. A CAN or Flexray connection is also available from the ECU.

Key fact

ABS ECUs contain two microcontrollers for safety.

Key fact

CAN, LIN and Flexray are different digital communication systems.

The hydraulic unit puts the ECU's commands into effect and regulates the pressure in the individual wheel brake cylinders by means of solenoid valves. It is in the engine compartment between the brake master cylinder and the wheel brake cylinders, so that the hydraulic lines to the brake master cylinder and the wheel brake cylinders can be kept short. The hydraulic unit has input and output solenoid valves for controlling the pressure in the individual wheel brakes. The ECU takes over all electrical and electronic tasks as well as the control functions of the system.

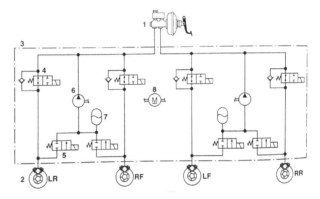

Figure 4.173 ABS hydraulic system diagram: 1, master cylinder; 2, wheel cylinders; 3, hydraulic unit; 4, inlet valves; 5, outlet valves; 6, return pump; 7, accumulator; 8, pump motor

Networking is what will provide the basis for further developments. ABS is already linked with (well, more accurately integrated with) stability control but is likely to now be linked with airbag control and other driver assistance systems. New functions then become possible that will improve accident avoidance but also protect occupants, pedestrians and other road users.

Key fact

Networking ABS is what will provide the basis for further developments.

The basic operating principle of ABS-9 has not changed (i.e. pressure holding, pressure reduction and pressure increase). However, version 9 has added several new options as well as incremental improvements meaning it will continue to be a popular choice by many manufacturers.

341

4.5.5 Predictive active ride control

Monroe (a subsidiary of Tenneco, well known for dampers and many other components) is developing a fully active suspension system with significant advantages over existing semi-active systems.

Current active suspension systems tend to rely on electronic actuation of the damper valves to change damping response. The Monroe system uses dampers that are fitted with two electronically activated valves. These valves control the rebound and compression strokes of the dampers. The aim of the project is to improve ride and handling.

The latest development of the ride-control system, known as ACOCAR, eliminates the need for anti-roll bars. Each damper is equipped with an integrated electrically powered pump (or in another variant, one pump for each axle). The pump continuously circulates oil through the damper, and damping valves control vehicle body movement. The result is a system that can react very rapidly to changes in surface condition.

The system can contain roll while cornering at speed. The controlled wheel movements improve vehicle handling by maintaining tyre–road contact. Further improvements are possible if the system incorporates other sensors. A camera scanning the road surface ahead of the vehicle, for example, would allow the system to anticipate factors such as bumps and potholes. Other sensors, such as radar and lidar, are also under evaluation as camera-based systems would not work properly in low light conditions.

Definition
Radar: Radio detection and ranging.
Lidar: Light detection and ranging.

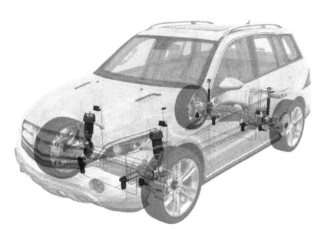

Figure 4.174 Predicting active ride control system. (Source: Tenneco/Monroe ACOCAR)

Figure 4.175 Damper for the active ride control system. (Source: Tenneco/Monroe ACOCAR)

Transmission systems

5.1 Manual transmission clutch

5.1.1 Clutch operation

A clutch is a device for disconnecting and connecting rotating shafts. In a vehicle with a manual gearbox, the driver pushes down the clutch when changing gear, thus disconnecting the engine from the gearbox. It allows a temporary neutral position for gear changes and also a gradual way of taking up drive from rest.

Drivetrain location varies with vehicle design. However, the clutch is always fitted between the engine and the transmission (Fig. 5.1). With few exceptions, the clutch and flywheel are bolted to the rear of the engine crankshaft.

Key fact

The clutch is always fitted between the engine and the transmission.

The driver operates the clutch by pushing down a pedal. This movement has to be transferred to the release mechanism. The two main methods used are cable and hydraulic. The cable method is the most common (Fig. 5.2). Electrohydraulic or electric clutch operation is used on direct shift gearboxes (DSGs).

A steel cable, which runs inside a plastic-coated steel tube, is used on most cars. The cable 'outer' must be fixed at each end. The cable 'inner' transfers the movement. One problem with cable clutches is that

Figure 5.1 The clutch cover is bolted to the engine flywheel

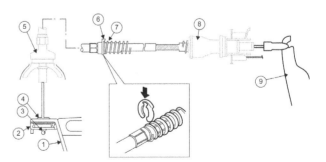

Figure 5.2 Clutch cable: 1, clutch release lever; 2, retaining clip; 3, seating plate; 4, rubber pad; 5, cable end fixed to gearbox or clutch housing; 6, spring retaining clip; 7, self-adjusting spring; 8, cable end fixed to bulkhead; 9, clutch pedal

Automobile Mechanical and Electrical Systems. 978-0-415-72578-1 © 2018 Tom Denton.
Published by Taylor & Francis. All rights reserved.

DOI: 10.1201/9781003299066-5

movement of the engine, with respect to the vehicle body, can cause the length to change. This results in a judder when the clutch is used. This problem has been almost eliminated by careful positioning and quality engine mountings.

This clutch cable works on a simple lever principle. The clutch pedal is the first lever. Movement is transferred from the pedal to the second lever, which is the release fork. The fork, in turn, moves the release bearing to operate the clutch (Fig. 5.3).

A hydraulic mechanism involves two cylinders (Fig. 5.4). These are termed the master and slave cylinders. The master cylinder is connected to the clutch pedal. The slave cylinder is connected to the release lever. The clutch pedal moves the master cylinder piston. This pushes fluid through a pipe, which in turn forces a piston out of the slave cylinder. The movement ratio can be set by the cylinder diameters and the lever ratios.

An electronic clutch was developed for racing vehicles to improve the getaway performance. For production vehicles (Fig. 5.5), a strategy has been developed to interpret the driver's intention. With greater throttle openings, the strategy changes to prevent abuse and driveline damage. Electrical control of the clutch

release bearing position is by a solenoid actuator, which can be modulated by signals from the electronic control unit (ECU). This allows the time to reach the ideal take-off position to be reduced and the ability of the clutch to transmit torque to be improved. The efficiency of the whole system can therefore be increased.

Figure 5.3 Clutch release bearing

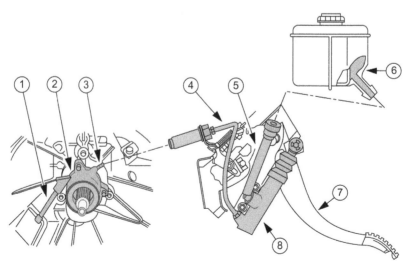

Figure 5.4 Clutch hydraulic components: 1, bleed valve; 2, slave cylinder; 3, pressure pipe connection; 4, pressure pipe; 5, replenishing pipe; 6, fluid reservoir; 7, clutch pedal; 8, master cylinder

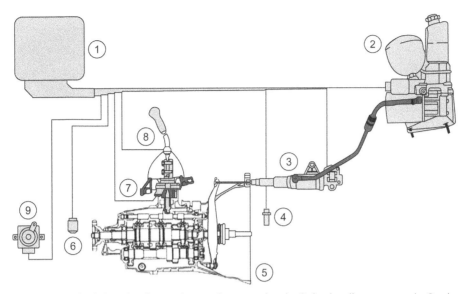

Figure 5.5 Electronic control of the clutch: 1, electronic control unit; 2, hydraulic power unit; 3, slave cylinder with position sensor; 4, engine speed sensor; 5, transmission; 6, vehicle speed sensor; 7, gear selection sensor; 8, gear lever; 9, accelerator pedal position sensor

5.1.2 Types of clutch

The clutch is made of two main parts: a pressure plate and a driven plate (Fig. 5.6). The driven plate, often termed the clutch disc, is fitted on the shaft, which takes the drive into the gearbox.

When the clutch is engaged, the pressure plate, inside the cover, presses the driven plate against the engine flywheel. This allows drive to be passed to the gearbox. Pushing down the clutch pedal moves the pressure plate away, which frees the driven plate.

The movement of the diaphragm during clutch operation is represented by Figures 5.7 and 5.8. The method of controlling the clutch is quite simple. The mechanism consists of either a cable or hydraulic system.

The clutch shaft, or gearbox input shaft, projects from the front of the gearbox (Fig. 5.9). Most shafts have a smaller section or spigot, which projects from its outer end. This rides in a spigot bearing in the engine crankshaft flange. The splined area of the shaft allows the clutch disc to move along the splines. When the clutch is engaged, the disc drives the gearbox input shaft through these splines.

The clutch disc is a steel plate covered with frictional material (Fig. 5.10). It fits between the flywheel face and the pressure plate. In the centre of the disc is the hub, which is splined to fit over the splines of the input shaft. As the clutch is engaged, the disc is firmly squeezed between the flywheel and pressure

Figure 5.6 Driven plate and pressure plate

plate. Power from the engine is transmitted by the hub to the gearbox input shaft. The width of the hub prevents the disc from rocking on the shaft as it moves along the shaft.

Key fact

The splined area of the gearbox input shaft allows the clutch disc to move but still transmit drive.

345

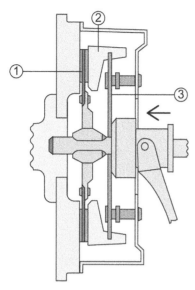

Figure 5.7 Clutch engaged: 1, driven (friction) plate; 2, pressure plate; 3, release levers

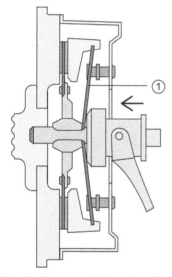

Figure 5.8 Clutch disengaged: 1, release levers (fingers)

Figure 5.9 Gearbox input shaft around which is the release bearing and its operating lever

Figure 5.10 Clutch disc or driven plate

The clutch disc has frictional material riveted or bonded on both sides. These frictional facings are either woven or moulded. Moulded facings are preferred because they can withstand high-pressure plate loading forces. Grooves are cut across the face of the friction facings to allow for smooth clutch action and increased cooling. The cuts also make a place for the facing dust to go as the clutch lining material wears.

Key fact

The clutch disc has frictional material riveted or bonded on both sides.

The frictional material wears as the clutch is engaged. At one time asbestos was in common use. Owing to awareness of the health hazards resulting from asbestos (asbestosis and mesothelioma, for example) new lining materials have been developed. The most commonly used types are paper-based and ceramic materials. They are strengthened by the addition of cotton and brass particles, and wire. These additives increase the torsional strength of the facings and prolong the life of the clutch.

The facings are attached to wave springs, which cause the contact pressure on the facings to rise gradually. This is because the springs flatten out when the clutch is engaged. These springs eliminate chatter when the clutch is engaged (Fig. 5.11). They also help to move the disc away from the flywheel, when it is disengaged. The wave springs and facings are attached to the steel disc.

There are two types of clutch discs: rigid and flexible (Fig. 5.12). A rigid clutch disc is a solid circular disc fastened directly to a centre splined hub. The flexible clutch disc has torsional dampener springs that circle the centre hub.

Figure 5.11 Wave springs eliminate chatter

Figure 5.13 Damping springs

Figure 5.12 Solid and flexible discs

Figure 5.14 Pressure plate

The dampener is a shock-absorbing feature built into a flexible clutch disc (Fig. 5.13). The primary purpose of the flexible disc is to absorb power impulses from the engine that would otherwise be transmitted directly to the gears in the transmission. A flexible clutch disc has torsion springs and friction discs between the plate and hub of the clutch.

When the clutch is engaged, these springs cushion the sudden loading by flexing and allowing twist between the hub and plate. When the loading is over, the springs release and the disc transmits power normally. The number and tension of these springs are determined by the amount of engine torque and the weight of the vehicle. Stop pins limit the torsional movement to a few millimetres.

The pressure plate and cover assembly squeezes the clutch disc onto the flywheel when the clutch is engaged. It moves away from the disc when the clutch is disengaged. These actions allow the clutch disc to transmit, or not transmit, the engine's torque to the gearbox.

Key fact
The pressure plate and cover assembly squeezes the clutch disc onto the flywheel when the clutch is engaged.

A pressure plate is a large spring-loaded clamp, which is bolted to, and rotates with, the flywheel (Fig. 5.14). The assembly includes a metal cover, heavy release springs and a metal pressure ring that provides a friction surface for the clutch disc. It also includes a thrust ring or fingers for the release bearing, and release levers.

The release levers release the holding force of the springs when the clutch is disengaged. Some pressure

347

plates are of a 'semi-centrifugal' design. They use centrifugal weights, which increase the clamping force on the thrust springs as engine speed increases.

The diaphragm spring assembly is a cone-shaped diaphragm spring between the pressure plate and the cover. Its purpose is to clamp the pressure plate against the clutch disc. This spring is normally secured to the cover by rivets. When pressure is exerted on the centre of the spring, the outer diameter of the spring tends to straighten out. When pressure is released, the spring resumes its normal cone shape.

The centre portion of the spring is slit into a number of fingers that act as release levers (Fig. 5.15). When the clutch is disengaged, these fingers are depressed by the release bearing. The diaphragm spring pivots over a fulcrum ring. This makes its outer rim move away from the flywheel. The retracting springs pull the pressure plate away from the clutch disc, to disengage the clutch.

As the clutch is engaged, the release bearing is moved away from the release fingers. As the spring pivots over the fulcrum ring, its outer rim forces the pressure plate tightly against the clutch disc. At this point, the clutch disc is clamped between the flywheel and pressure plate.

The individual parts of a pressure plate assembly are contained in the cover. Most covers are vented to allow heat to escape and air to enter. Other covers are designed to provide a fan action to force air circulation around the clutch assembly. The effectiveness of the clutch is affected by heat. Therefore, by allowing the assembly to cool, it works better.

Key fact

As the clutch is engaged, the release bearing is moved away from the release fingers.

Figure 5.15 Fingers

Figure 5.16 Coil spring clutch assembly

Earlier clutches, and some heavy-duty types, used coil springs instead of a diaphragm (Fig. 5.16). However, the diaphragm clutch has replaced the coil spring type because the diaphragm type has the following advantages:

▷ It is not affected by high speeds (coil springs can be thrown outwards).
▷ The low pedal force makes for easy operation.
▷ It is light and compact.
▷ The clamping force increases or at least remains constant as the friction lining on the plate wears.

5.2 Manual transmission gearbox

5.2.1 Gearbox operation

A transmission system gearbox is required because the power of an engine consists of speed and torque. Torque is the twisting force of the engine's crankshaft and speed is its rate of rotation. The transmission can adjust the proportion of torque and speed that is delivered from the engine to the drive shafts. When torque is increased, speed decreases, and when speed is increased, the torque decreases. The transmission also reverses the drive and provides a neutral position when required.

Key fact

A transmission system gearbox is required because torque from an engine changes at different speeds.

Helical gears are used for almost all modern gearboxes (Fig. 5.17). They run more smoothly and are quieter in operation. Earlier sliding-mesh gearboxes used straight-cut gears, as these were easier to manufacture. Helical gears do produce some sideways force when operating, but this is dealt with using thrust bearings.

Figure 5.18 Front-wheel drive gearbox. (Source: Ford Media)

Figure 5.17 Straight cut and helical gears and at the top, the principle of gear operation

For most light vehicles, a gearbox has five forward gears and one reverse gear (Fig. 5.18). It is used to allow operation of the vehicle through a suitable range of speeds and torque. A manual gearbox needs a clutch to disconnect the engine crankshaft from the gearbox while changing gears. The driver changes gears by moving a lever, which is connected to the box by a mechanical linkage. Alternatively, gears can be changed by electrohydraulic methods in response to paddle switches located behind the steering wheel.

The gearbox converts the engine power by a system of gears, providing different ratios between the engine and the wheels. When the vehicle is moving off from rest, the gearbox is placed in first, or low gear. This produces a high torque but low wheel speed. As the car speeds up, the next higher gear is selected. With each higher gear, the output turns more quickly but with less torque.

Fourth gear on most rear-wheel drive light vehicles is called direct drive, because there is no gear reduction in the gearbox (Fig. 5.20). In other words, the gear ratio is 1:1. The output of the gearbox therefore turns at the same speed as the crankshaft. For front-wheel drive vehicles, the ratio can be 1:1 or slightly different. Most modern light vehicles now have a fifth gear. This can be thought of as a kind of overdrive because the output turns more quickly than the engine crankshaft.

Power comes into the gearbox via the input shaft. A gear at the end of this shaft drives a gear on another shaft called the countershaft or layshaft. A number of gears of various sizes are mounted on the layshaft. These gears drive other gears on a third motion shaft, also known as the output shaft.

Figure 5.19 Pontiac six-speed gear selector

Key fact

Power comes into the gearbox via the input shaft.

Older vehicles used sliding-mesh gearboxes. With these gearboxes, the cogs moved in and out of contact with each other. Gear changing was, therefore, a skill that took time to master. These have now been replaced by constant mesh gearboxes. The modern gearbox still produces various gear ratios by engaging different combinations of gears. However, the gears are constantly in mesh. For reverse, an extra gear called an idler operates between the countershaft and the output shaft. It turns the output shaft in the opposite direction to the input shaft.

349

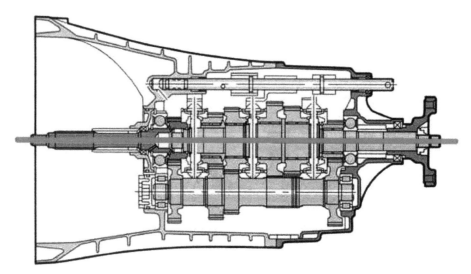

Figure 5.20 Fourth gear is often 'straight through'

Figure 5.21 Lincoln five-speed rear-wheel drive gearbox. (Source: Ford Media)

Figure 5.22 Front-wheel drive transaxle gearbox with cable shift. (Source: Ford Media)

With the exception of reverse, the gears do not move. This is why this type of gearbox has become known as constant mesh. In other words, the gears are running in mesh with each other at all times. In constant mesh boxes, dog clutches are used to select which gears will be locked to the output shaft. These clutches, which are moved by selector levers, incorporate synchromesh mechanisms.

A manual gearbox allows the driver to select the gear appropriate to the driving conditions. Low gears produce low speed but high torque; high gears produce higher speed but lower torque.

Key fact

Low gears produce low speed but high torque; high gears produce higher speed but lower torque.

5.2.2 Gear change mechanisms

On all modern manual change (shift) gearboxes, the selection of different ratios is achieved by locking gears to the mainshaft. A synchromesh and clutch mechanism does this when moved by a selector fork (Fig. 5.23). The selector fork is moved by a rod, or rail, which in turn is moved by the external mechanism and the gearstick.

To save space, some manufacturers use a single selector shaft. This means the shaft has to twist and move lengthways. The twisting allows a finger to contact with different selector forks. The lengthways movement pushes the synchronizers into position. All selector forks are fitted on the same shaft.

Key fact

On all modern manual lever change (shift) gearboxes, the selection of different ratios is achieved by locking gears to the mainshaft.

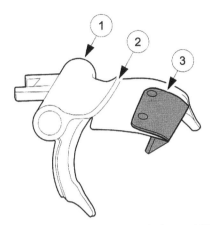

Figure 5.23 Gear shift or selector fork: 1, shift fork; 2, fluid guide edge (for lubrication); 3, fluid baffle

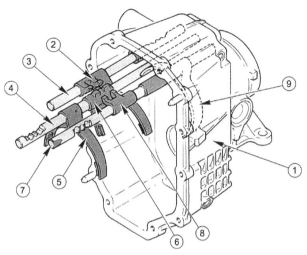

Figure 5.25 Triple selector shaft: 1, housing; 2, 1st/2nd gear selector; 3, 1st/2nd selector shaft; 4, 3rd/4th selector shaft; 5, 3rd/4th selector fork; 6, 3rd/4th gear driver; 7, 5th/reverse gear selector shaft; 8, 5th/reverse gear driver; 9, 5th/reverse gear selector fork. (Source: Ford Motor Company)

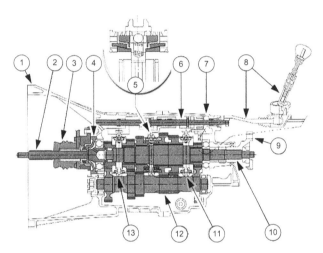

Figure 5.24 Double selector shaft: 1, clutch (bell) housing; 2, input shaft; 3, clutch mechanism; 4, seal housing; 5, interlock; 6, two shift shafts (rails) one behind the other; 7, housing; 8, external change mechanism; 9, output drive flange; 10, output shaft; 11, shaft, 5th gear and reverse gear; 12, countershaft (layshaft); 13, 3rd and 4th gears. (Source: Ford Motor Company)

On a two-shaft system, the main selector shaft often operates the first/second gear selector fork. An auxiliary shaft operates the third/fourth selector fork (Fig. 5.24).

The three-rail, or three-shaft system, is similar to the two-shaft type (Fig. 5.25). However, each individual shaft can be moved lengthways. In turn, the shafts will move the first/second, third/fourth or fifth/reverse forks.

A common external linkage is shown in Figure 5.26. Movement of the lever is transferred to the gearbox by a shift rod. The rod will only move to select reverse gear when the lock sleeve is lifted. This prevents accidental selection of reverse gear.

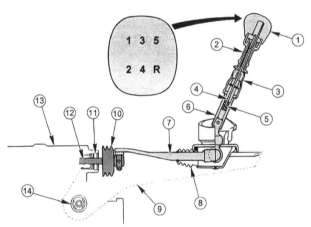

Figure 5.26 Rod-operated shift mechanism: 1, gear knob; 2, reverse lock mechanism; 3, gaiter; 4, reverse lock cable; 5, spring; 6, gear lever; 7, external shift rod; 8, gaiter; 9, housing gaiter; 10, bellows gaiter; 11, selector shaft radial lip oil seal; 12, internal selector shaft; 13, transmission housing; 14, insulator (noise vibration and harshness)

A more recent development is the cable shift mechanism. The advantage of this system is that the shift lever does not have to be fixed to the gearbox or in a set position. This allows designers more freedom (see Fig. 5.22).

Definition
NVH: Noise vibration and harshness.

351

A detent mechanism is necessary to hold the selected gear in mesh. In most cases, this is just a simple ball and spring acting on the selector shaft(s). Figure 5.27 shows a gearbox with the detent mechanisms highlighted.

Gear selection interlocks are a vital part of a gearbox. They prevent more than one gear from being engaged at any one time. When any selector clutch is in mesh, the interlock will not allow the remaining selectors to change position. As the main selector shaft is turned by side-to-side movement of the gear stick, the gate restricts the movement. The locking plate, shown in Figure 5.28, will only allow one shaft to be moved at a time. Because the gate restricts the movement, selection of more than one gear is prevented.

Key fact

A detent system holds the selected gear in place.

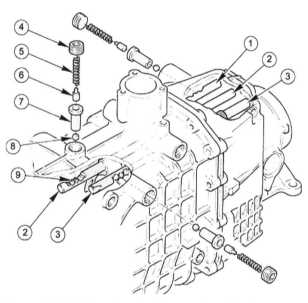

Figure 5.27 Ball and spring detent: 1, 1st/2nd gear selector shaft; 2, 3rd/4th gear selector shaft; 3, 5th/reverse gear selector shaft; 4, threaded plug; 5, spring; 6, pin; 7, sleeve; 8, ball; 9, detent notches

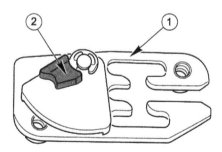

Figure 5.28 Gate and reverse gear lock: 1, gate; 2, reverse gear lock

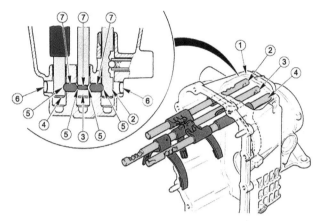

Figure 5.29 Plunger type interlock: 1, housing; 2, 1st/2nd gear selector shaft; 3, 3rd/4th gear selector shaft; 4, 5th/reverse gear selector shaft; 5, locking notch; 6, plug; 7, locking pin

When three rails are used to select the gears, plungers or locking pins can be used (Fig. 5.29). These lock the two remaining rails when one has moved. In the neutral position, each of the rails is free to move. When one rail (rod or shaft) has moved, the pins move into the locking notch, preventing the other rails from moving.

Key fact

A gearbox mechanism must only allow selection of one gear at a time by the use of an interlock.

Gear selection must be a simple process for the driver. To facilitate changing, several mechanical components are needed. The external shift mechanism must transfer movement to the internal components. The internal mechanism must only allow selection of one gear at a time by the use of an interlock. A detent system helps to hold the selected gear in place.

5.2.3 Gears and components

There is a wide range of gearboxes in use. However, although the internal components differ, the principles remain the same. The examples in this section are, therefore, useful for learning the way in which any gearbox works.

The input shaft transmits the torque from the clutch, via the countershaft (or layshaft) to the transmission output shaft. It runs inside a bearing at the front and has an internal bearing, which runs on the mainshaft, at the rear. The input shaft carries the countershaft driving gear and the synchronizer teeth and cone for fourth gear (Fig. 5.30).

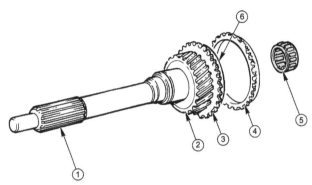

Figure 5.30 Input shaft components: 1, input shaft; 2, countershaft drive gear; 3, 4th gear synchronizing teeth; 4, 4th gear synchronizing ring; 5, mainshaft pilot bearing; 6, synchronizing cone

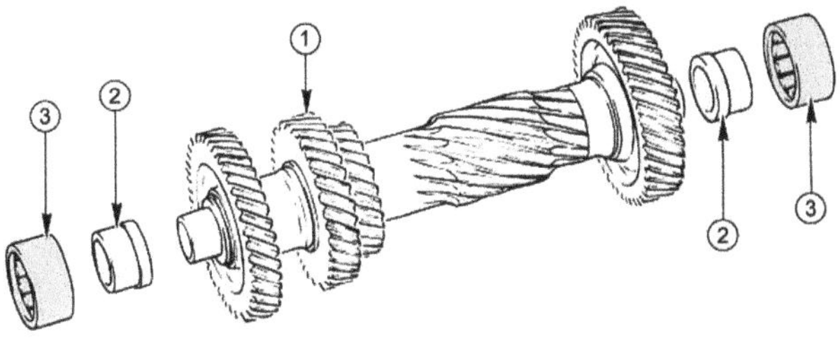

Figure 5.31 Countershaft/layshaft: 1, gears fixed to the countershaft; 2, inner race; 3, roller bearing

Figure 5.32 Reverse gear idler: 1, shaft; 2, needle roller bearing; 3, reverse gear idler; 4, roll pin; 5, mounting

Key fact

The input shaft transmits the torque from the clutch, via the countershaft (or layshaft) to the transmission output shaft.

The mainshaft is mounted in the transmission housing at the rear and the input shaft at the front. This shaft carries all the main forward gears, the selectors and clutches. All the gears run on needle roller bearings. The gears run freely unless selected by one of the synchronizer clutches.

Definition

Countershaft: The layshaft.

The countershaft is sometimes called a layshaft (Fig. 5.31). It is usually a solid shaft containing four or more gears. Drive is passed from here to the output shaft, in all gears except fourth. The countershaft runs in bearings, fitted in the transmission case, at the front and rear.

An extra gear has to be engaged to reverse the direction of the drive. A low ratio is used for reverse, even lower than first gear in many cases. The reverse idler connects the reverse gear to the countershaft (Fig. 5.32).

Power travels into the gearbox via the input shaft. A gear at the end of this shaft drives a gear on the countershaft (layshaft). Gears of various sizes are mounted on the layshaft. These gears drive other gears on a third motion shaft, also known as the output shaft.

The gearbox produces various gear ratios by engaging different combinations of gears. For reverse, an extra gear called an idler operates between the countershaft

and the output shaft. It turns the output shaft in the opposite direction to the input shaft. Figure 5.33 shows a front-wheel drive gearbox and the power flows through it in each of the different gears. Note how in each case (with the exception of reverse) the gears do not move. This is why this type of gearbox has become known as constant mesh. In other words, the gears are running in mesh with each other at all times. Dog clutches are used to select which gears will be locked to the output shaft. These clutches are moved by selector levers and incorporate synchromesh mechanisms.

A synchromesh mechanism is needed because the teeth of dog clutches clash if they meet at different speeds. A synchromesh system synchronizes the speed of two shafts before the dog clutches are meshed; hence the name.

There are three stages of engagement:

▷ neutral
▷ synchronizing
▷ shift position.

Key fact

A synchromesh mechanism is needed because the teeth of dog clutches clash if they meet at different speeds.

The system works like a friction-type cone clutch. The collar is in two parts and contains an outer

353

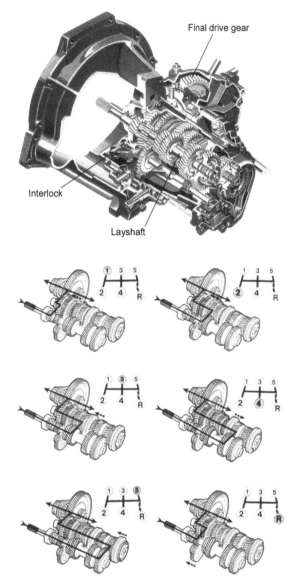

Figure 5.33 Gearbox operation and power flows

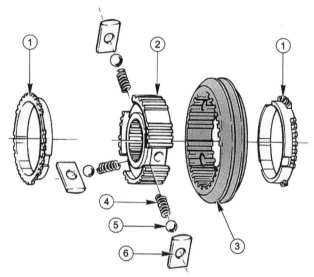

Figure 5.34 Synchronizer components: 1, synchronizer; 2, hub; 3, selector sleeve; 4, spring; 5, ball; 6, blocker bar

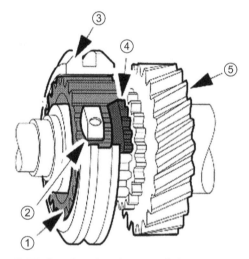

Figure 5.35 Synchronizer in neutral: 1, synchronizer body; 2, blocker bars; 3, shift/selector ring; 4, synchronizer ring; 5, gearwheel

toothed ring, which is spring-loaded to sit centrally on the synchromesh hub. When the outer ring, or synchronizer sleeve, is made to move by the action of the selector mechanism, the cone clutch is also moved because of the blocker bars (Fig. 5.34).

In the neutral position the shift ring and blocker bars are centralized (Fig. 5.35). There is no connection between the shift ring and the gear wheel. The gear wheel can turn freely on the shaft.

When the shift fork is moved by the driver, the shift ring is slid towards the gear wheel. In the process, the shift ring carries three blocker bars, which move the synchronizer ring axially and press it onto the friction surface (cone clutch) of the gear wheel. As long as there is a difference in speed, the shift ring cannot move any further. This is because the frictional force

turns the synchronizer ring, causing the tooth flanks to rest on the side of the synchronizer body (Fig. 5.36).

Once the shift ring and gear are turning at the same speed, circumferential force no longer acts on them. The force still acting on the shift ring turns it until it slides onto the teeth of the gear wheel. The gear wheel is now locked to its shaft (Fig. 5.37).

For two rotating shafts to mesh using a dog clutch, they should ideally be rotating at the same speed. Early motorists had to be skilled in achieving this through a process known as double-declutching. However, all modern gearboxes make life much easier for us by the use of synchromesh systems (Fig. 5.38).

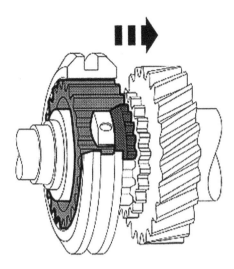

Figure 5.36 Synchronizer synchronizing

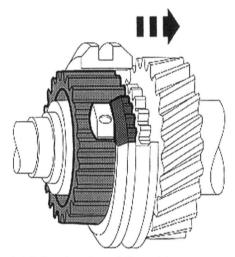

Figure 5.37 Synchronizer shift position

Transmission fluid is an important component of any gearbox and must meet the following requirements:

- Viscosity must be largely unaffected by temperature.
- It must have a high ageing resistance (gearboxes are usually filled for life).
- It must have a minimal tendency to foaming.
- It must be compatible with different sealing materials.

Only the specified transmission fluid should be used when topping up or filling after dismantling and reassembly. Bearing and tooth flank damage can occur if this is disregarded.

On earlier vehicles, a four-speed gearbox was the norm. Further improvements in operation could be gained by fitting an overdrive. This was mounted on

Figure 5.38 Synchromesh components

the output of the gearbox (rear-wheel drive). In fourth gear, the drive ratio is usually 1:1. Overdrive would allow the output to rotate more quickly than the input, hence the name. Most gearboxes now incorporate a fifth gear, which is effectively an overdrive but does not form a separate unit.

The transmission gearbox on all modern cars is a sophisticated component. However, the principle of operation does not change because it is based on simple gear ratios and clutch operation. Most current gearboxes are five-speed, constant mesh, and use helical gears.

5.3 Automatic transmission

5.3.1 Introduction and torque converter

An automatic gearbox contains special devices that automatically provide various gear ratios as they are needed. Most automatic gearboxes have three or four forward gears and one reverse gear. Instead of a gearstick, the driver moves a lever called a selector. Some automatic gearboxes have selector positions for park, neutral, reverse, drive, 2 and 1 (or 3, 2 and 1 in some cases). Some more sophisticated types with electronic control just have drive, park and reverse positions. The fluid flywheel or torque converter is the component that makes automatic operation possible.

> **Key fact**
> A torque converter is the component that makes automatic operation possible.

The engine will only start if the selector is in either the park or neutral position. In park, the drive shaft is locked so that the drive wheels cannot move (Fig. 5.39). It is also now common, when the engine is running, to only be able to move the selector out of park if you are pressing the brake pedal. This is

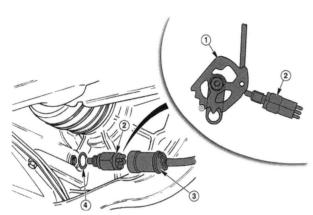

Figure 5.39 Starter circuit with inhibitor switch: 1, selector mechanism; 2, interlock switch; 3, connector; 4, sealing washer

Figure 5.40 Modern auto-box. (Source: GM Media)

a very good safety feature as it prevents sudden, uncontrolled movement of the vehicle.

For ordinary driving, the driver moves the selector to the 'drive' position. The transmission starts out in the lowest gear and automatically shifts into higher gears as the car picks up speed. The driver can use the lower positions of the gearbox for going up or down steep hills or driving through mud or snow. When in position 3, 2 or 1, the gearbox will not change above the lowest gear specified (Fig. 5.40).

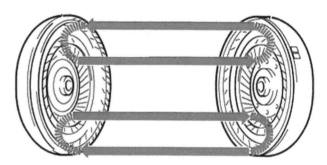

Figure 5.41 Fluid flywheel principle

A fluid flywheel consists of an impeller and turbine, which are immersed in oil (Fig. 5.41). They transmit drive from the engine to the gearbox. The engine-driven impeller faces the turbine, which is connected to the gearbox. Each of the parts, which are bowl-shaped, contains a number of vanes. They are both a little like half of a hollowed-out orange facing each other. When the engine is running at idle speed oil is flung from the impeller into the turbine, but not with enough force to turn the turbine.

As engine speed increases so does the energy of the oil. This increasing force begins to move the turbine and hence the vehicle. The oil gives up its energy to the turbine and then recirculates into the impeller at the centre, starting the cycle over again. As the vehicle accelerates the difference in speed between the impeller and turbine reduces until the slip is about 2%.

Figure 5.42 Impeller

A good analogy for a fluid flywheel that you can try for yourself is to place two desktop-type cooling fans facing each other. Switch one on (the impeller) and the air it blows will drive the blades of the other (the turbine).

Definition

Analogy: Using an analogy means drawing a comparison in order to show a similarity.

A problem with a basic fluid flywheel is that it is slow to react when the vehicle is moving off from rest. This can be improved by fitting a reactor or stator between the impeller and turbine. We now know this device as a torque converter. All modern cars fitted with automatic transmission use a torque converter.

The torque converter delivers power from the engine to the gearbox like a basic fluid flywheel, but also increases the torque when the car begins to move. Similar to a fluid flywheel, the torque converter resembles a large doughnut sliced in half. One half, called the pump impeller (Fig. 5.42), is bolted to the drive plate or flywheel. The other half, called the turbine, is connected to the gearbox-input shaft. Each half is lined with vanes or blades. The pump and the turbine face each other in a case filled with oil. A bladed wheel called a stator is fitted between them. The components and operation of the torque converter are shown in Figures 5.43–5.46.

The engine causes the pump (impeller) to rotate and throw oil against the vanes of the turbine. The force of the oil makes the turbine rotate and send power to

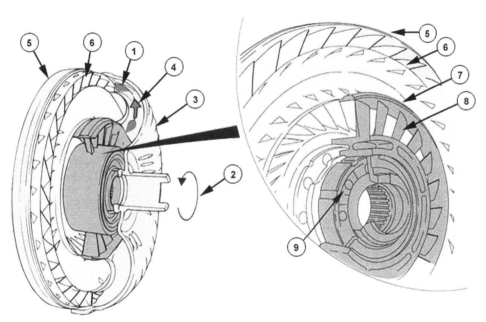

Figure 5.43 Torque converter components and operation – idle speed: 1, automatic transmission fluid (red, ATF); 2, direction of engine rotation; 3, impeller; 4, impeller blade; 5, turbine; 6, turbine blade; 7, reactor; 8, reactor blade; 9, one-way clutch

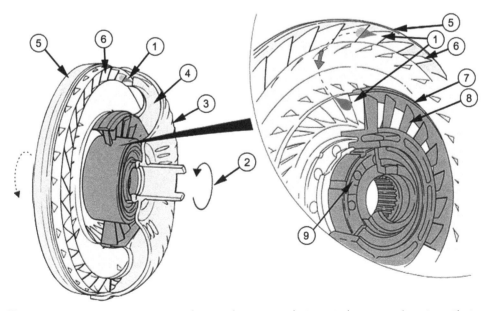

Figure 5.44 Torque converter components and operation – speed starts to increase: 1, automatic transmission fluid (red, ATF); 2, direction of engine rotation; 3, impeller; 4, impeller blade; 5, turbine; 6, turbine blade; 7, reactor; 8, reactor blade; 9, one-way clutch

357

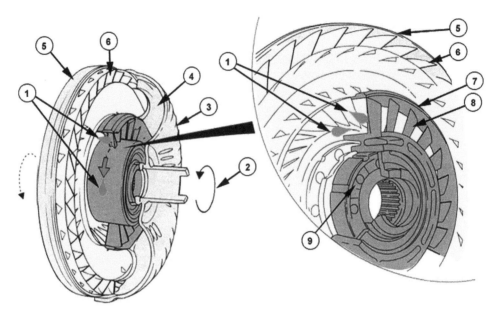

Figure 5.45 Torque converter components and operation – stator comes into use: 1, automatic transmission fluid (red, ATF); 2, direction of engine rotation; 3, impeller; 4, impeller blade; 5, turbine; 6, turbine blade; 7, reactor; 8, reactor blade; 9, one-way clutch

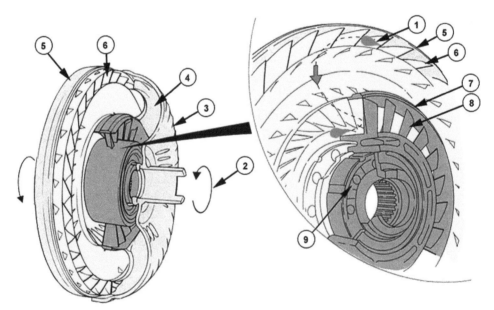

Figure 5.46 Torque converter components and operation – drive is transmitted: 1, automatic transmission fluid (red, ATF); 2, direction of engine rotation; 3, impeller; 4, impeller blade; 5, turbine; 6, turbine blade; 7, reactor; 8, reactor blade; 9, one-way clutch

the transmission. After striking the turbine vanes, the oil passes through the stator and returns to the pump. When the pump reaches a specific rate of rotation, a reaction between the oil and the stator increases the torque. In a fluid flywheel, oil returning to the impeller tends to slow it down. In a torque converter, the stator or reactor diverts the oil towards the centre of the impeller for extra thrust.

When the engine is running slowly, the oil may not have enough force to rotate the turbine. However, when the driver presses the accelerator pedal, the engine runs more quickly and so does the impeller. The action of the impeller increases the force of the oil. This force gradually becomes strong enough to rotate the turbine and moves the vehicle. Torque converters can double the applied torque when moving off from rest. As engine speed increases, the torque multiplication tapers off until at cruising speed there is no increase in torque. The reactor or stator then freewheels on its one-way clutch at the same speed as the turbine.

The converter housing is bolted to the crankshaft and driven directly. It is welded to the impeller and filled with automatic transmission fluid. The impeller:

- is welded to the converter housing
- has blades arranged radially
- turns at the same speed as the engine
- conveys fluid to the turbine blades and, as a result, produces a radial force at the turbine.

The turbine is splined to, and drives, the transmission input shaft. It has blades arranged in a curved pattern, which allows fluid to flow inwards owing to the reduced centrifugal force compared with the impeller. The fluid is then passed to the stator.

The purpose of the stator is to deflect the stream of fluid into the impeller until the coupling speed ratio is reached. The stator and one-way clutch assembly is located between the impeller and the turbine. It is splined on the stator support, which is locked to the fluid pump housing and hence to the transmission housing. The stator has blades arranged in a curved pattern. It locks, counter to the normal direction of rotation of the engine, and runs freely in the normal direction of rotation of the engine. The purpose is to boost torque, through ram pressure, up to the coupling point. It is exposed to flow from the rear until a turbine to impeller speed ratio of 85% is reached. The stator now rotates with the converter.

The fluid flywheel action of a torque converter or fluid flywheel reduces efficiency because the pump tends to rotate more quickly than the turbine. In other words, some slip will occur. This is usually about 2%. To improve efficiency, many transmissions now include a lock-up facility. When the pump reaches a specific rate of rotation, the pump and turbine are locked together, allowing them to rotate as one.

Key fact

The purpose of the stator is to deflect the stream of fluid into the impeller until the coupling speed ratio is reached.

A converter lock-up clutch allows slip-free and hence loss-free transmission of the engine torque to the automatic transmission. When engaged, it creates a frictional connection between the converter housing and the turbine. It consists of a clutch pressure plate with a friction lining, and a torsional vibration damper

to damp the crankshaft torsional vibrations. It is connected positively to the turbine and is exposed to fluid pressure from one side for clutch disengagement and engagement. A modulating valve is often used to allow controlled pressure build-up and reduction. This is to ensure smooth opening and closing. The valve is controlled electronically by means of the transmission ECU.

Key fact

A converter lock-up clutch allows loss-free transmission of engine torque.

5.3.2 Automatic transmission components

All standard automatic transmission gearboxes (Fig. 5.47) use epicyclic gearing. Epicyclic gears are a special set of gears that are part of most automatic gearboxes. In their basic form they consist of three main elements:

- a sun gear, located in the centre
- a carrier that holds two, three or four planet gears, which mesh with the sun gear and revolve around it
- an internal gear or annulus, which is a ring with internal teeth; it surrounds the planetary gears and meshes with them.

Any part of a set of planetary gears can be held stationary or locked to one of the others. This will produce different gear ratios. Most automatic gearboxes have two or more sets of planetary gears that are arranged in line. This provides the necessary

Figure 5.47 Cutaway transmission mounted on an engine. (Source: Ford Media)

number of gear ratios. As the gear selector is moved (automatically) into different positions, the power flow through the gearbox changes. Figures 5.48–5.51 show four power flows; in each case the drive comes in on the sun gear shaft and is taken out on the annulus.

> **Key fact**
>
> Most automatic gearboxes have two or more sets of planetary gears that are arranged in line.

A special feature of some transmission systems is a Ravigneaux planetary gearset (Fig. 5.52). This gearset has the following features:

▷ It offers the possibility of more gears than a conventional planetary gearset.
▷ It has a compact design (space-saving) in relation to the available transmission ratios.
▷ It has two sun gears with different diameters.
▷ It has two sets of planetary gears, inner and outer. The inner planetary gears are in constant mesh with the smaller, rear sun gear and with the outer planetary gears. The outer planetary gears are in constant mesh with the ring gear and the larger, front sun gear.
▷ The planetary gears are all fitted on a joint planetary gear holder.

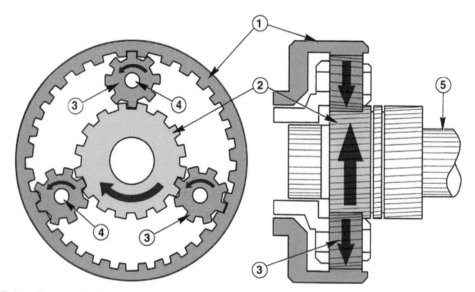

Figure 5.48 Epicyclic gears in direct drive: 1, annulus; 2, sun gear; 3, planet gears; 4, planet carrier; 5, sun gear shaft

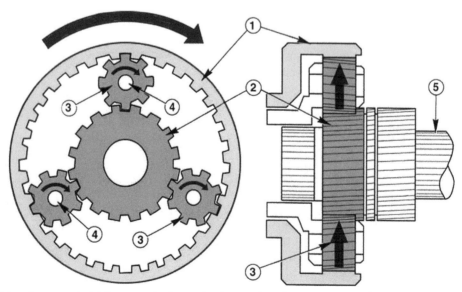

Figure 5.49 Epicyclic gears in high transmission ratio: 1, annulus; 2, sun gear; 3, planet gears; 4, planet carrier; 5, sun gear shaft

360

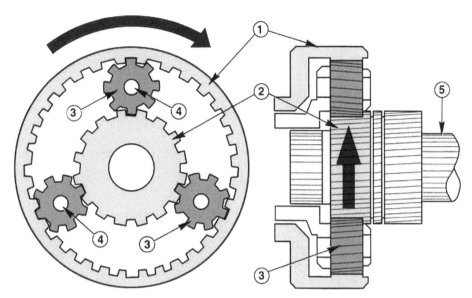

Figure 5.50 Epicyclic gears in low transmission ratio: 1, annulus; 2, sun gear; 3, planet gears; 4, planet carrier; 5, sun gear shaft

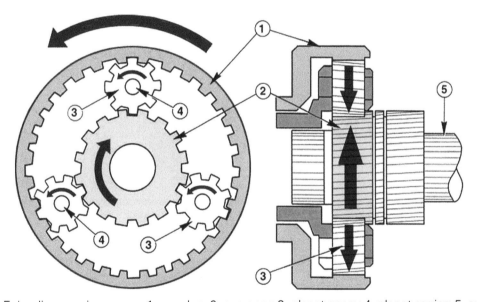

Figure 5.51 Epicyclic gears in reverse: 1, annulus; 2, sun gear; 3, planet gears; 4, planet carrier; 5, sun gear shaft

The transmission ratios are achieved via the combinations of locked and/or coupled components listed in Table 5.1. On many transmission systems the Ravigneaux planetary gearset is connected in series with a conventional epicyclic gearset. The conventional gearset can generate a speed reduction from the two sun gears of the Ravigneaux planetary gearset for all gears.

The power flows shown in Figures 5.53–5.61 are a representation of what occurs in one type of automatic gearbox. Note in particular that only the top half is shown in section. In other words, the complete picture would include a reflection of what is represented here.

The appropriate elements in the gear train of an automatic gearbox are held stationary by a system of hydraulically operated brake bands and clutches (Fig. 5.62). These are worked by a series of hydraulically operated valves, usually in the lower part of the gearbox. In most systems these are now electrically controlled.

Oil pressure to operate the clutches and brake bands is supplied by a pump (Fig. 5.63). The supply for this is the oil in the sump (pan) of the gearbox.

A cable from the throttle also allows a facility known as 'kick down'. This allows the driver to change down a gear such as for overtaking, by pressing the throttle

361

Table 5.1 Gear ratios in the Ravigneaux gearset shown in Figure 5.52

Gear/ratio	Input torque (output is always on the ring gear)	Locked
1	Small sun gear	Planetary gear holder
2	Small sun gear	Large sun gear
3	Small sun gear and large sun gear	No component
4	Small sun gear and planetary gear holder	No component
5	Large sun gear and planetary gear holder	No component
6	Planetary gear holder	Large sun gear
Reverse	Large sun gear	Planetary gear holder

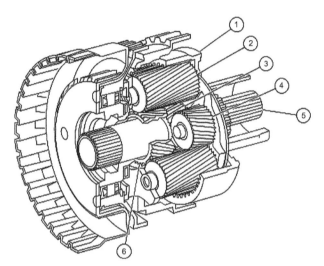

Figure 5.52 Ravigneaux gearset: 1, ring gear; 2, long planet gear; 3, small planet gear; 4, short planet gear; 5, transmission input shaft; 6, large sun gear. (Source: Ford Motor Company)

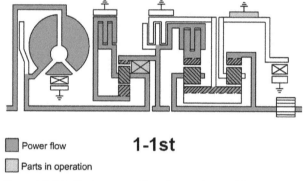

■ Power flow
□ Parts in operation

1-1st

Figure 5.54 First gear with selector in position 1

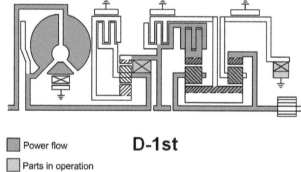

■ Power flow
□ Parts in operation

D-1st

Figure 5.55 First gear with selector in position D

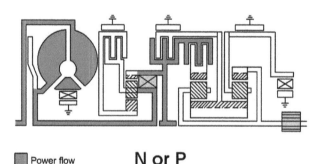

■ Power flow
□ Parts in operation

N or P

Figure 5.53 Neutral or park with selector in position N or P

all the way down. Alternatively, electronic controls determine when this downshift is needed.

> **Key fact**
>
> Different parts of a gear train in an automatic gearbox are held stationary by a system of hydraulically operated brake bands and clutches.

The main aim of electronically controlled automatic transmission (ECAT) is to improve on conventional automatic transmission in the following ways:

▷ smoother and quieter gear changes
▷ improved performance
▷ reduced fuel consumption
▷ reduction of characteristic changes over system life
▷ increased reliability.

Gear changes and lock-up of the torque converter are caused by hydraulic pressure, but under electronic control. In an ECAT system, electrically controlled solenoid valves can influence this hydraulic pressure. Most ECAT systems now have a transmission ECU that is in communication with the engine control

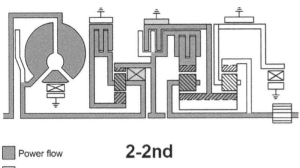

2-2nd

■ Power flow
□ Parts in operation

Figure 5.56 Second gear with selector in position 2

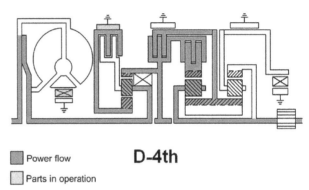

D-3rd

■ Power flow
□ Parts in operation

Figure 5.59 Third gear with selector in position D

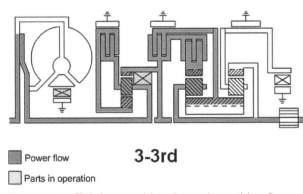

D-2nd

■ Power flow
□ Parts in operation

Figure 5.57 Second gear with selector in position D

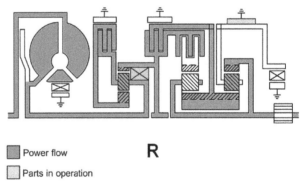

D-4th

■ Power flow
□ Parts in operation

Figure 5.60 Fourth gear with selector in position D

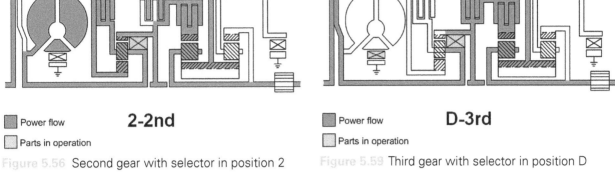

3-3rd

■ Power flow
□ Parts in operation

Figure 5.58 Third gear with selector in position 3

R

■ Power flow
□ Parts in operation

Figure 5.61 Reverse gear with selector in position R

system. Control of gearshift and torque converter lock-up is determined by the ECU. With an ECAT system, the actual point of gearshift is determined from preprogrammed memory within the ECU. Data from other sensors is also taken into consideration. Actual gearshifts are initiated by changes in hydraulic pressure, which are controlled by solenoid valves.

Typical components of a modern ECAT system are listed here:

▷ transmission control module (TCM), which controls activation and deactivation of the solenoid valves by processing the input signals from, among others, the speed sensor and rotational speed sensor of the transmission, as well as the temperature sensor; it also saves adaptive data, diagnostic trouble codes (DTCs) and frozen values for diagnosis

▷ gearshift position sensor, integrated in the TCM

▷ solenoid valves to control engine braking in first gear

▷ solenoid valve to control the lock-up function, and also used during certain gearshift processes

▷ solenoid valve to control the transmission system pressure

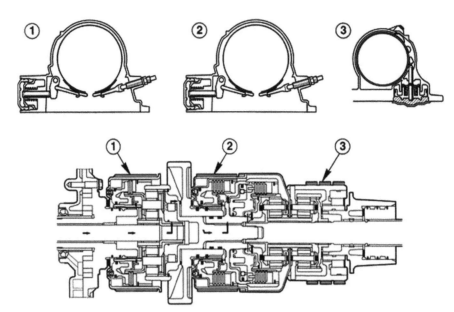

Figure 5.62 Brake bands lock different components in the gear train: 1, front band; 2, centre band; 3, rear band

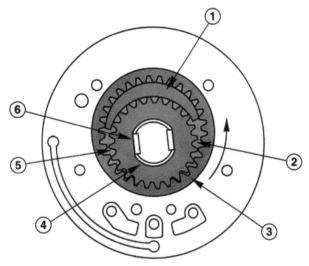

Figure 5.63 Oil pump and governor: 1, pump crescent; 2, intake side; 3, driven gear; 4, driving gear; 5, delivery side; 6, pump drive

Figure 5.64 Porsche Carrera four-wheel drive auto-gearbox. (Source: Porsche Media)

▷ solenoid valves to control which clutches are used
▷ solenoid valve to control the use of a brake band
▷ rotational speed sensor, input shaft
▷ output shaft rotational speed sensor
▷ oil temperature sensor
▷ gear selector module (GSM) in the gear selector housing, which provides information to the TCM about the desired driving mode, etc.; shift select mode and sport mode
▷ hydraulic system with oil pump, torque converter, hydraulic control housing and oil cooler

▷ a front planetary gearset and a rear planetary gearset, the transmission ratios of which are controlled by clutches and brakes.

The two main control functions of this system are hydraulic pressure and engine torque. A temporary reduction in engine torque during gear shifting allows smooth operation. This is because the peaks of gearbox output torque, which cause the characteristic surge during gear changes on conventional automatics, are suppressed. Because of control functions smooth gearshifts are possible, and because of the learning ability of some ECUs, the characteristics remain throughout the life of the system.

The ability to lock up the torque converter has been used for some time even on vehicles with more conventional automatic transmission. This gives better fuel economy and quietness and improved driveability. Lock-up is carried out using a hydraulic valve, which can be operated gradually to produce a smooth transition. The timing of lock-up is determined from ECU memory in terms of the vehicle speed and acceleration (Fig. 5.65).

Figure 5.66 shows a modern electronically controlled valve block. The ECU is built in to the system. A connection to other ECUs is made via a controller area network (CAN) connection.

Oil pressure is used to actuate clutches and brake bands to change the ratio of the epicyclic gears, just like on a system with no electronic control.

There is a wide range of electronic and hydraulic control systems. All types, however, serve to operate brake bands and/or clutches. Electronically controlled system operation is determined by ECU programming. Other systems work by sensing a combination of road speed (governor pressure) and throttle (throttle

Figure 5.66 Electrohydraulic valve block. (Source: Bosch Media)

pressure). The transmission (main line pressure) is then used to operate clutches and bands, which are under the control of valves.

5.3.3 Constantly variable transmission

The constantly variable transmission (CVT) uses a pair of cone-shaped pulleys connected by a metal belt (Fig. 5.67). The key to the operation of a CVT

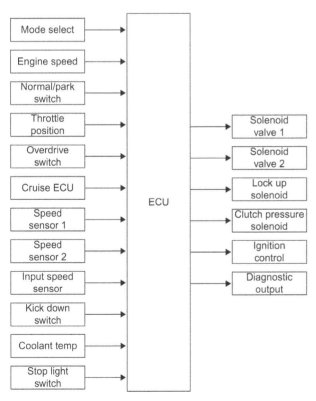

Figure 5.65 Generic electronically controlled automatic transmission (ECAT) system block diagram

Figure 5.67 Constantly variable transmission (CVT) cutaway. (Source: Ford Media)

Figure 5.68 Constantly variable transmission (CVT) components. (Source: Bosch Media)

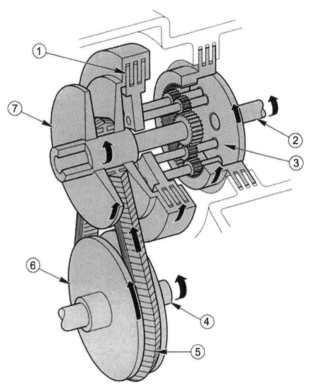

Figure 5.69 Drive belt and cone pulleys: 1, forward and reverse clutch assembly; 2, input shaft; 3, planetary gearset; 4, output shaft; 5, drive belt; 6, secondary cone pulley; 7, primary cone pulley. (Source: Ford Motor Company)

system is a high-friction drive belt (Fig. 5.68). The belt, made from high-performance steel, transmits drive by thrust rather than tension. The ratio of the rotations, or the gear ratio, is determined by how far the belt rides from the centres of two pulleys. The transmission can produce an unlimited number of ratios. As the car changes speed, the ratio is continuously adjusted.

> **Definition**
> **CVT:** Constantly variable transmission.
> **CTX:** Constantly variable transaxle.

> **Key fact**
> A key component of a CVT system is a high-friction drive belt.

Cars with this system are said to use fuel more efficiently than cars with set gear ratios. In the gearbox, hydraulic control is used to move the pulleys and hence change the ratio. To achieve forward and reverse, a standard epicyclic gear set is used.

The drive belt transmits torque from the primary cone pulley to the secondary cone pulley unit (Fig. 5.69). The belt is V-shaped and consists of several hundred steel elements held together by steel strips.

5.3.4 Direct shift gearbox

The DSG (Fig. 5.70) is an interesting development as it could be described as a manual gearbox that can change gear automatically. It can be operated by 'paddles' behind the steering wheel, by a lever in the

Figure 5.70 Direct shift gearbox (DSG). (Source: Volkswagen Media)

centre console or in a fully automatic mode. The gear train and synchronizing components are similar to those of a normal manual change gearbox.

> **Definition**
> **DSG:** Direct shift gearbox.

The DSG is made of two transmission units that are independent of each other. Each transmission unit is constructed in the same way as a manual gearbox and is connected by a multiplate clutch. They are regulated, opened and closed by a mechatronics system. On the system outlined in this section:

- first, third, fifth and reverse gears are selected via multiplate clutch 1
- second, fourth and sixth gears are selected via multiplate clutch 2.

One transmission unit is always in gear and the other transmission unit has the next gear selected ready for the next change, but with its clutch still in the open position.

Torque is transmitted from the crankshaft to a dual-mass flywheel. The splines of the flywheel, on the input hub of the double clutch, transmit the torque to the drive plate of the multiplate clutch. This is joined to the outer plate carrier of clutch 1 with the main hub of the multiplate clutch. The outer plate carrier of clutch 2 is also positively joined to the main hub.

Torque is transmitted into the relevant clutch through the outer plate carrier. When the clutch closes, the torque is transmitted further into the inner plate carrier and then into the relevant gearbox input shaft. One multiplate clutch is always engaged.

Clutch 1 is the outer clutch and transmits torque into input shaft 1 for first, third, fifth and reverse gears (Fig. 5.71). To close the clutch, oil is forced into the pressure chamber. Plunger 1 is therefore pushed along its axis and the plates of clutch 1 are pressed together. Torque is then transmitted via the plates of the inner plate carrier to input shaft 1. When the clutch opens, a diaphragm spring pushes plunger 1 back into its start position.

Clutch 2 is the inner clutch and transmits torque into input shaft 2 for second, fourth and sixth gears. As with clutch 1, oil is forced into the pressure chamber so that plunger 2 then joins the drive via the plates to input shaft 2. The coil springs press plunger 2 back to its start position when the clutch is opened.

Input shaft 2 is shown in relation to the installation position of input shaft 1. It is hollow and is joined via splines to multiplate clutch 2. The helical gear wheels for sixth, fourth and second gear can be found on input shaft 2. For sixth, fourth and second gear, a common gear wheel is used. A pulse wheel is used to measure the speed of input shaft 2. The sender is adjacent to the gear wheel for second gear.

Input shaft 1 rotates inside shaft 2 and it is joined to multiplate clutch 1 via splines. Located on input shaft 1 are the helical gear wheels for fifth gear, the common

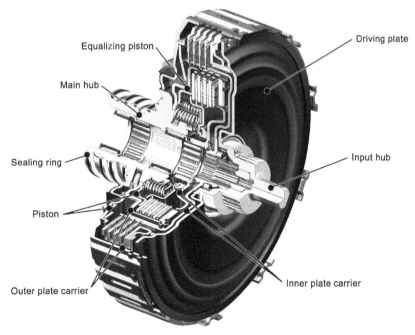

Figure 5.71 Multiplate twin clutch: clutch 1 is the outer. (Source: Volkswagen Media)

gear wheel for first and reverse gear and the gear wheel for third gear. A second pulse wheel is used to measure the speed of input shaft 1. The sender is between the gear wheels for first/reverse gear and third gear.

In line with the two input shafts, the DSG also has two output shafts (Figs 5.72 and 5.73). Because the gear wheels for first and reverse gear are the same, and fourth and sixth gear are on the input shafts, it was possible to reduce the length of the gearbox.

The reverse shaft changes the direction of rotation of output shaft 2 and therefore the direction of rotation of the final drive in the differential (Fig. 5.74). It engages in the common gear wheel for first gear and reverse gear on input shaft 1 and the selector gear for reverse gear on output shaft 2. Both output shafts transmit the torque to the input shaft of the differential (Fig. 5.75). The differential transmits the torque via the drive shafts to the road wheels.

A parking brake is integrated in the differential to secure the vehicle in the parked position and to prevent the vehicle from creeping forwards or backwards unintentionally, when the handbrake is not applied. Engagement of the locking pawl

is by mechanical means via a cable between the selector lever and the parking brake lever on the gearbox.

The mechatronics are housed in the gearbox, surrounded by oil. They comprise an ECU and an electrohydraulic control unit (Fig. 5.76). The mechatronics form the central control unit in the gearbox. Housed in this compact unit are 12 sensors. Only two sensors are located outside the mechatronics system.

The mechatronic control unit uses hydraulics to control or regulate eight gear actuators via six pressure valves. It also controls the pressure and flow of cooling oil from both clutches. The mechatronics control unit learns and remembers (adapts) the position of the clutches, the positions of the gear actuators, when a gear is engaged, and the main pressure.

A variation on the system described above is shown in Figure 5.77. I am unsure whether to describe the DSG as a manual or an automatic box, so the only thing to do is to make up a new word. This section has therefore outlined the operation of an 'auto-man gearbox' (AMG)!

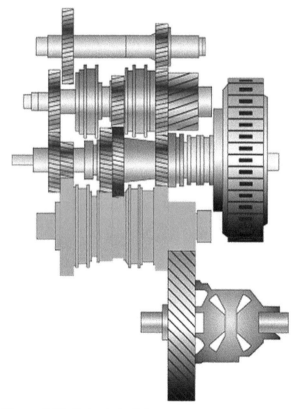

Figure 5.72 Output shaft 1

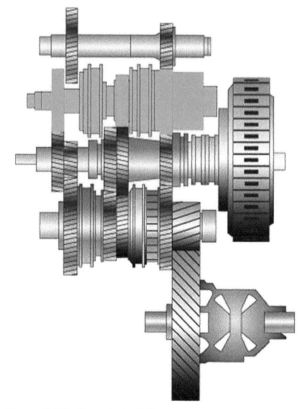

Figure 5.73 Output shaft 2

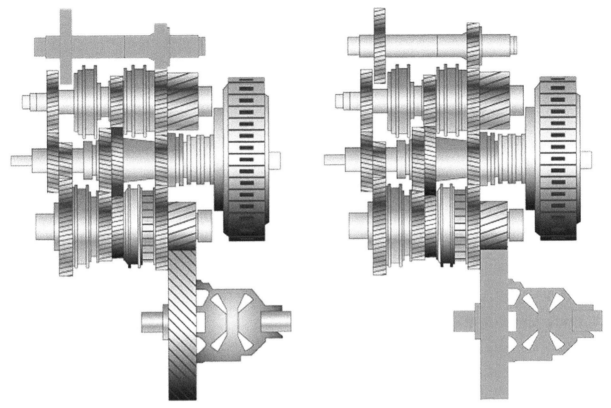

Figure 5.74 Reverse shaft   Figure 5.75 Final drive and differential

Figure 5.76 Electronic control. (Source: Volkswagen Media)

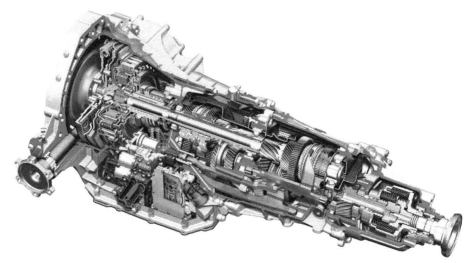

Figure 5.77 Audi four-wheel drive direct shift gearbox. (Source: Audi Media)

5.4 Transmission driveline

5.4.1 Propshafts and driveshafts

Propshafts, with universal joints (UJs), are used on rear- or four-wheel drive vehicles (Fig. 5.78). They transmit drive from the gearbox output to the final drive in the rear axle. Drive then continues through the final drive and differential, via two half shafts to each rear wheel. A hollow steel tube is used for the main shaft (Fig. 5.79). This is lightweight, but will still transfer considerable turning forces and resist bending forces.

Universal joints allow for the movement of the rear axle with the suspension, while the gearbox remains fixed. Two joints are used on most systems and must always be aligned correctly (Fig. 5.80).

Figure 5.78 Propshaft

Key fact

A hollow steel tube is used for a propshaft because it is lightweight, but will still transfer considerable turning forces and resist bending.

Key fact

Two joints are used on most propshafts and must always be aligned correctly.

Because of the angle through which the drive is turned by UJs, a speed variation results. This is caused because two arms of the UJ rotate in one plane and two in another. The cross of the UJ, therefore, has to change position twice on each revolution. This problem can be overcome by making sure the two UJs are aligned correctly (Fig. 5.81). If the two UJs on a propshaft are aligned correctly, the variation in speed caused by the first can be cancelled out by the second. However, the angles through which the shaft works must be equal. The main body of the propshaft will run with variable velocity but the output drive will be constant.

The simplest and most common type of UJ consists of a four-point cross, which is sometimes called a spider. Four needle roller bearings are fitted, one on each arm of the cross. Two bearings are held in the driver yoke and two in the driven yoke (Fig. 5.82).

Several types of UJ have been used on vehicles (Figs 5.83–5.84). These developed from the simple 'Hooke'-type joint to the later cross-type, sometimes known as a Hardy Spicer. Rubber joints are also used on some vehicles.

The 'doughnut' coupling has the advantage that it is flexible and absorbs torsional shocks. It will also tend to reduce vibrations caused by other joints. Its other

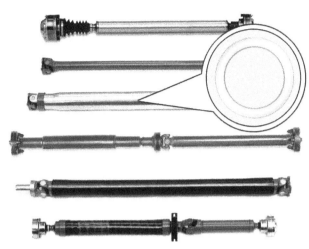

Figure 5.79 Section of a propshaft

Figure 5.82 Detail of a universal joint

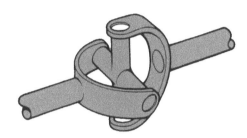

Figure 5.83 Hooke-type joint

Figure 5.80 Universal joint in position

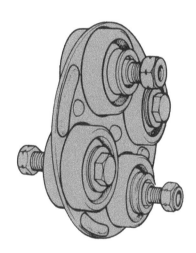

Figure 5.84 Layrub joint

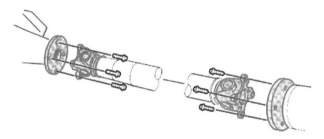

Figure 5.81 These joints are aligned correctly

Key fact

As a car's suspension moves up and down, the length of the driveline changes.

main advantage is that it allows some axial (back and forth) movement.

As the suspension moves up and down, the length of the driveline changes slightly. As the rear wheels hit a bump, the axle moves upwards. This tends to shorten the driveline. The splined sliding joint allows for this movement. However, it will also transfer the rotational drive. Internal splines are used on the propshaft so that the external surface is smooth (Fig. 5.87). This allows an oil seal to be fitted into the gearbox output casing.

When long propshafts are used, there is a danger of vibration. This is because the weight of the propshaft can cause it to sag slightly and therefore 'whip' (like a skipping rope) as it rotates. Most centre bearings are standard ball bearings mounted in rubber (Fig. 5.88).

Propshafts are used on rear- or four-wheel drive vehicles. They transmit drive from the gearbox output to the rear axle. Most propshafts contain two UJs. A single joint produces rotational velocity variations, but this can be cancelled out if the second joint is aligned

371

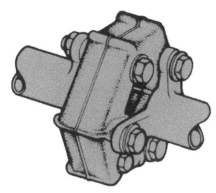

Figure 5.85 Rubber 'doughnut' joint

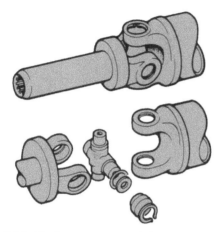

Figure 5.86 Modern cross-type joint

Figure 5.87 A splined joint connects to the gearbox

Figure 5.88 This rubber mounted bearing prevents propshaft whip

Figure 5.89 Inner constant velocity (CV) joint

Figure 5.90 Outer constant velocity (CV) joint

A CV joint is a universal joint. However, it is constructed so that the output rotational speed is the same as the input speed. The speed of rotation remains constant even as the suspension and steering move the joint.

The inner and outer joints (Figs 5.89 and 5.90) have to perform different tasks. The inner joint has to plunge in and out, to take up the change in length as the suspension moves. The outer joint has to allow suspension and steering movement up to about 45°. A solid steel shaft transmits the drive.

When a normal UJ operates, the operating angle of the cross changes as described above. This is what causes the speed variations. A CV joint spider (or cross) operates in one plane because balls or rollers are free to move in slots. The cross therefore bisects the driving and driven planes.

The rubber boot or gaiter keeps out dirt and water, and keeps in lubricant. Usually, a graphite or molybdenum grease is used, but check the manufacturer's specifications to be sure.

There are a number of different types of CV joint (Fig. 5.91). The most common is the Rzeppa

correctly. Centre bearings are used to prevent vibration due to propshaft whip.

On front-wheel drive cars, driveshafts with constant velocity (CV) joints transmit drive from the output of the final drive and differential, to each front wheel. They must also allow for suspension and steering movements.

Key fact

Constant velocity (CV) joints allow for suspension and steering movements.

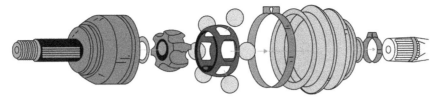

Figure 5.91 Details of a constant velocity (CV) joint (Rzeppa type)

Figure 5.92 Constant velocity (CV) joint: double offset type

(pronounced reh-ZEP-ah). The inner joint must allow for axial movement due to changes in length as the suspension moves. It has six steel balls held in a cage between an inner and outer race inside the joint housing. Each ball rides in its own track on the inner and outer races. The tracks are manufactured into an arch shape so that the balls stay in the mid-point at all times, ensuring that the angle of the drive is bisected. This joint is used on the outer end of a driveshaft and can handle steering angles of up to 45°.

The other common CV joint is the double offset joint, which is a variation of the Rzeppa type (Fig. 5.92). The main difference is that the outer race has long straight tracks. This allows a plunge (axial movement) of about 50 mm (2 in.) and a steering angle of up to 24°. This makes it ideal as the inner joint on the driveshaft.

Driveshafts with CV joints are used on front-wheel drive vehicles. They transmit drive from the differential to each front wheel. They must also allow for suspension and steering movements. Inner joints must 'plunge' to allow for changes in length of the shaft.

Key fact

The two main types of bearing are ball and roller.

5.4.2 Wheel bearings

Two main types of bearing are used in rear wheel hubs: ball and roller (or tapered roller) bearings (Figs 5.93 and 5.94).

Key fact

Different bearing layouts are used on the fixed live axle of rear-wheel drive vehicles.

Figure 5.93 Ball bearing

Figure 5.94 Tapered roller bearing

Axle shafts transmit drive from the differential to the rear wheel hubs. An axle shaft has to withstand:

▷ torsional stress due to driving and braking forces
▷ shear and bending stress due to the weight of the vehicle
▷ tensile and compressive stress due to cornering forces.

A number of bearing layouts are used on the fixed live axle (driven) of a rear-wheel drive car, to handle these stresses.

Figure 5.95 shows a typical axle mounting used on many rear-wheel drive cars. A single bearing is used,

373

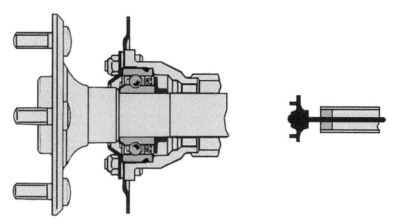

Figure 5.95 Wheel bearing: semi-floating

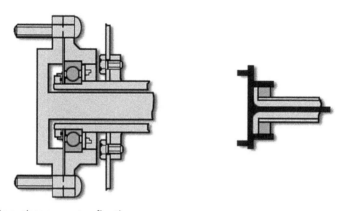

Figure 5.96 Wheel bearing: three-quarter floating

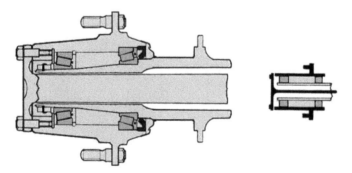

Figure 5.97 Wheel bearing: fully floating

which is mounted in the axle casing. With this design, the axle shaft has to withstand all the operating forces. The shaft is therefore strengthened and designed to do this. An oil seal is incorporated because oil from the final drive can work its way along the shaft. The seal prevents the brakes being contaminated.

The three-quarter floating bearing shown in Figure 5.96 reduces the main shear stresses on the axle shaft but the other stresses remain. The bearing is mounted on the outside of the axle tube. An oil seal is included to prevent the brake linings from being contaminated.

Fully floating systems are generally used on heavy or off-road vehicles (Fig. 5.97). This is because the stresses on these applications are greater. Two widely spaced bearings are used, which take all the loads, other than torque, off the axle shaft. Bolts or studs are used to connect the shaft to the wheel hub. When these are removed, the shaft can be taken out without jacking up the vehicle.

The front hubs on rear-wheel drive cars consist of two bearings. These are either ball or tapered roller types (Figs 5.98 and 5.99). The roller types are generally

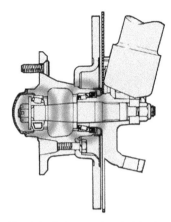

Figure 5.98 Front hub (rear-wheel drive) with tapered roller bearings

Figure 5.100 Rear hub (dead axle) with taper bearing

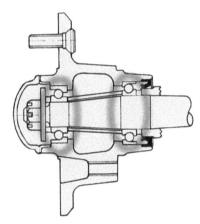

Figure 5.99 Front hub with ball bearings

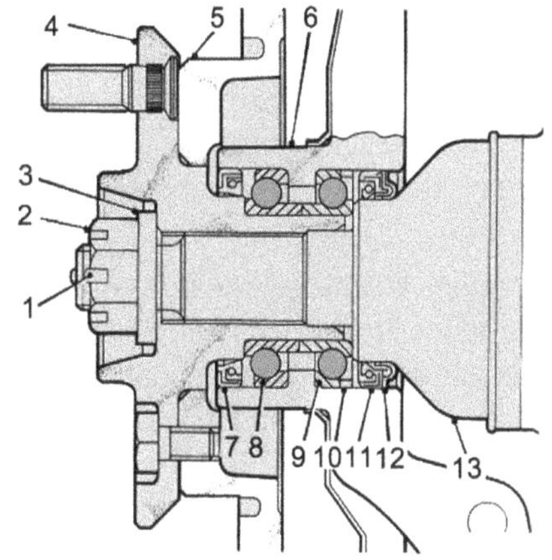

Figure 5.101 Front hub and bearings: 1, split pin; 2, hub nut; 3, flat washer; 4, drive flange; 5, brake disc; 6, hub; 7, outlet oil seal; 8, outer bearing; 9, inner bearing; 10, spacer; 11, inner oil seal; 12, bearing water shield; 13, drive shaft

used on earlier vehicles. They have to be adjusted by tightening the hub nut and then backing it off by about half a turn. The more modern hub bearings, known as contact-type ball races, do not need adjusting. This is because the hub nut tightens against a rigid spacer. This nut must always be set at a torque specified by the manufacturer.

The most common systems for rear-wheel drive cars are semi-floating rear bearings at the rear and twin ball bearings at the front. The front bearings are designed to withstand side forces as well as vertical loads. Front-wheel drive cars have a dead axle at the rear so, typically, two taper bearings are used, or a double-race ball bearing (Fig. 5.100).

Key fact

Wheel bearings must allow smooth rotation of the wheel and withstand high suspension and steering stresses.

Wheel bearings must allow smooth rotation of the wheel but also be able to withstand high stresses such as when cornering. In addition, front-wheel drive arrangements must allow the drive to be transmitted via the driveshafts.

A front hub works as an attachment for the suspension and steering as well as supporting the bearings (Fig. 5.101). It supports the weight of the vehicle at the front, when still or moving. Ball or roller bearings are used for most vehicles with specially shaped tracks. This is so that the bearings can stand side loads when cornering. The bearings support the driveshaft as well as the hub.

The stub axle, which is solid-mounted to the suspension arm, fits in the centre of two bearings.

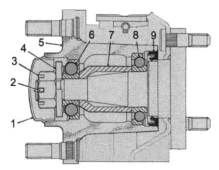

Figure 5.102 Rear hub and bearings: 1, grease retaining cap; 2, split pin; 3, hub nut; 4, flat washer; 5, hub and brake drum assembly; 6, outer bearing; 7, spacer; 8, inner bearing; 9, oil seal

The axle supports the weight of the vehicle at the rear, when still or moving. Ball bearings are used for most vehicles, with specially shaped tracks for the balls, so that the bearings can stand side loads when cornering. A spacer is used to ensure the correct distance between, and pressure on, the two bearings.

On front-wheel drive vehicles, the hub and bearing arrangement on the front must bear weight, withstand driving forces and support the driveshaft. The rear hub and bearings (Fig. 5.102) must support the vehicle and withstand side forces.

5.4.3 Four-wheel drive

Four-wheel drive (4WD) systems can be described as part-time or full-time. All 4WD systems must include some type of transfer gearbox. The main components of a 4WD system are shown in Figure 5.103. Each axle must be fitted with a differential. A transfer box takes drive from the output of the normal gearbox and distributes it to the front and rear. The transfer box may also include gears to allow the selection of a low ratio. High ratio is a straight-through drive.

> **Key fact**
> Four-wheel drive systems can be described as part-time or full-time.

A 4WD system, when described as part-time, means that the driver selects 4WD only when the vehicle needs more traction. When the need no longer exists, the driver reverts to the normal two-wheel drive. This keeps driveline friction, and therefore the wear rate, to a minimum.

When a 4WD system is described as full-time, it means that the drive is engaged all the time. To

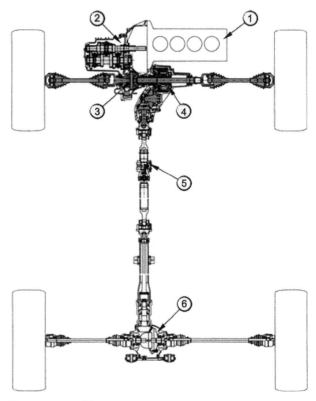

Figure 5.103 The main components of a four-wheel drive system: 1, engine; 2, transmission; 3, front axle differential and final drive; 4, transfer box with longitudinal differential; 5, two-piece driveshaft; 6, rear final drive and differential

prevent 'wind-up', which occurs if the front and rear axles rotate at different speeds, a centre differential or viscous drive is used.

An all-wheel drive (AWD) system automatically transfers drive to the axle with better traction (Fig. 5.104). It is designed for normal road use. The drive, on full-time systems, is passed to the rear via a viscous coupling. When the front wheels spin, the viscous coupling locks and transfers drive to the rear.

> **Key fact**
> The drive, on full-time four-wheel drive systems, is passed to the rear via a viscous coupling to prevent wind-up.

If used, a transfer box on a part-time 4WD system may allow the driver to choose from four options:

▷ neutral
▷ 2WD high
▷ 4WD high
▷ 4WD low (Fig. 5.105).

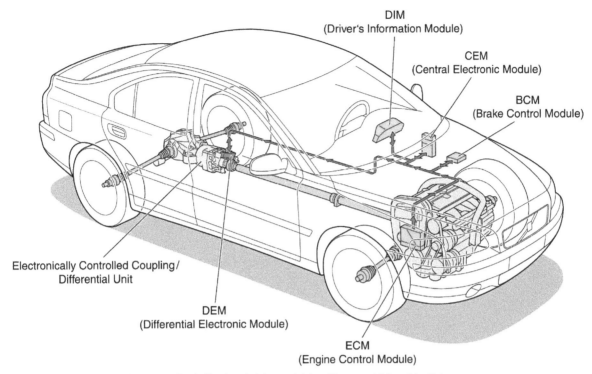

Figure 5.104 Electronically controlled all-wheel drive vehicle. (Source: Volvo Media)

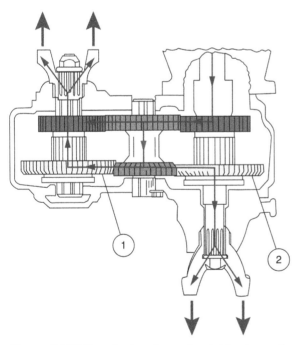

Figure 5.105 Transfer box four-wheel drive low ratio selected: 1, low ratio gear selector; 2, high ratio gear selector

A typical system will have the transfer box, attached to the normal rear-wheel drive gearbox, in place of the extension housing. This arrangement is more often used for off-road vehicles.

A differential in a transfer box allows its two outputs to be driven at different speeds. In 4WD systems, it is possible that, for example, the front axle could rotate more quickly than the rear axle. This would produce driveline wind-up, so centre differentials or viscous couplings are used. On modern vehicles, they often consist of planetary gears.

A viscous coupling is designed to transmit drive when the axle speeds differ (Fig. 5.106). This occurs because the difference in speed of the two axles increases the friction in the coupling. This results in greater torque transmission, which in turn reduces the speed difference. As the speed difference reduces, less torque is transmitted. In this way, the torque is shared proportionally between the two axles.

> **Key fact**
>
> A differential in a transfer box allows its two outputs to be driven at different speeds.

A 'silent' drive chain is used on many vehicles to pass the drive to the auxiliary output shaft (Fig. 5.107). The chain takes up less space than do gears. It is designed to last the life of the vehicle and adjustment is not normally possible.

Four-wheel drive systems use a combination of propshafts and driveshafts together with viscous couplings and transfer boxes. A number of variations are possible and may be described as full- or part-time.

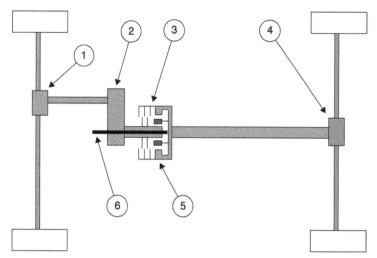

Figure 5.106 A viscous coupling transfers torque when axle speeds differ: 1, front differential; 2, drive chain transfer box; 3, viscous coupling; 4, rear differential; 5, planetary gear train; 6, transmission gearbox output shaft

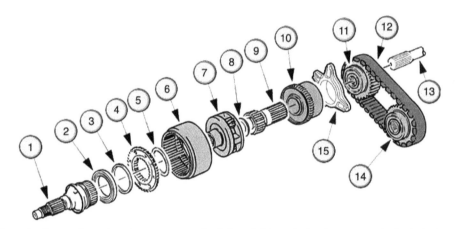

Figure 5.107 Transfer box using planetary gears and a 'silent' drive chain: 1, output shaft; 2, ball bearing; 3, spacer; 4, drive plate; 5, spacer; 6, annulus; 7, planet carrier and planet gears; 8, thrust washer; 9, sun wheel shaft; 10, viscomatic locking unit; 11, driving sprocket and bearing; 12, drive chain; 13, transmission gearbox output shaft; 14, driven sprocket and bearing; 15, bracket

5.5 Final drive and differential

5.5.1 Final drive

Because of the speed at which an engine runs, and in order to produce enough torque at the road wheels, a fixed gear reduction is required. This is known as the final drive and it consists of just two gears. The final drive is fitted on the output of the gearbox on front-wheel drive vehicles, and in the rear axle after the propshaft on rear-wheel drive vehicles. The ratio is normally a speed reduction of between about 2:1 and about 4:1. For example, at 4:1, when the gearbox output is turning at 4000 rpm, the wheels will turn at 1000 rpm.

Key fact

The final drive consists of two gears and provides a fixed speed reduction.

Many cars now have a transverse engine, which drives the front wheels. The final drive contains ordinary helical gears. When a transaxle system is used as in Figure 5.108, it always consists of the final drive and two driveshafts.

On rear-wheel drive vehicles, the final drive gears also turn the drive through 90° (Fig. 5.109). This is done using bevel gears known as a crown wheel and pinion. Four-wheel drive vehicles also have a similar

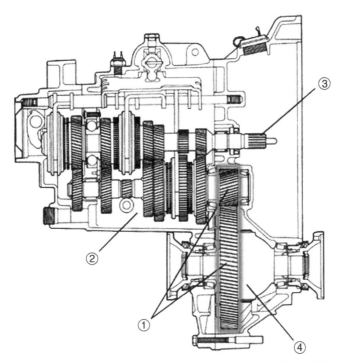

Figure 5.108 Transaxle final drive gears: 1, gearbox output shaft; 2, input shaft; 3, differential; 4, final drive gears (highlighted)

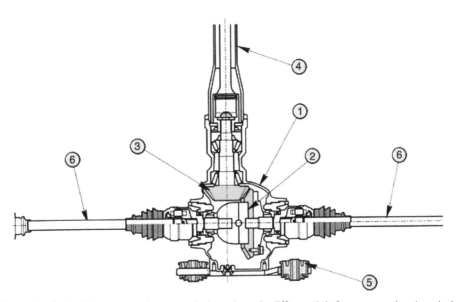

Figure 5.109 Rear axle final drive gears: 1, rear axle housing; 2, differential; 3, crown wheel and pinion final drive gears; 4, extension tube; 5, mounting; 6, driveshafts

arrangement as part of the rear axle. The crown wheel and pinion are special types of bevel gears, which mesh at right angles to each other (Fig. 5.110). They carry power through a right angle to the drive wheels. The crown wheel is driven by the pinion, which receives power from the propeller shaft.

Final drive gears reduce the speed from the propeller shaft and increase the torque. The reduction in the final drive multiplies any reduction that has already taken place in the transmission.

The crown wheel gear (and pinion) of a rear-wheel drive system is usually a hypoid type, and is named

379

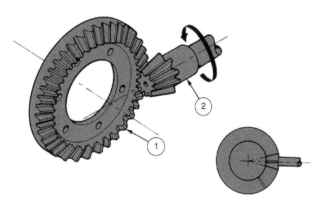

Figure 5.110 Bevel gears change ratio and drive angle: 1, crown wheel; 2, pinion (both on centre line)

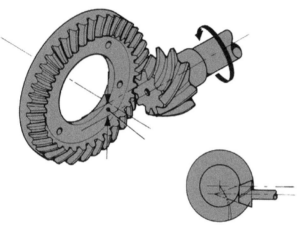

Figure 5.111 The hypoid design allows a lower propshaft to be used (the offset is shown by the arrows)

after the way the teeth are cut (Fig. 5.111). As well as quiet operation, this allows the pinion to be set lower than the crown wheel centre, thus saving space in the vehicle because a smaller transmission tunnel can be used. Some early vehicles used spiral-cut gears, which had the advantage of running more quietly but still had to be set at the centre line.

Because the teeth of hypoid gears cause 'extreme pressure' on the lubrication oil, a special type is used. This oil may be described as 'hypoid gear oil' or 'EP', which stands for extreme pressure. As usual, refer to manufacturers' recommendations when topping up or changing oil.

> **Key fact**
> A crown and pinion hypoid gear set allows the propshaft to be set lower.

The rigid rear axle assembly consists of other components as well as the final drive gears. The other main components are the differential, halfshafts and bearings. On split axle types, as in Figure 5.109, the final drive is mounted to the chassis, and driveshafts are used to connect to the wheels.

> **Key fact**
> Because the teeth of hypoid gears cause 'extreme pressure', special oil is used.

To produce enough torque at the road wheels, a fixed gear reduction is required. This is known as the final drive. It consists of just two gears. On rear-wheel drive systems, the gears are bevelled to turn the drive through 90°. On front-wheel drive systems, this is not necessary. The drive ratio is similar for front- and rear-wheel drive cars.

5.5.2 Differential

The differential is a set of gears that divides the torque evenly between the two drive wheels. The differential allows one wheel to rotate more quickly than the other. As a car goes around a corner, the outside driven wheel travels further than the inside one (Fig. 5.112). The outside wheel must therefore rotate faster than the inside one to cover the greater distance in the same time. Tyre scrub and poor handling would be the result if a fixed axle were used.

> **Key fact**
> A differential is a set of gears that divides torque evenly between two drive wheels.

The differential consists of sets of bevel gears, and pinions within a cage, attached to the large final drive gear. The bevel gears can be described as sun and

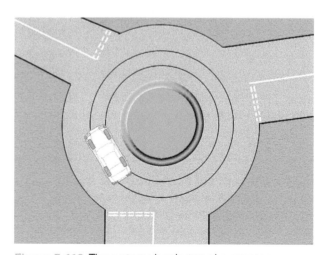

Figure 5.112 The outer wheels travel a greater distance

planet gears. The sun gears provide the drive to the wheels via halfshafts or driveshafts. The planet gears either rotate with the sun gears or rotate around them, depending on whether the car is cornering or not.

Bearings support the differential casing, which is in turn bolted to the final drive gear. The casing transmits the drive from the final drive gear, to the planet gear pinion shaft. The planet gears are pushed round by this shaft. The sun gear pinions, which are splined to the drive shafts, take their drive from the planet gears. The sun gears always rotate at the same speed as the road wheels.

The planet shaft is secured in the differential casing so that it pushes the planet gears. If the sun gears, which are attached to the road wheels via the driveshafts, are moving at the same speed, the planet gears do not spin on their shaft. However, when the vehicle is cornering, the sun gears need to move at different speeds. In this case, the planet gears rotate on their shaft to make up for the different wheel speeds.

When the vehicle is travelling in a straight line, the planet gears turn with the sun gears, but do not rotate on their shaft. This occurs because the two sun gears attached to the driveshafts are revolving at the same speed.

A standard differential can be described as a torque equalizer. This is because the same torque is always provided to each wheel, even if they are revolving at different speeds. At greater speeds, more power is applied to the wheel, so the torque remains the same.

A good way to understand the differential action is to consider the extreme situation. This is when a corner is so sharp that the inner wheel would not move at all! Although this is impossible, it can be simulated by jacking up one wheel of the car. All the drive is transferred to the free wheel (Fig. 5.113). The planets roll around the stationary sun wheel but drive the free wheel because they are rotating on their shaft.

The previous extreme example highlights the problem with a differential. If one of the driven wheels is stuck in the mud, all the drive is transferred to that wheel and it normally spins. Of course, in this case, drive to the wheel on the hard ground would be more useful. The solution to this problem is the limited slip differential (LSD).

Some higher-performance vehicles use an LSD to improve traction. Clutch plates, or similar, are connected to the two output shafts and can, therefore, control the amount of slip. This can be used to counteract the effect of one wheel losing traction when high power is applied.

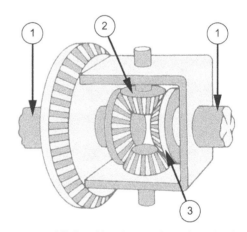

Figure 5.113 All the drive is transferred to the free wheel: 1, driveshaft; 2, planet gears; 3, sun gears

Figure 5.114 This differential is described as an open type

381

A standard differential always applies the same amount of torque to each wheel. Two factors determine how much torque can be applied to a wheel. In dry conditions, when there is plenty of traction, the amount of torque applied to the wheels is limited by the engine and gearing. When the conditions are slippery, such as on ice, the torque is limited by the available grip. LSDs allow more torque to be transferred to the non-slipping wheel.

The clutch-type LSD is the most common (Fig. 5.115). It is the same as a standard differential, except that it also has a spring pack and a multiplate clutch. The spring pack pushes the sun gears against the clutch plates, which are attached to the cage and both sun gears spin with the cage. When both wheels are moving at the same speed, the clutches have little or no effect. However, the clutch plates try to prevent either wheel from spinning more quickly than the other. The stiffness of the springs and the friction of the clutch plates determine how much torque it takes to make the clutch slip.

If one drive wheel is on a slippery surface and the other one has good traction, drive can be transmitted to the wheel with good traction. The torque supplied to the wheel not on the slippery surface is equal to the amount of torque it takes to overpower the clutches. The result is that the car will move, but not with all the available power.

The viscous coupling often found in all-wheel-drive vehicles acts like an LSD (see Fig. 5.106). It is commonly used to link the back wheels to the front wheels so that when one set of wheels starts to slip, torque will be transferred to the other set. The viscous coupling has two sets of plates inside a sealed housing that is filled with a thick fluid. One set of plates is connected to each output shaft. Under normal conditions, both sets of plates and the viscous fluid spin at the same speed. However, when one set of wheels spins more quickly, there will be a difference in speed between the two sets of plates.

The viscous fluid between the plates tries to catch up with the faster discs, dragging the slower discs along. This transfers more torque to the slower wheels. When a vehicle is cornering, the difference in speed between the wheels is not as large as when one wheel is slipping. The faster the plates spin, relative to each other, the more torque the coupling transfers. This effect can be demonstrated by spinning an egg. Spin the egg on a table top and then stop it. Let go, and it will start to spin again because the viscous fluid inside is still spinning and drags the shell around with it.

Conventional LSDs cannot be designed for optimum performance because of the effect on the vehicle when cornering and on the steering. These issues prompted the development of electronic control. The slip limiting action is controlled by a multidisc clutch, as discussed previously. The pressure on the clutch plates is controlled by hydraulic pressure, which in turn is controlled by a solenoid valve under the influence of an ECU. The ECU is able, if required, to fully lock the axle. Data is provided to the ECU from standard ABS-type wheel sensors.

Differential locks are used on many off-road type vehicles. A simple dog clutch or similar device

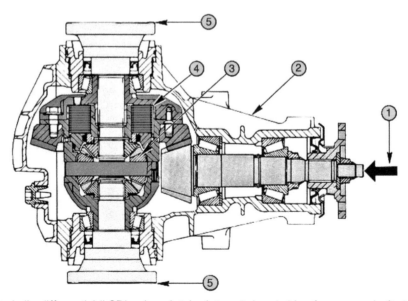

Figure 5.115 Limited slip differential (LSD) using clutch plates: 1, input drive from propshaft; 2, housing; 3, differential; 4, clutch plates; 5, output drive flanges. (Source: Ford Motor Company)

Figure 5.116 Differential locking mechanism [highlighted]. (Source: Ford Media)

Figure 5.117 This electronically controlled clutch closes the gap between automatic and manual transmission. (Source: Bosch Media)

prevents the differential action. This allows far better traction on slippery surfaces. An electric, hydraulic or pneumatic mechanism is used to lock the two output pinions together (Fig. 5.116).

This mechanism is usually activated manually by switch and, when activated, both wheels will spin at the same speed. If one wheel ends up off the ground, the other wheel will continue to spin at the same speed.

5.6 Transmission systems: new developments and updates

5.6.1 Electronic clutch

An electronic clutch (eClutch) can save fuel and make driving easier. Traffic congestion is bad enough at the best of times, but worse without automatic transmission. The reason is the constant switching between clutch, throttle and brake. In stop-and-go traffic, the eClutch allows drivers using manual transmission to use first gear without using the clutch. They can simply use the brake and throttle pedal, just like in an automatic transmission, without accidentally stalling the engine.

There are several benefits and features that can be achieved with an electronically controlled clutch actuator. Fitting an eClutch costs significantly less than a conventional automatic transmission, and is thus an attractive alternative in the compact car segment. Sport mode settings are also available on some systems. In addition, the eClutch makes a coasting function possible, which saves fuel. Independently of the driver, the clutch decouples the engine from the

transmission if the driver is no longer accelerating, and the engine is then stopped. The result is a real fuel saving in the region of 10%.

This clutch system is described as drive-by-wire as there is no direct connection between the pedal and the clutch actuator. Under normal operating conditions the actuator position will just follow the driver's clutch pedal position. However, on some systems it will intervene if the pedal was released too quickly, for example. The operation of different types of actuator varies but in general an electric motor is used that drives a segment of a wheel via a worm gear. The worm wheel presses the release pin forwards, thus preventing the flow of power through the clutch. The large tensioning forces to separate the clutch are balanced by a spring, which means that the electric motor only supplies the torque to overcome friction.

Definition

Drive-by-wire: When there is no direct connection between the control and actuator.

As well as the stop-and-go feature and the possibility of saving fuel, the eClutch offers several other functions. For example, it can be used to support gear shifts, making them smoother. A special sensor detects the start of a gear shift and adjusts engine speed to make the gear change smoother.

The cost of an eClutch system is a lot more than a clutch cable. However, the potential benefits may soon outweigh the extra initial expense, as manufacturers work to meet ever lower emissions targets.

Figure 5.118 Clutch actuator sectioned view. (Source: www.technolab.org)

Figure 5.119 Front transverse engine, nine-speed transmission. (Source: ZF)

5.6.2 Automatic transmission with nine speeds

ZF Friedrichshafen's 9HP nine-speed automatic transmission was designed for passenger cars with front-transverse engine configurations. Model ranges cover 200 to 480 Nm. Four individual gear sets are nested within the transmission, with six shifting elements and nine gear ratios. The 9HP also features interlocking dog clutches for power shifting; hydraulically operated constant-mesh elements reduce overall transmission length and optimize efficiency.

With nine speeds, the 9HP improves fuel economy and achieves small gear ratios. The rpm is lowered by approximately 700 at a speed of 75 mph (120 km/h), reducing noise. A torque converter is the 9HP's standard starting element, and a multilevel torsion damper system minimizes hydraulic losses.

> **Key fact**
> Nine-speed transmission improves fuel economy with lower rpm, and small gear ratios result in smoother operation.

Key features of this transmission are:

- ▶ wide transmission ratio spread with small ratio steps

Figure 5.120 Gearsets producing nine speeds. (Source: ZF)

- ▶ interlocking dog clutches
- ▶ torque converter with excellent vibration and oscillation isolation for optimal comfort during drive-off and shifting
- ▶ proven ZF control systems technology to provide excellent shift quality.

The design makes the most of the available space and creates the best possible conditions for versatile front transverse applications.

CHAPTER 6

Emerging technologies

6.1 Safety

6.1.1 Risks of working with EVs and other technologies

EVs introduce hazards into the workplace in addition to those normally associated with the repair and maintenance of vehicles, roadside recovery and other vehicle-related activities. These include:

> **Safety First**
>
> Remember, to work on high voltage systems, you must be suitably qualified and use correct equipment and PPE!

- the presence of high-voltage components and cabling capable of delivering a fatal electric shock
- the storage of electrical energy with the potential to cause explosion or fire
- components that may retain a dangerous voltage even when a vehicle is switched off
- electric motors or the vehicle itself that may move unexpectedly due to magnetic forces within the motors
- manual handling risks associated with battery replacement
- the potential for the release of explosive gases and harmful liquids if batteries are damaged or incorrectly modified
- the possibility of people being unaware of vehicles moving, because when electrically driven they are silent in operation
- the potential for the electrical systems on the vehicle to affect medical devices such as pacemakers and insulin controllers.

Figure 6.1 Orange high-voltage cables are easy to spot (Source: Renault Media)

Most of the high-voltage components are combined in a power unit. This is often located behind the rear seats or under the luggage compartment floor (or the whole floor in a Tesla). The unit is a metal box that is completely closed with bolts. A battery module switch, if used, may be located under a small secure cover on the power unit. The electric motor is located between the engine and the transmission or as part of the transmission on a hybrid or on a pure-EV; it is the main driving component. A few vehicles use wheel motors too. High-voltage wires are always brightly coloured. They have an orange outer casing with a yellow inner casing.

> **Safety First**
>
> Always refer to and follow the manufacturer's instructions – it is not possible to outline all the variations.

DOI: 10.1201/9781003299066-6

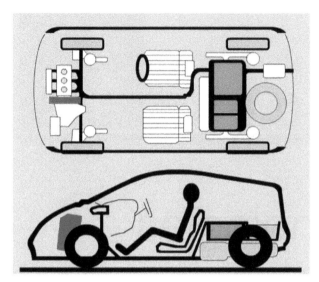

Figure 6.2 Motor (blue) and power pack (orange) locations on a typical hybrid

The electrical energy is conducted to or from the motor by thick orange or brightly coloured wires on most vehicles. If these wires must be disconnected, SWITCH OFF or DE-ENERGIZE the high-voltage system. This will prevent the risk of electric shock or short-circuit of the high-voltage system.

The following table lists some identified risks involved with working on ALL vehicles – and these become even more of a risk on EVs. The table is by no means exhaustive but serves as a good guide.

6.1.2 IMI TechSafe™

The IMI TechSafe™ professional registration scheme is designed to ensure complex automotive technologies are repaired safely and that technicians work safely – particularly in the UK, but it is just as effective internationally.

Figure 6.3 Risks and their reduction

Table 6.1 Risks and their reduction

Identified risk	Reducing the risk
Electric shock 1	Voltages and the potential for electric shock when working on an EV mean a high risk level.
Electric shock 2	Ignition HT is the most likely place to suffer a shock when working on an ICE vehicle; up to 40,000 V is quite normal. Use insulated tools if it is necessary to work on HT circuits with the engine running. Note that high voltages are also present on circuits containing windings due to back EMF as they are switched off; a few hundred volts is common. Mains-supplied power tools and their leads should be in good condition and using an earth leakage trip is highly recommended. Only work on HEVs and EVs if you are trained in the high-voltage systems.
Battery acid	Sulphuric acid is corrosive, so always use good personal protective equipment. In this case, overalls and if necessary, rubber gloves. A rubber apron is ideal, as are goggles if working with batteries a lot.
Raising or lifting vehicles	Apply brakes and/or chock the wheels when raising a vehicle on a jack or drive-on lift. Only jack under substantial chassis and suspension structures. Use axle stands in case the jack fails.
Running engines	Do not wear loose clothing; good overalls are ideal. Keep the keys in your possession when working on an engine to prevent others starting it. Take extra care if working near running drive belts.
Exhaust gases	Suitable extraction must be used if the engine is running indoors. Remember it is not just the carbon monoxide that might make you ill or even kill you; other exhaust components could cause asthma or even cancer.
Moving loads	Only lift what is comfortable for you; ask for help if necessary and/or use lifting equipment. As a general guide, do not lift on your own if it feels too heavy!
Short-circuits	Use a jump lead with an in-line fuse to prevent damage due to a short when testing. Disconnect the battery (earth lead off first and back on last) if any danger of a short exists. A very high current can flow from a vehicle battery, it will burn you as well as the vehicle.
Fire	Do not smoke when working on a vehicle. Fuel leaks must be attended to immediately. Remember the triangle of fire: heat–fuel–oxygen. Don't let the three sides come together.
Skin problems	Use a good barrier cream and/or latex gloves. Wash skin and clothes regularly.

Figure 6.4 Compact hatchback (Source: Citroën Media)

To be added to the register, a technician must successfully complete a specified qualification (e.g. Electric/Hybrid Vehicles Level 2, 3, or 4) and join the IMI Professional Register. To retain IMI TechSafe™ recognition, you must complete a set amount of continuing professional development (CPD) over 3 years. This must be aligned to a CPD subject framework. Not everything has to be covered and it is of course sensible to tailor it to vehicles you work on or are likely to do so.

For EVs this will fully meet the requirements that anyone working on high voltages must be competent (Electricity at Work Regulations 1989). ADAS and other areas are also covered in a similar way by the IMI. Technology safe means technician safe, means customer safe. (www.theimi.org.uk)

Safety First

For EVs IMI TechSafe™ fully meets the requirement that anyone working on high voltages must be competent (Electricity at Work Regulations 1989).

6.2 Electric and hybrid vehicles overview

6.2.1 Electric vehicle market

The EV market has increased year on year, and the number of battery technology native drivers (meaning that this technology has been easily available to them from their first car) also rises. Toyota and Honda first entered the hybrid market in the 1990s, with Toyota dominating the market sector with the Prius. Nissan brought a cost-effective hatchback to the market with the LEAF in 2010, a market which now enjoys an offering from almost every manufacturer thus bringing increased choice to the consumer. 'Range anxiety' is the term applied to a fear that the vehicle could not complete a journey, and this can be a barrier to uptake of this technology. This problem, however, is becoming less apparent as the charging network increases together with motorist knowledge.

The government's intention to disallow new internal combustion engine (ICE) car sales from 2030 (correct at time of press) has been met with mixed reaction. There is speculation that this will age the UK car parc while others have embraced the target. In addition, the 2020 pandemic may bring changes in employment frameworks that might impact how cars are used. The pandemic resulted in a greater reliance on the private motor car particularly with concerns for safety using public travel.

Key fact

The UK government's intention is to disallow new internal combustion engine (ICE) car sales from 2030.

As more city centres create restrictions and/or financial penalties for using ICE cars in areas of high congestion, this may also impact uptake of electric car ownership, particularly small and cheap quadricycle models and/or shared ownership or rental solutions.

Figure 6.6 IMI TechSafe™ logo

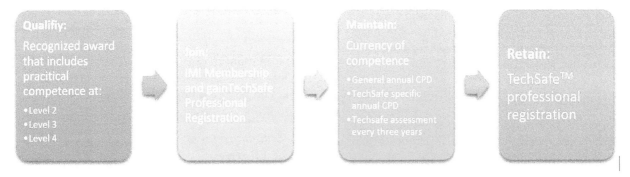

Figure 6.5 IMI TechSafe™ process

Figure 6.7 Citroën AMI - electric quadricycle (Source: Citroën Media)

6.2.2 The EV experience

Living with an electric car is not that different to ownership of an ICE; any motorist with an appropriate license can purchase one and drive away with no special training. The standout difference is how the car receives power to run. An ICE vehicle has an advantage of a short refuelling time, the fuel is transferred from one place to another quickly and pumps are easily found owing to the large number of refilling stations. ICE vehicles are not as impacted in changes of environmental conditions, whereas an electric car can use more energy when the weather is extreme. A motorist used to the convenience of fuelling an ICE may find the transition to electric more difficult than an electric car native.

The electric car can be charged at home, once the necessary equipment is safely fitted, and uptake for charging availability at the workplace is increasing. Rapid charging can dump a lot of energy into a battery in a much shorter time in a more similar fashion to the ICE but this can have a derogatory effect on the battery if this is the primary method of charging. The advantage of charging the car at home and work not only removes the inconvenient visit to the fuel station entirely, but it also allows the development of bi-directional charging, the CHAdeMO protocol enables the use of a vehicle as temporary energy storage device, allowing it to charge during the day (whilst renewable energy is easy to make) and then when connected to the home, used to power the home using a home battery system, reducing the demand on the grid.

Key fact

Rapid charging can dump a lot of energy into a battery in a much shorter time in a more similar fashion to the ICE but this can have a derogatory effect on the battery if this is the primary method of charging.

EVs deliver power faster than the ICE counterpart, giving swift acceleration, and there are no gears to slow the progress. EVs are quite heavy, but that weight is mainly made up by the batteries, which can be placed to give the car a low centre of gravity and a planted driving experience. With no engine noise, the compartment is quiet and without vibration, in fact, in can be difficult to tell if the car is on without looking at the dash! Due to their quiet nature, EVs create an artificial sound when driving to alert pedestrians of their presence.

The maintenance and repair of an EV requires specialist equipment and qualified technicians, although you do not need a special driving licence to drive one. Roles such as valeting currently do not require formal qualifications to work with cars with high-voltage systems, although it is an excellent idea to have a knowledge of the technology and how it is different to an ICE vehicle.

Safety First

The maintenance and repair of an EV requires specialist equipment and qualified technicians.

Maintenance and repairs of hybrid technologies need both the requirements of the EV and the ICE technology that are all packaged into the same engine bay as either a pure-EV or standalone ICE – potentially increasing labour times and expense to maintain two systems that are tightly fitted into the same space.

For the motorist, maintenance of the pure EV is different as the vehicle does not require the fluids and filters of an ICE car. It is important to note that they do still require maintenance, and some of those maintenance tasks can be of a different technical level to that of an ICE. EVs still share a lot of components with their ICE cousins: steering, suspension and infotainment will still look remarkably familiar in their functionality, whereas brakes and air conditioning may be integrated into the high-voltage systems. Some high-voltage batteries have sophisticated cooling systems that are now adopting filtration.

Key fact

It is important to note that EVs do still require maintenance, and some of those maintenance tasks can be of a different technical level to that of an ICE.

6.2.3 History

It may be surprising to learn that EVs were first invented in the early 1800s. Popular for replacing the

Figure 6.8 Supermarket delivery vehicle (Source: Waitrose & Partners Media)

horse and cart, particularly for doorstop deliveries, uptake in the UK of EVs was booming in the late 1800s to the early 1900s. Many of these vehicles were developed and built in the UK and shipped all over the world. World War 2 interrupted production but their popularity returned with around 40,000 of them in service in the 1970s. Although termed 'milk floats' the service sold anything that could be useful for a doorstep delivery including petfood, other dairy products and compost. One milkman was even convicted of selling cannabis to his elderly neighbours in 2009.

> **Key fact**
>
> The land speed record of 105.882 kilometres per hour (65.792 mph) was established by a car called La Jamais Contente (The Never Contented) on 29 April or 1 May 1899 at Achères, Yvelines near Paris, France. The vehicle had two Postel-Vinay 25 kW motors, each driving the rear axle via a chain, running at 200 V.

The 2020/1 pandemic saw an unprecedented surge in demand for doorstep delivery and it is interesting to note there is a return to electric vehicles; the regular short run and stop use of the vehicle makes this technology an excellent choice once again.

6.3 Electric vehicle systems and components

6.3.1 Types of electric vehicles

Electric vehicles (EVs) usually refers to any vehicle that is powered, in part or in full, by a battery that may or

may not be directly plugged into the mains. For this reason, the term EV can be generally used to cover all types of electric vehicle.

Pure-electric vehicles are those powered only by a battery. They are also referred to as a battery electric vehicle (BEV); this is commonly being dropped in favour of just EV as a term. Table 6.2 outlines some of the variations.

6.3.2 Identifying electric and hybrid vehicles

Identifying electric and hybrid technology can sometimes be tricky if unfamiliar. Manufacturers are aware of this, and they wish to demonstrate the distinction of this new technology and so will brand a vehicle to distinguish it from the ICE counterparts. The government also announced in June 2020 that zero emission vehicles would be further identified by a green flash on the registration plate. The purpose of the registration plate flash is to make for easy identification, this could be important for areas that only allow zero emissions vehicles entry, and it is also hoped it may incentivize zero emission car ownership.

Six areas to check when identifying an EV, hybrid, or alternative fuel vehicle:

▷ construction
▷ badging
▷ model
▷ procedure to follow if initial identification is not possible
▷ cable colouring
▷ registration plate (post 2020).

Examples of where high-voltage technology is available

▷ cars that use this technology
  • EVs (or BEVs)
  • HEVs
  • PHEVs
  • micro hybrid (cars that use start-stop technology)
▷ fuel cell vehicle/alternative fuel source vehicle
▷ two-wheel vehicles (currently only full electric or ICE are offered)
▷ commercial vehicles
▷ passenger transport
▷ mechanical handling equipment (MHE)
▷ plant.

6.3.3 High-voltage components

High-voltage components are designed by manufacturers to be easily identifiable and

Table 6.2 Summary of EVs and HEVs and their alternative names

Electric Vehicle/Car (EV), Electrically Chargeable Vehicle/Car	Generic terms for a vehicle powered, in part or in full, by a battery that can be plugged into the mains
Pure-EV, Pure-Electric Car/Vehicle, All Electric, Battery Electric Vehicle (BEV), Fully Electric	A vehicle powered only by a battery charged from mains electricity. Currently, typical pure-electric cars have a range of 100–200 miles. Top of the range vehicles will cover more miles.
Plug-In Hybrid Electric Vehicle (PHEV), Plug-In Hybrid Vehicle (PHV)	A vehicle with a plug-in battery and an internal combustion engine (ICE). Typical PHEVs will have a pure-electric range of 10–30 miles (a few do more). After the pure-electric range is used up, the vehicle reverts to the benefits of full hybrid capability.
Extended-Range Electric Vehicle (E-REV), Range-Extended Electric Vehicle (RE-EV)	A vehicle powered by a battery with an ICE-powered generator on board. E-REVs are like pure-EVs but may have a shorter battery range of around 50–100 miles. Range is extended by an on-board generator providing additional miles of mobility. With an E-REV the vehicle is still always electrically driven and is known as a series hybrid.
Hybrid Electric Vehicles (HEV), Full/Normal/Parallel/Standard hybrid	A hybrid vehicle is powered by a battery and/or an ICE. The power source is selected automatically by the vehicle, depending on speed, engine load, and battery charge. This battery cannot be plugged in, so charge is maintained by regenerative braking supplemented by ICE-generated power.
Micro hybrid/Stop-start	A stop-start system shuts off the engine when the vehicle is stationary. An enhanced starter motor is used to support the increased number of engine starts.
Mild hybrid	A mild hybrid vehicle cannot be plugged in or driven solely on battery power. However, it does harvest power during regenerative braking and uses this during acceleration.
Strong hybrid	A hybrid that can drive on electric only as well as by the internal combustion engine. A PHEV is always a strong hybrid.
Alternative Fuel Vehicle (AFV)	Any vehicle that is not solely powered by traditional fuels (i.e. petrol or diesel) is referred to as alternative fuel.
Internal Combustion Engine (ICE)	Petrol or diesel engine, as well as those adapted to operate on alternative fuels.
Electric quadricycles	This is a four-wheeled vehicle that is categorized and tested in a similar way to a moped or three-wheeled motorcycle.
Electric motorcycles	Battery only, so some full electric drive motorcycles may have a limited range. However, the Volt 220, which takes its name from its range of 220 km, and will do up to 60 mph according to the manufacturer. Other manufacturers are also developing products, the Harley Davidson shown in Fig. 6.11 has a 146-mile range.

Figure 6.9 Nissan LEAF – pure-EV (Source: Nissan Media)

Figure 6.10 Manufacturer's badge (Source: Toyota Media)

high-voltage cables are brightly coloured with an orange outer casing and a yellow inner casing. High-voltage cables also require greater insulation, so will appear more heavy duty than the cabling used in lower voltage systems.

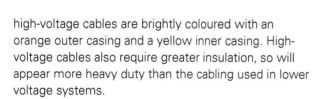

Safety First

High-voltage cables are brightly coloured with an orange outer casing and a yellow inner casing.

Figure 6.11 Electric motorcycle (Source: Harley-Davidson Media)

Figure 6.12 Under bonnet components of a pure EV

Manufacturers also use stickers, labels, and tags to identify components that may not be easily brightly coloured or could be discoloured over time. It is important to be able to identify EV components in order to work with technology safely. In many cases, manufacturer's information will be required to assist with this task. Names of components can differ slightly between brands, but generally the main components are:

▷ high-voltage battery
▷ motor
▷ high-voltage cables and connectors
▷ relays (switching components)
▷ control units (power electronics)
▷ charging points
▷ isolators (safety devices)
▷ interlock devices (safety devices)
▷ inverter
▷ DC to DC converter
▷ battery management controller
▷ ignition key/key on control switch
▷ driver display panel/interface.

There are additions to these systems, such as braking, steering, and even air conditioning that integrate into the high-voltage system and are covered in later qualifications. The high-voltage battery is not easily accessed, and most high-voltage systems will still carry a 12V battery, one just like an ICE would use, to run ancillary components and to activate the critical relays that bring the high-voltage battery online and allow for it to be charged. A flat 12V battery will disable the high-voltage system and will therefore not allow the vehicle to operate.

6.3.4 Motors and batteries overview

The electronically commutated motor (ECM) is, in effect, half way between an AC and a DC motor. The rotor contains permanent magnets and hence no slip rings. It is sometimes known as a brushless motor. The rotor operates a sensor, which provides feedback to the control and power electronics. This control system produces a rotating field, the frequency of which determines motor speed.

Brushless motors are also described as brushless DC motors (BLDC) and they are effectively AC motors because the current through them alternates. However, because the supply frequency is variable, it has to be derived from DC and its speed/torque characteristics are similar to a brushed DC motor, it is called a DC motor. Two typical motors are shown in Figures 6.13 and 6.14, one is integrated with the engine flywheel and the other a separate unit. Both are DC brushless motors and are water cooled.

Lithium-ion (Li-ion) batteries are the technology of choice, and they still have plenty of potential. Today's batteries have an energy density of up to 140 Wh/kg or more in some cases, but have the potential to go as high as 280 Wh/kg. Much research in cell optimization is taking place to create a battery with a higher energy density and increased range. Lithium-ion technology is currently considered the safest.

Key fact
Almost all modern EVs use lithium-ion (Li-ion) batteries.

The Li-ion battery works as follows. A negative pole (anode) and a positive pole (cathode) are part of the individual cells of a lithium-ion battery together with the electrolyte and a separator. The anode is a graphite structure and the cathode is layered metal oxide. Lithium-ions are deposited between these layers. When the battery is charging, the lithium-ions move from the anode to the cathode and take on electrons.

Figure 6.13 Bosch integrated motor generator (IMG) also called integrated motor assist (IMA) by some manufacturers

Figure 6.14 Separate motor unit showing the coolant connections on the side and the three main electrical connections on top

The number of ions therefore determines the energy density. When the battery is discharging, the lithium-ions release the electrons to the anode and move back to the cathode.

6.3.5 Charging

Most electric cars will be charged at home, but national infrastructures are developing. There are, however, competing organizations and commercial companies, so it is necessary to register with a few different organizations to access their charging points. Many businesses now also provide charging stations for staff and visitors. Some are pay in advance, some are pay as you go, and others require a monthly subscription.

Key fact

Most electric cars will be charged at home, but national infrastructures are developing.

Many apps and websites are available for locating charge points. An example of this can be found at https://www.zap-map.com.

There are different plugs, types of supply and methods of connection. It is important to use the correct plug as it is difficult to insert a plug into an incorrect socket due to the shapes of the available plugs. Some power supplies are 'tethered' meaning that the cable is fixed to the supply and is plugged into the car. Other supplies are simply a socket that requires the motorist to select their own cable and connect themselves, and there is a lock facility to ensure that the cable cannot be removed by passers-by. Once connected, before electricity begins to flow, a 'handshake' is completed to determine the situation is safe to operate the relays and begin charging. A supply can either be DC or AC. Most vehicles will provide status updates, such as unexpectedly becoming unplugged or reaching a full charge, and inform the motorist via an app.

The uptake of this technology is now prompting large fuel sellers to incorporate charging facilities into their existing business model and deploying through their established infrastructure, making the availability of charging along main routes more accessible.

Safety note: Although the rechargeable electric vehicles and equipment can be recharged from a domestic wall socket, a charging station has additional current or connection sensing mechanisms to disconnect the power when the EV is not charging. There are two main types of safety sensor:

▶ Current sensors monitor the power consumed and only maintain the connection if the demand is within a predetermined range.

Figure 6.15 Battery packs are several hundred volts

Figure 6.16 Charging point on the roadside (Source: Citroën Media)

Figure 6.17 A tethered domestic charging point (Source: EDF Media)

▷ Additional sensor wires provide a feedback signal that requires special power plug fittings.

It is safe to charge in wet weather. When you plug in the charge lead, the connection to the supply is not made until the plug is completely in position. Circuit breaker devices are also used for additional safety. Clearly some common sense is necessary, but EV charging is very safe.

Safety First

When you plug in the charge lead, the connection to the supply is not made until the plug is completely in position.

It is strongly recommended that home charging sockets and wiring are installed and approved by a qualified electrician. A home charge point with its own dedicated circuit is the best way of charging an EV safely. For rapid charging, special equipment and an upgraded electrical supply would be required and is therefore unlikely to be installed at home, where most consumers will charge overnight.

How long it takes to charge an EV depends on the type of vehicle, how discharged the battery is and the type of charge point used. Typically, pure-electric cars using standard charging will take between six and eight hours to charge fully and can be 'opportunity charged' whenever possible to keep the battery topped up.

Pure-EVs capable of using rapid charge points could be fully charged in around 30 minutes and can be topped up in around 20 minutes, depending on the type of charge point and available power. PHEVs take approximately two hours to charge from a standard electricity supply. E-REVs take approximately four hours to charge from a standard electricity supply. PHEVs and E-REVs require less time to charge because their batteries are usually smaller.

The cost of charging an EV depends on the size of the battery and how much charge is left in the battery before charging. As a guide, charging an electric car from flat to full will cost from as little as £1 to £5 (2022). This is for a pure-electric car with a 24 kWh battery that will offer around a 100-mile range. This results in an average cost of a few pence per mile.

The cost of charging from public points will vary; many will offer free electricity in the short term. It is also possible to register with supply companies who concentrate on energy from renewable sources.

Table 6.3 Estimated charging times

Charging time for 100 km range	Power supply	Power	Voltage	Max. current
6–8 hours	Single phase	3.3 kW	230 V AC	16 A
3–4 hours	Single phase	7.4 kW	230 V AC	32 A
2–3 hours	Three phase	10 kW	400 V AC	16 A
1–2 hours	Three phase	22 kW	400 V AC	32 A
20–30 minutes	Three phase	43 kW	400 V AC	63 A
20–30 minutes	Direct current	50 kW	400–500 V DC	100–125 A
10 minutes	Direct current	120 kW	300–500 V DC	300–350 A

6.4 Advanced driver assistance systems overview

6.4.1 Introduction

Advanced driver assistance systems (ADAS) are, as the name suggests, designed to help the driver. This improves safety because most road accidents occur due to human error. Automated systems help to minimize human error, which has led to reduced road fatalities. These are also the enabling technologies for full automated driving.

ADAS relies on inputs from several sources such as LiDAR, radar, cameras and vehicle CAN data. Safety features are designed to alert the driver to potential problems, or to avoid collisions by implementing safeguards. In some cases, this means taking control of the vehicle.

ADAS features can, for example, switch on lights, provide adaptive cruise control and collision avoidance, incorporate traffic warnings, alert the driver to other vehicles and dangers, warn if lane departure is detected or even initiate automated lane guidance. Cameras are used to see in blind spots. Many modern vehicles now have systems such as electronic stability control, anti-lock brakes, lane departure warning, adaptive cruise control and traction control. All these systems can be affected by mechanical alignment. For this reason, correct repairs, adjustments and servicing are essential.

Figure 6.18 Sensing activity

Figure 6.19 Automatic parking (Source: Park4U)

Some example advanced driver assistance systems (ADAS) are listed here:

- adaptive cruise control
- adaptive light control
- anti-lock braking system
- automatic parking
- blind spot monitor
- collision avoidance
- collision warning
- driver drowsiness detection
- electric vehicle warning sounds
- emergency driver assistant
- glare-free high beam and pixel light
- lane change assistance
- lane departure warning
- navigation system with traffic information
- night vision
- parking sensor
- pedestrian protection
- rain sensor
- surround view
- traffic sign recognition
- tyre pressure monitoring.

6.4.2 Systems

Conventional cruise control is not always practical because the speed of the general traffic varies constantly, and traffic is often very heavy. The driver must take over from a standard cruise control system on many occasions to speed up or slow down. Adaptive cruise control (ACC) can automatically adjust the vehicle speed to the current traffic situation. The system has three main features:

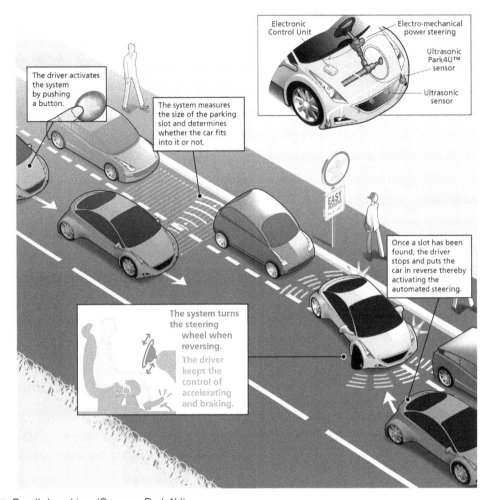

Figure 6.20 Parallel parking (Source: Park4U)

- maintain a speed as set by the driver
- adapt this speed and maintain a safe distance from the vehicles in front
- provide a warning if there is a risk of collision.

The main extra components, compared to standard cruise control, are the headway sensor and the steering angle sensor; the first of these is clearly the most important. Information on steering angle is used to further enhance the data from the headway sensor by allowing greater discrimination between hazards and spurious signals. Two types of headway sensor are in use: radar and LiDAR. Both contain transmitter and receiver units. The radar system uses microwave signals of up to 80 GHz, and the reflection time of these gives the distance to the object in front. LiDAR uses a laser diode to produce infrared

Figure 6.21 Camera used for night vision
(Source: First Sensor AG)

Figure 6.22 Adaptive cruise control prevents accidents

Figure 6.23 Lidar sensor

Figure 6.24 Headway sensor is fitted at the front of a vehicle (Source: Bosch Media)

light signals, the reflections of which are detected by a photodiode.

These two types of sensors have advantages and disadvantages. The radar system is not affected by rain and fog, but the LiDAR can be more selective by recognizing the standard reflectors on the rear of the vehicle in front. Radar can produce strong reflections from bridges, trees, posts and other normal roadside items. It can also suffer loss of signal return due to multipath reflections. Under ideal weather conditions, the LiDAR system appears to be the best, but it becomes very unreliable when the weather changes. A beam divergence of about 2.5° vertically and horizontally has been found to be the most suitable whatever headway sensor is used. An important consideration is that signals from other vehicles fitted with this system must not produce erroneous results. Fig. 6.24 shows a typical headway sensor and control electronics.

Key fact

Radar is not affected by rain and fog, but LiDAR can be more selective.

Fundamentally, the operation of an adaptive cruise system is the same as a conventional system, except when a signal from the headway sensor detects an obstruction. In this case, the vehicle speed is decreased. If the optimum stopping distance cannot be achieved by just backing off the throttle, a warning is supplied to the driver. Later systems also take control of the vehicle transmission and brakes.

This system, sometimes called collision avoidance radar, can be looked at in two ways. First, it is an aid

to reversing, which gives the driver some indication as to how much space is behind the car. Second, it can be used as a vision enhancement system. The principle of radar as a reversing aid is illustrated. This technique is, in effect, a range-finding system. The output can be audio but visual is perhaps more appropriate, as the driver is likely to be looking backwards. The audible signal is a 'pip-pip-pip' type sound, the repetition frequency of which increases as the car comes nearer to the obstruction and becomes almost continuous as impact is imminent. Many systems now also make the noise come from the appropriate speaker(s) to indicate direction.

A reverse sensing system is a reverse-only parking aid system that uses sensors mounted in the rear bumper. Parking aid systems feature both front and rear sensors. Low-cost, high-performance ultrasonic range sensors are fitted to the vehicle. Generally, four sensors are used to form a detection zone as wide as the vehicle. A microprocessor monitors the sensors and emits

Figure 6.25 Forward sensing

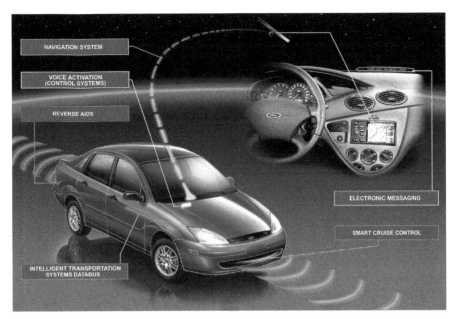

Figure 6.26 Reversing aid as part of a control system (Source: Ford)

Figure 6.27 Rear sensors

audible beeps during slow reverse parking to help the driver reverse or park the vehicle. The technique is relatively simple as the level of discrimination required is low and the system only has to operate over short distances.

Drivers are taught to assess surrounding traffic before changing lanes by checking their rear-view and side mirrors and looking over each shoulder. However, the area alongside and just behind the vehicle is a constant source of danger and often the cause of serious accidents. Drivers are not able to see into this area using either the rear view or side mirrors, but it is big enough for a vehicle to be missed by a cursory glance.

To help minimize this risk, a lane-changing assistant receives the information it needs from a mid-range

radar sensor for rear-end applications. This means drivers are effectively looking over their shoulders all the time, because it reliably and accurately recognizes other road users in their vehicle's blind spot.

A typical installation is to have two sensors in the rear bumper, one on the left, one on the right. These two rear sensors monitor the area alongside and behind the car. Powerful control software collates the sensor information to produce a complete picture of all traffic in the area behind the vehicle. Whenever another vehicle approaches at speed from behind or is already present in the blind spot, a signal such as a warning light in the side mirror alerts the driver to the hazard. Should the driver still activate the turn signal with the intention of changing lanes, the lane-changing assistant issues an additional acoustic and/or haptic warning.

Key fact

Powerful control software collates sensor information.

The rear radar system can do more than just assist with lane changing. These sensors also form part of a cross-traffic alert system, which supports drivers reversing out of perpendicular parking spaces when their rear view is obstructed. Able to recognize cars, cyclists and pedestrians crossing behind the reversing vehicle at up to 50 m, the system alerts the driver to the imminent danger of collision by issuing an audible or visible signal.

Figure 6.28 Audi ultrasonic rear sensor

Figure 6.30 Mid-range radar (MRR) sensor
(Source: Bosch Media)

In this example it is 1 microsecond.

$$t = \frac{2 \times 150}{3 \times 10^8} = 1\mu S$$

If the measured time for the round trip is 1 microsecond, as in the above example, then the distance must be 150 m. Relative closing speed can be calculated from the current vehicle speed. The type of display or output that may be used on a motor vehicle will vary from an audible warning to a warning light or series of lights or more likely a display screen.

6.4.4 Camera

Bosch has developed a stereo video camera such that an emergency braking system can function based solely on camera data. Normally, this would require a radar sensor or a combination of radar and video sensors.

Emergency braking systems are among the most effective assistance systems in the car. It is estimated that something like 70% of all rear-end collisions, resulting in personal injury, could be avoided if all vehicles were equipped with them.

Land Rover uses a stereo video camera together with the Bosch emergency braking system as standard in its Discovery Sport. When the camera recognizes another vehicle ahead in the lane as an obstruction, the emergency braking system prepares for action. If the driver does not react, then the system initiates maximum braking.

Other driver assistance functions can also be based on the stereo video camera. One such function is road-sign recognition, which keeps the driver informed about the current speed limit. Another is a lane-departure warning. This vibrates the steering wheel to warn drivers before they unintentionally drift out of lane.

Figure 6.29 Sensors monitor all traffic in the area behind the vehicle

6.4.3 Radar

The operation of a basic radar system is as follows: a radio transmitter generates radio wave pulses, which are then radiated from an antenna. A target, such as another vehicle, scatters a small portion of the radio energy back to a receiving antenna. This weak signal is amplified and displayed on a screen. To determine its position, the distance (range) and bearing must be measured. Because radio waves travel at the speed of light (3 x 10⁸ m/s), the range may be found by measuring the time taken for a radio wave to travel from transmitter to obstacle, and back to the receiver.

If the range were 150 m, the time for the round trip would be:

$$t = \frac{2d}{C}$$

Where: t = time, d = distance to object, and C = speed of light.

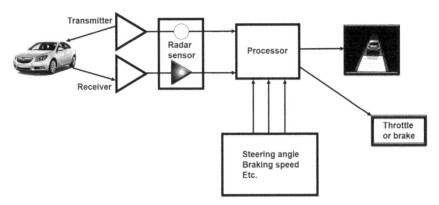

Figure 6.31 Block diagram animation of obstacle avoidance radar

Figure 6.32 Land Rover using the camera system (Source: Bosch Media)

Figure 6.33 Stereo video camera for ADAS (Source: Bosch Media)

Figure 6.34 The camera recognizes another vehicle ahead in the same lane (Source: Bosch Media)

With its light-sensitive lenses and video sensors, the camera covers a 50° horizontal field of vision and can take measurements in 3D at over 50 m. Thanks to these spatial measurements, the video signal alone provides enough data to calculate, for example, the distance to vehicles ahead. Its pair of highly sensitive video sensors are equipped with colour recognition and CMOS (complementary metal oxide semiconductor) technology. They have a resolution of 1,280 by 960 pixels and can also process high-contrast images.

6.5 Alternative fuels

The use of an alternative fuel can lessen dependence upon oil and reduce greenhouse gas emissions. There are a number of alternative fuels and each of these is outlined briefly in this section.

Key fact

Alternative fuels can lessen dependence upon oil and reduce greenhouse gas emissions.

6.5.1 Ethanol

Ethanol is an alcohol-based fuel made by fermenting and distilling starch crops, such as corn. It can also be made from plants such as trees and grasses. E10 is a blend of 10% ethanol and 90% petrol/gasoline. Almost all manufacturers approve the use of E10 in their vehicles. E85 is a blend of 85% ethanol and 15% petrol/gasoline and can be used in flexible fuel vehicles (FFVs).

Safety First

Almost all manufacturers approve the use of E10 in their vehicles – but double check if in doubt!

399

Figure 6.35 E85 Vehicle

Figure 6.36 E85 logo

FFVs are specially designed to run on petrol/gasoline, E85, or any mixture of the two. These vehicles are offered by several manufacturers. There is no noticeable difference in vehicle performance when E85 is used. However, FFVs operating on E85 can experience a 20–30% drop in miles per gallon due to ethanol's lower energy content.

Here are some advantages and disadvantages of this alternative fuel:

Advantages:

▷ lower emissions of air pollutants
▷ more resistant to engine knock
▷ added vehicle cost is very small.

Disadvantages:

▷ can only be used in flex-fuel vehicles
▷ lower energy content, resulting in fewer miles per gallon
▷ limited availability.

6.5.2 Biodiesel

Biodiesel is a form of diesel fuel manufactured from vegetable oils, animal fats or recycled restaurant oils. It is safe, biodegradable, and produces fewer air pollutants than petroleum-based diesel.

Biodiesel can be used in its pure form (B100) or blended with petroleum diesel. Common blends include B2 (2% biodiesel), B5 and B20. B2 and B5 can be used safely in most diesel engines. However, most vehicle manufacturers do not recommend using blends greater than B5, and engine damage caused by higher blends is not covered by some manufacturer warranties.

Here are some advantages and disadvantages of this alternative fuel:

Figure 6.37 Biodiesel engine

Advantages:

▷ can be used in most diesel engines, especially newer ones
▷ fewer air pollutants (other than NO_x) and less greenhouse gases
▷ biodegradable
▷ non-toxic
▷ safer to handle.

Disadvantages:

▷ use of blends above B5 may not yet be approved by manufacturers
▷ lower fuel economy and power (10% lower for B100, 2% for B20)
▷ more nitrogen oxide emissions
▷ B100 generally not suitable for use in low temperatures
▷ concerns about B100's impact on engine durability.

6.5.3 Natural gas

Natural gas is a fossil fuel made up mostly of methane. It is one of the cleanest burning alternative fuels. It can be used in the form of compressed natural gas (CNG) or liquefied natural gas (LNG) to fuel cars and trucks.

Figure 6.38 Biodiesel logo

Figure 6.39 CNG logo

Figure 6.40 Propane tank (Source: http://www.
rasoenterprises.com)

Dedicated natural gas vehicles are designed to run on natural gas only, while dual-fuel or bi-fuel vehicles can also run on petrol/gasoline or diesel. Dual-fuel vehicles take advantage of the widespread availability of conventional fuels but use a cleaner, more economical alternative when natural gas is available. Natural gas is stored in high-pressure fuel tanks so dual-fuel vehicles require two separate fuelling systems, which takes up extra space.

Natural gas vehicles are not produced commercially in large numbers. However, conventional vehicles can be retrofitted for CNG.

Here are some advantages and disadvantages of this alternative fuel:

Advantages:

- 60–90% less smog-producing pollutants
- 30–40% less greenhouse gas emissions
- less expensive than petroleum fuels.

Disadvantages:

- limited vehicle availability
- less readily available
- fewer miles on a tank of fuel.

6.5.4 Propane or liquified petroleum gas

Propane or liquefied petroleum gas is a clean-burning fossil fuel that can be used to power internal combustion engines. LPG-fuelled vehicles produce fewer toxic and smog-forming air pollutants.

Petrol/gasoline and diesel vehicles can be retrofitted to run on LPG in addition to conventional fuel. The LPG

Figure 6.41 LPG logo

is stored in high-pressure fuel tanks, so separate fuel systems are needed in vehicles powered by both LPG and a conventional fuel.

Here are some advantages and disadvantages of this alternative fuel:

Advantages:

- fewer toxic and smog-forming air pollutants
- less expensive than petrol/gasoline.

Disadvantages:

- no new passenger cars or trucks commercially available, but vehicles can be retrofitted for LPG
- less readily available than conventional fuels
- fewer miles on a tank of fuel.

6.5.5 Hydrogen

Hydrogen (H_2) can be produced from fossil fuels (such as coal), nuclear power, or renewable resources, such as hydropower. Fuel cell vehicles powered by pure hydrogen emit no harmful air pollutants. Hydrogen is being aggressively explored as a fuel for passenger vehicles. It can be used in fuel cells to power electric

401

motors or burned in internal combustion engines. It is stored under very high pressure.

Safety First

Hydrogen is stored under very high pressure.

It is an environmentally friendly fuel that has the potential to dramatically reduce dependence on oil, but several significant challenges must be overcome before it can be widely used.

Key fact

Fuel cell vehicles powered by pure hydrogen emit no harmful air pollutants.

Figure 6.42 Fuel cell vehicle (Source: Honda Media)

Figure 6.43 Hydrogen logo

Here are some advantages and disadvantages of this alternative fuel:

Advantages:

▷ can be produced from several sources, reducing dependence on petroleum
▷ no air pollutants or greenhouse gases when used in fuel cells
▷ it produces only NO_x when burned in internal combustion engines.

Disadvantages:

▷ expensive to produce and is only available at a few locations
▷ fuel cell vehicles are currently too expensive for most consumers
▷ hydrogen has a lower energy density than conventional petroleum fuels
▷ a hydrogen-powered car uses about three times as much electricity as a pure-EV.

Key fact

A hydrogen powered car uses about three times as much electricity as a pure-EV.

6.5.6 Summary

This section has given an overview of some alternative fuels. All of them offer some significant advantages either commercially, environmentally, or both. There are also some disadvantages, not least of which is that the cost of production can be high. This is likely to change, however, if the use of alternative fuels becomes more widespread.

CHAPTER 7

Automotive Technology Academy

7.1 Introduction

The online Automotive Technology Academy has been created by the author (Tom Denton), who has over 40 years of relevant automotive experience, and over 30 published textbooks that are used by students and technicians worldwide.

The aims of the academy are to:

▷ Improve automotive technology **skills** and **knowledge.**
▷ Provide **free** access to study resources to support the textbooks.
▷ Create a worldwide **community** of automotive learners.

▷ Freely **share** automotive related information and ideas.
▷ **Reach** out to learners who are not able to attend school or college.
▷ Improve automotive training **standards** and **quality.**
▷ Provide online access to **certification** for a range of automotive subjects.

To access the academy visit: www.automotive-technology.org and create an account for yourself. To access the free courses that work in conjunction with the textbooks you will need to enter an enrolment key. This will be described something like this:

'The third word on the last line of page ## of the associated textbook'

You will therefore need to own the book!

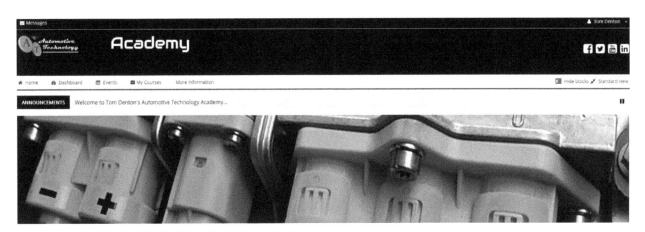

Figure 7.1 Automotive Technology Academy website

DOI: 10.1201/9781003299066-7

Key fact

To access the free courses, that work in conjunction with the textbooks, you will need to enter an enrolment key.

All you will need to do is enter the word in a box and you will have full unrestricted access to the course and its associated resources.

7.2 Resources

The following is a list of some of the resources that will be available to you:

- images
- videos
- activities
- 3D models
- hyperlinks
- assignments
- quizzes
- forums
- chat features
- social media
- interactive features, games, and much more.

A progress bar is used in the courses so you can see at a glance how you are getting on, if you are working to cover all the content. Alternatively, you can just dip in and out to find what you need.

Figure 7.2 Still image from one of our 3D models

More formal assessments will also be available for those who are not able to attend traditional training centres. It will be possible to obtain certification relating to theory and practical work. A charge will apply to this aspect but all other resources are free.

Updates and interesting new articles will also be available, so what are you waiting for? Come and visit and join in!

www.automotive-technology.org

Index

Page numbers in *italics* denote an illustration, **bold** indicates a table

9 781032 289083